LONDON MATHEMATICAL SOCIETY LECTURE

Managing Editor: Professor M. Reid, Mathematics Institute, University of

The titles below are available from booksellers, or from Cambridge Unive

214 Generalised Euler–Jacobi inversion formula and asymptotics beyo
215 Number theory 1992–93, S. DAVID (ed.)
216 Stochastic partial differential equations, A. ETHERIDGE (ed.)
217 Quadratic forms with applications to algebraic geometry and topol
218 Surveys in combinatorics, 1995, P. ROWLINSON (ed.)
220 Algebraic set theory, A. JOYAL & I. MOERDIJK
221 Harmonic approximation, S. J. GARDINER
222 Advances in linear logic, J.-Y. GIRARD, Y. LAFONT & L. REGNIER (eds)
223 Analytic semigroups and semilinear initial boundary value problems, KAZUAKI TAIRA
224 Computability, enumerability, unsolvability, S. B. COOPER, T. A. SLAMAN & S. S. WAINER (eds)
225 A mathematical introduction to string theory, S. ALBEVERIO et al.
226 Novikov conjectures, index theorems and rigidity I, S. FERRY, A. RANICKI & J. ROSENBERG (eds)
227 Novikov conjectures, index theorems and rigidity II, S. FERRY, A. RANICKI & J. ROSENBERG (eds)
228 Ergodic theory of Z^d actions, M. POLLICOTT & K. SCHMIDT (eds)
229 Ergodicity for infinite dimensional systems, G. DA PRATO & J. ZABCZYK
230 Prolegomena to a middlebrow arithmetic of curves of genus 2, J. W. S. CASSELS & E. V. FLYNN
231 Semigroup theory and its applications, K. H. HOFMANN & M. W. MISLOVE (eds)
232 The descriptive set theory of Polish group actions, H. BECKER & A. S. KECHRIS
233 Finite fields and applications, S. COHEN & H. NIEDERREITER (eds)
234 Introduction to subfactors, V. JONES & V. S. SUNDER
235 Number theory 1993–94, S. DAVID (ed.)
236 The James forest, H. FETTER & B. G. DE BUEN
237 Sieve methods, exponential sums, and their applications in number theory, G. R. H. GREAVES et al.
238 Representation theory and algebraic geometry, A. MARTSINKOVSKY & G. TODOROV (eds)
240 Stable groups, F. O. WAGNER
241 Surveys in combinatorics, 1997, R. A. BAILEY (ed.)
242 Geometric Galois actions I, L. SCHNEPS & P. LOCHAK (eds)
243 Geometric Galois actions II, L. SCHNEPS & P. LOCHAK (eds)
244 Model theory of groups and automorphism groups, D. EVANS (ed.)
245 Geometry, combinatorial designs and related structures, J. W. P. HIRSCHFELD et al.
246 p-Automorphisms of finite p-groups, E. I. KHUKHRO
247 Analytic number theory, Y. MOTOHASHI (ed.)
248 Tame topology and o-minimal structures, L. VAN DEN DRIES
249 The atlas of finite groups: ten years on, R. CURTIS & R. WILSON (eds)
250 Characters and blocks of finite groups, G. NAVARRO
251 Gröbner bases and applications, B. BUCHBERGER & F. WINKLER (eds)
252 Geometry and cohomology in group theory, P. KROPHOLLER, G. NIBLO, R. STÖHR (eds)
253 The q-Schur algebra, S. DONKIN
254 Galois representations in arithmetic algebraic geometry, A. J. SCHOLL & R. L. TAYLOR (eds)
255 Symmetries and integrability of difference equations, P. A. CLARKSON & F. W. NIJHOFF (eds)
256 Aspects of Galois theory, H. VÖLKLEIN et al.
257 An introduction to noncommutative differential geometry and its physical applications 2ed, J. MADORE
258 Sets and proofs, S. B. COOPER & J. TRUSS (eds)
259 Models and computability, S. B. COOPER & J. TRUSS (eds)
260 Groups St Andrews 1997 in Bath, I, C. M. CAMPBELL et al.
261 Groups St Andrews 1997 in Bath, II, C. M. CAMPBELL et al.
262 Analysis and logic, C. W. HENSON, J. IOVINO, A. S. KECHRIS & E. ODELL
263 Singularity theory, B. BRUCE & D. MOND (eds)
264 New trends in algebraic geometry, K. HULEK, F. CATANESE, C. PETERS & M. REID (eds)
265 Elliptic curves in cryptography, I. BLAKE, G. SEROUSSI & N. SMART
267 Surveys in combinatorics, 1999, J. D. LAMB & D. A. PREECE (eds)
268 Spectral asymptotics in the semi-classical limit, M. DIMASSI & J. SJÖSTRAND
269 Ergodic theory and topological dynamics, M. B. BEKKA & M. MAYER
271 Singular perturbations of differential operators, S. ALBEVERIO & P. KURASOV
272 Character theory for the odd order theorem, T. PETERFALVI
273 Spectral theory and geometry, E. B. DAVIES & Y. SAFAROV (eds)
274 The Mandelbrot set, theme and variations, T. LEI (ed.)
275 Descriptive set theory and dynamical systems, M. FOREMAN et al.
276 Singularities of plane curves, E. CASAS-ALVERO
277 Computational and geometric aspects of modern algebra, M. D. ATKINSON et al.
278 Global attractors in abstract parabolic problems, J. W. CHOLEWA & T. DLOTKO
279 Topics in symbolic dynamics and applications, F. BLANCHARD, A. MAASS & A. NOGUEIRA (eds)
280 Characters and automorphism groups of compact Riemann surfaces, T. BREUER
281 Explicit birational geometry of 3-folds, A. CORTI & M. REID (eds)
282 Auslander–Buchweitz approximations of equivariant modules, M. HASHIMOTO
283 Nonlinear elasticity, Y. FU & R. OGDEN (eds)
284 Foundations of computational mathematics, R. DEVORE, A. ISERLES & E. SÜLI (eds)
285 Rational points on curves over finite fields, H. NIEDERREITER & C. XING
286 Clifford algebras and spinors 2ed, P. LOUNESTO
287 Topics on Riemann surfaces and Fuchsian groups, E. BUJALANCE et al.
288 Surveys in combinatorics, 2001, J. HIRSCHFELD (ed.)

289	Aspects of Sobolev-type inequalities, L. SALOFF-COSTE
290	Quantum groups and Lie theory, A. PRESSLEY (ed.)
291	Tits buildings and the model theory of groups, K. TENT (ed.)
292	A quantum groups primer, S. MAJID
293	Second order partial differential equations in Hilbert spaces, G. DA PRATO & J. ZABCZYK
294	Introduction to operator space theory, G. PISIER
295	Geometry and integrability, L. MASON & Y. NUTKU (eds)
296	Lectures on invariant theory, I. DOLGACHEV
297	The homotopy category of simply connected 4-manifolds, H.-J. BAUES
298	Higher operads, higher categories, T. LEINSTER
299	Kleinian groups and hyperbolic 3-manifolds, Y. KOMORI, V. MARKOVIC & C. SERIES (eds)
300	Introduction to Möbius differential geometry, U. HERTRICH-JEROMIN
301	Stable modules and the D(2)-problem, F. E. A. JOHNSON
302	Discrete and continuous nonlinear Schrödinger systems, M. J. ABLOWITZ, B. PRINARI & A. D. TRUBATCH
303	Number theory and algebraic geometry, M. REID & A. SKOROBOGATOV (eds)
304	Groups St Andrews 2001 in Oxford Vol. 1, C. M. CAMPBELL, E. F. ROBERTSON & G. C. SMITH (eds)
305	Groups St Andrews 2001 in Oxford Vol. 2, C. M. CAMPBELL, E. F. ROBERTSON & G. C. SMITH (eds)
306	Peyresq lectures on geometric mechanics and symmetry, J. MONTALDI & T. RATIU (eds)
307	Surveys in combinatorics 2003, C. D. WENSLEY (ed.)
308	Topology, geometry and quantum field theory, U. L. TILLMANN (ed.)
309	Corings and comodules, T. BRZEZINSKI & R. WISBAUER
310	Topics in dynamics and ergodic theory, S. BEZUGLYI & S. KOLYADA (eds)
311	Groups: topological, combinatorial and arithmetic aspects, T. W. MÜLLER (ed.)
312	Foundations of computational mathematics, Minneapolis 2002, FELIPE CUCKER et al. (eds)
313	Transcendantal aspects of algebraic cycles, S. MÜLLER-STACH & C. PETERS (eds)
314	Spectral generalizations of line graphs, D. CVETKOVIC, P. ROWLINSON & S. SIMIC
315	Structured ring spectra, A. BAKER & B. RICHTER (eds)
316	Linear logic in computer science, T. EHRHARD et al. (eds)
317	Advances in elliptic curve cryptography, I. F. BLAKE, G. SEROUSSI & N. SMART
318	Perturbation of the boundary in boundary-value problems of partial differential equations, DAN HENRY
319	Double affine Hecke algebras, I. CHEREDNIK
320	L-functions and Galois representations, D. BURNS, K. BUZZARD & J. NEKOVÁŘ (eds)
321	Surveys in modern mathematics, V. PRASOLOV & Y. ILYASHENKO (eds)
322	Recent perspectives in random matrix theory and number theory, F. MEZZADRI & N. C. SNAITH (eds)
323	Poisson geometry, deformation quantisation and group representations, S. GUTT et al. (eds)
324	Singularities and computer algebra, C. LOSSEN & G. PFISTER (eds)
325	Lectures on the Ricci flow, P. TOPPING
326	Modular representations of finite groups of Lie type, J. E. HUMPHREYS
328	Fundamentals of hyperbolic manifolds, R. D. CANARY, A. MARDEN & D. B. A. EPSTEIN (eds)
329	Spaces of Kleinian groups, Y. MINSKY, M. SAKUMA & C. SERIES (eds)
330	Noncommutative localization in algebra and topology, A. RANICKI (ed.)
331	Foundations of computational mathematics, Santander 2005, L. PARDO, A. PINKUS, E. SULI & M. TODD (eds)
332	Handbook of tilting theory, L. ANGELERI HÜGEL, D. HAPPEL & H. KRAUSE (eds)
333	Synthetic differential geometry 2ed, A. KOCK
334	The Navier–Stokes equations, P. G. DRAZIN & N. RILEY
335	Lectures on the combinatorics of free probability, A. NICA & R. SPEICHER
336	Integral closure of ideals, rings, and modules, I. SWANSON & C. HUNEKE
337	Methods in Banach space theory, J. M. F. CASTILLO & W. B. JOHNSON (eds)
338	Surveys in geometry and number theory, N. YOUNG (ed.)
339	Groups St Andrews 2005 Vol. 1, C. M. CAMPBELL, M. R. QUICK, E. F. ROBERTSON & G. C. SMITH (eds)
340	Groups St Andrews 2005 Vol. 2, C. M. CAMPBELL, M. R. QUICK, E. F. ROBERTSON & G. C. SMITH (eds)
341	Ranks of elliptic curves and random matrix theory, J. B. CONREY, D. W. FARMER, F. MEZZADRI & N. C. SNAITH (eds)
342	Elliptic cohomology, H. R. MILLER & D. C. RAVENEL (eds)
343	Algebraic cycles and motives Vol. 1, J. NAGEL & C. PETERS (eds)
344	Algebraic cycles and motives Vol. 2, J. NAGEL & C. PETERS (eds)
345	Algebraic and analytic geometry, A. NEEMAN
346	Surveys in combinatorics, 2007, A. HILTON & J. TALBOT (eds)
347	Surveys in contemporary mathematics, N. YOUNG & Y. CHOI (eds)
348	Transcendental dynamics and complex analysis, P. RIPPON & G. STALLARD (eds)
349	Model theory with applications to algebra and analysis Vol 1, Z. CHATZIDAKIS, D. MACPHERSON, A. PILLAY & A. WILKIE (eds)
350	Model theory with applications to algebra and analysis Vol 2, Z. CHATZIDAKIS, D. MACPHERSON, A. PILLAY & A. WILKIE (eds)
351	Finite von Neumann algebras and masas, A. SINCLAIR & R. SMITH
352	Number theory and polynomials, J. MCKEE & C. SMYTH (eds)
353	Trends in stochastic analysis, J. BLATH, P. MÖRTERS & M. SCHEUTZOW (eds)
354	Groups and analysis, K. TENT (ed)
355	Non-equilibrium statistical mechanics and turbulence, J. CARDY, G. FALKOVICH & K. GAWEDZKI, S. NAZARENKO & O. V. ZABORONSKI (eds)
356	Elliptic curves and big Galois representations, D. DELBOURGO
357	Algebraic theory of differential equations, M. A. H. MACCALLUM & A. V. MIKHAILOV (eds)
358	Geometric and cohomological methods in group theory, M. BRIDSON, P. KROPHOLLER & I. LEARY (eds)
359	Moduli spaces and vector bundles, L. BRAMBILA-PAZ, S. B. BRADLOW, O. GARCÍA-PRADA & S. RAMANAN (eds)
360	Zariski geometries, B. ZILBER
361	Words: notes on verbal width in groups, D. SEGAL
362	Differential tensor algebras and their module categories, R. BAUTISTA, L. SALMERÓN & R. ZUAZUA
363	Foundations of computational mathematics, Hong Kong 2008, F. CUCKER, A. PINKUS & M. J. TODD (eds)
364	Partial differential equations and fluid mechanics, J. C. ROBINSON & J. L. RODRIGO (eds)

London Mathematical Society Lecture Note Series. 365

Surveys in Combinatorics 2009

Edited by

SOPHIE HUCZYNSKA

JAMES D. MITCHELL

COLVA M. RONEY-DOUGAL

University of St Andrews

CAMBRIDGE UNIVERSITY PRESS
Cambridge, New York, Melbourne, Madrid, Cape Town, Singapore,
São Paulo, Delhi, Dubai, Tokyo

Cambridge University Press
The Edinburgh Building, Cambridge CB2 8RU, UK

Published in the United States of America by Cambridge University Press, New York

www.cambridge.org
Information on this title: www.cambridge.org/9780521741736

© Cambridge University Press 2009

This publication is in copyright. Subject to statutory exception
and to the provisions of relevant collective licensing agreements,
no reproduction of any part may take place without the written
permission of Cambridge University Press.

First published 2009

A catalogue record for this publication is available from the British Library

ISBN 978-0-521-74173-6 Paperback

Transferred to digital printing 2010

Cambridge University Press has no responsibility for the persistence or
accuracy of URLs for external or third-party internet websites referred to in
this publication, and does not guarantee that any content on such websites is,
or will remain, accurate or appropriate. Information regarding prices, travel
timetables and other factual information given in this work are correct at
the time of first printing but Cambridge University Press does not guarantee
the accuracy of such information thereafter.

Contents

	Preface	*page* vii
1	Graph decompositions and symmetry *Arrigo Bonisoli*	1
2	Combinatorics of optimal designs *R. A. Bailey and Peter J. Cameron*	19
3	Regularity and the spectra of graphs *W. H. Haemers*	75
4	Trades and t-designs *G. B. Khosrovshahi and B. Tayfeh-Rezaie*	91
5	Extremal graph packing problems: Ore-type versus Dirac-type *H. A. Kierstead, A. V. Kostochka and G. Yu*	113
6	Embedding large subgraphs into dense graphs *Daniela Kühn and Deryk Osthus*	137
7	Counting planar graphs and related families of graphs *Omer Giménez and Marc Noy*	169
8	Metrics for sparse graphs *B. Bollobás and O. Riordan*	211
9	Recent results on chromatic and flow roots of graphs and matroids *G. F. Royle*	289

Preface

The Twenty-Second British Combinatorial Conference was organised by the University of St Andrews. It was held in St Andrews in July 2009. The British Combinatorial Committee had invited nine distinguished combinatorialists to give survey lectures in areas of their expertise, and this volume contains the survey articles on which these lectures were based.

In compiling this volume we are indebted to the authors for preparing their articles so accurately and professionally, and to the referees for their rapid responses and keen eye for detail. We would also like to thank Roger Astley, Diana Gillooly and Peter Thompson at Cambridge University Press, and Max Neunhöffer at the University of St Andrews for their advice and assistance. Finally, without the previous efforts of editors of earlier *Surveys* and the guidance of the British Combinatorial Committee, the preparation of this volume would have been somewhat daunting.

We gratefully acknowledge the financial support provided by the London Mathematical Society, the Edinburgh Mathematical Society, the Glasgow Mathematical Journal Trust and the EPSRC.

Sophie Huczynska
James D. Mitchell
Colva M. Roney-Dougal

The University of St Andrews

January 2009

Graph decompositions and symmetry

Arrigo Bonisoli

To the memory of Adriano Barlotti and Lucia Gionfriddo

Abstract

In this paper I shall try to review some results which were obtained in the area of factorizations and decompositions of complete graphs admitting an automorphism group with some specified properties. These properties primarily involve the action of the group on the objects of the decomposition, most often vertices, but also edges, subgraphs of the decomposition or factors of the factorization.

Classification theorems were obtained in highly symmetric situations, for example when the group acts doubly transitively on vertices, and it is often the case that all examples arise from geometry in this context.

A "less" symmetric situation involves a group acting sharply transitively on vertices, which means for any two given vertices there exists precisely one group element mapping the first vertex to the second one. The vertices of the complete graph can be identified with group elements in this case, and the decomposition or factorization can be described entirely within the group by techniques which are generally known as "difference" or "starter-like" methods. Existence may be a non-trivial question and generally depends on the isomorphism type of the chosen group.

1 The general point of view: automorphism groups of combinatorial structures

I tend to believe that mathematicians always used criteria which have to do with symmetry. In our context, I heard many times that it was Felix Klein in his "Erlanger Programm" [35] who set the basic framework for studying the properties of a geometric structure which are invariant under a given group of transformations, and so people have been speaking of "Geometry from Klein's point of view". E. Artin [4, p. 54] felt that the study of automorphisms "has always yielded the most powerful results" and this citation is also quoted in the introduction of P. Dembowski's monograph [28], in which one of the seven chapters is devoted to collineations of finite planes. Speaking of graphs, I will only mention the fact that the "Handbook of Graph Theory" [32] has an entire section devoted to algebraic graph theory, within which a chapter [50] on automorphisms is included.

The automorphism group of a finite structure is a finite group by definition, since it has to be a subgroup of the group of all permutations of the finitely many objects involved in the structure. The work which eventually led to the classification of finite simple groups produced a wide spectrum of properties that can be used in studying the combinatorial structures of a given family from Klein's point of view. These properties may relate to the algebraic structure of the automorphism group under consideration. They may also have to do with the action of the given automorphism group on the objects of the combinatorial structure. Both aspects are very often present simultaneously and may be related in some meaningful manner.

Two questions which are typical in this approach can be stated in the following general form.

- Does there exist a combinatorial structure in the given family admitting an automorphism group with the prescribed properties?
- Within the given family of combinatorial structures, is it possible to classify those which admit an automorphism group with the prescribed properties?

In this paper I shall try to review some of the answers which were given to these two questions in the case where the combinatorial structure involved is a decomposition or a factorization of a graph.

Let T be a graph. A *decomposition* of T is a collection of pairwise edge-disjoint subgraphs $\Gamma_1, \Gamma_2, \ldots, \Gamma_s$ of T satisfying

$$E(T) = E(\Gamma_1) \cup E(\Gamma_2) \cup \cdots \cup E(\Gamma_s)$$

If each of the subgraphs $\Gamma_1, \Gamma_2, \ldots, \Gamma_s$ is isomorphic to a given graph Γ then we speak of a *Γ-decomposition* of T. A Γ-decomposition of the complete graph K_v is also said to be a *Γ-design*. The reason for this lies in the circumstance that if Γ is a complete graph on k vertices then a Γ-decomposition of K_v is precisely a 2-$(v, k, 1)$ design. In a Γ-design we therefore think of the blocks as having the "shape" of the graph Γ.

If r is a positive integer, an r-regular spanning subgraph of T is said to be an *r-factor* of T. A collection of pairwise edge-disjoint r-factors $\Gamma_1, \Gamma_2, \ldots, \Gamma_s$ of T satisfying

$$E(T) = E(\Gamma_1) \cup E(\Gamma_2) \cup \cdots \cup E(\Gamma_s)$$

is said to form an *r-factorization* of T. For $r = 1$ an r-factor is a perfect matching, consequently all subgraphs $\Gamma_1, \Gamma_2, \ldots, \Gamma_s$ will be isomorphic. For $r > 1$ we cannot expect this to be the case in general (2-factorizations with non-isomorphic 2-factors can easily be exhibited). An r-factorization can thus be regarded as a Γ-decomposition only in the special case in which all of its r-factors are isomorphic to a given r-factor Γ.

Let $\mathcal{F}$ be a Γ-decomposition of the graph T. An automorphism of $\mathcal{F}$ is by definition an automorphism of the graph T which leaves $\mathcal{F}$ globally invariant. The set of all automorphisms of the Γ-decomposition $\mathcal{F}$ is the *full automorphism group* of $\mathcal{F}$, sometimes denoted by $\mathrm{Aut}(\mathcal{F})$. Any subgroup of $\mathrm{Aut}(\mathcal{F})$ will be referred to as *an* automorphism group of $\mathcal{F}$. If G is an automorphism group of $\mathcal{F}$, then G acts on three different sets: the vertex-set $V(T)$, the edge-set $E(T)$ and the set $\mathcal{F}$ itself, that is the subgraph-set $\{\Gamma_1, \Gamma_2, \ldots, \Gamma_s\}$. Each such action partitions the corresponding underlying set into G-orbits and we shall speak of vertex-orbits, edge-orbits and subgraph-orbits.

An assumption on one or more of these actions yields an instance for which it makes sense to consider one or both of the previous questions. In what follows I shall be primarily concerned with assumptions involving the action on vertices. As a rule of thumb, if this action is highly symmetric, say, multiply transitive, then there are few examples and there is some hope for classification. Otherwise the focus may be on construction methods once the nature of the group G is further specified, for example by assigning its isomorphism type.

2 Highly symmetric factorizations

Let $\mathcal{F}$ be a doubly transitive 1-factorization of the complete graph K_v, v even. That means $\mathcal{F}$ admits an an automorphism group G acting doubly transitively on the set of vertices $V(K_v)$, in other words transitively on the ordered pairs of distinct vertices. Can we say anything about $\mathcal{F}$ and G?

P. J. Cameron started working in this direction in [25], [26]. The complete classification was eventually achieved by P. J. Cameron and G. Korchmáros in [27, Theorem 3]. The outcome is a prototype for results of this kind: there is an infinite family of examples, forming somehow the "classical" cases; in addition there are finitely many "sporadic" examples.

Theorem 2.1 [27, Theorem 3] *If $\mathcal{F}$ is a 1-factorization of K_v, v even, admitting an automorphism group G acting doubly transitively on the set of vertices $V(K_v)$, then v is a power of 2 or one of the values 6, 12 or 28. Furthermore, when v is a power of 2, the 1-factorization $\mathcal{F}$ can always be described as arising from the affine line-parallelism of an affine space over the field of two elements, and the group G is a subgroup of the group of affine transformations.*

The corresponding classification of doubly transitive 2-factorizations of complete graphs was achieved more recently, see [46], [7]. The situation differs from that of 1-factorizations in the sense that there are no "exceptional cases". I had learned about this problem from a lecture of Alex Rosa [44] at the International Symposium on Graphs, Designs and Applications, held in Messina (Italy) from 30 September to 4 October 2003. In particular there was then no classification for doubly transitive hamiltonian decompositions of complete graphs, which are the 2-factorizations in which every 2-factor is a hamiltonian cycle.

One family of such hamiltonian decompositions was known: if p is an odd prime and $V(K_p) = \mathbf{Z}_p$ we can define $C^{(i)} = \bigl(0, i, 2i, \ldots, (p-1)i\bigr)$ for $i = 1, 2, \ldots, (p-1)/2$ and obtain the hamiltonian decomposition $\mathcal{H} = \{C^{(1)}, C^{(2)}, \ldots, C^{((p-1)/2)}\}$ of K_p. This hamiltonian decomposition is invariant under the group $AGL(1,p)$ in its natural doubly transitive permutation representation.

A further family of doubly transitive 2-factorizations was known [44, Section 3]: if we identify $V(K_{3^r})$ with the point-set of the affine space $AG(r,3)$, then each class of parallel lines yields a 2-factor which is a union of 3-cycles, the vertices in each 3-cycle being the three points on some line of the class. The 2-factors arising from all parallel classes actually form a 2-factorization of K_{3^r}, as any two points lie on a unique line of the affine space. The group $AGL(r,3)$ of affine transformations acts doubly transitively on points and clearly leaves this 2-factorization invariant.

The former examples are the unique doubly transitive hamiltonian decompositions of complete graphs, while the latter examples can be generalized to an arbitrary odd prime p, thus yielding doubly transitive 2-factorizations of K_{p^r}, r a positive integer. I presented these results at the conference "Giornate di Geometria" held at the Università "La Sapienza" in Rome, 4–6 December 2003.

These two families of examples yield all doubly transitive 2-factorizations, up to isomorphism. The way we achieved this classification in [7] was by proving that the 2-factorization arises from a 2-$(v, k, 1)$ design and that the group acting on the 2-factorization induces an automorphism group of the design.

The basic steps in the proof can be outlined as follows.

- The 2-factors are pairwise isomorphic, each of them consists of cycles of equal length k, a divisor of v.

- The k-cycles occurring in the 2-factors yield a C_k-design which is actually a Steiner k-cycle system of K_v, that means for any two vertices x, y and any positive integer $d \leq (k-1)/2$ there exists precisely one such k-cycle in which x and y occur at distance d.

- If two of the above k-cycles share at least two vertices, then they share all vertices.

- The integer k is a prime.

Doubly transitive 2-$(v, k, 1)$ designs are classified by the work of W. M. Kantor [34]: using the previous properties we could establish that affine spaces over prime fields are the only surviving items in his list in the situation under consideration and the 2-factorization can be reconstructed.

This result was announced by Marco Buratti at the "International Conference on Incidence Geometry" held in La Roche-en-Ardennes (Belgium) 23-29 May 2004.

Meanwhile the paper by T. Q. Sibley [46] had appeared, with a classification of edge colorings of the complete graph admitting an automorphism group acting doubly transitively on vertices: the case of 2-factorizations falls within that description if it is assumed that the color classes are 2-factors. We decided to submit our paper anyway because we felt our proof was somehow more "design theoretic". It should be remarked that Sibley's methods as well as ours require the classification of doubly transitive finite permutation groups, hence ultimately rest on the classification of finite simple groups, a circumstance which is certainly not unexpected.

Theorem 2.2 [46, Theorem 15], [7, Section 7] *Each doubly transitive 2-factorization arises from the affine line parallelism of $AG(r,p)$, where p is an odd prime.*

It is probably worth mentioning how 2-factors arise from the classes of parallel lines. Each such class yields $(p-1)/2$ such 2-factors. A class of parallel lines in the affine space $AG(r,p)$ is uniquely determined by a non-zero vector $\mathbf{w}$. Each line in this class can be written in parametric form as $P = P_0 + \alpha \mathbf{w}$ where P_0 is an arbitrary point on the line, $\alpha \in GF(p)$.

For $\alpha = 1, 2, \ldots, (p-1)/2$ form a p-cycle Γ_α on the given line as

$$(P_0, \ P_0 + \alpha\mathbf{w}, \ P_0 + 2\alpha\mathbf{w}, \ \ldots, \ P_0 + (p-1)\alpha\mathbf{w}).$$

A different choice of the initial point P_0 on the given line only alters the representation of Γ_α by a cyclic reordering of its points. Choosing the opposite vector $-\mathbf{w}$ rather than $\mathbf{w}$ amounts to walking along Γ_α in the reverse order. A different choice of the non-zero vector $\mathbf{w}$ in the 1-dimensional vector subspace spanned by $\mathbf{w}$ still yields the same cycles $\Gamma_1, \Gamma_2, \ldots, \Gamma_{(p-1)/2}$, possibly in a different order.

For a fixed choice of the non-zero vector $\mathbf{w}$ for all the lines of the given parallel class, form a 2-factor F_α where α is a fixed integer between 1 and $(p-1)/2$: this

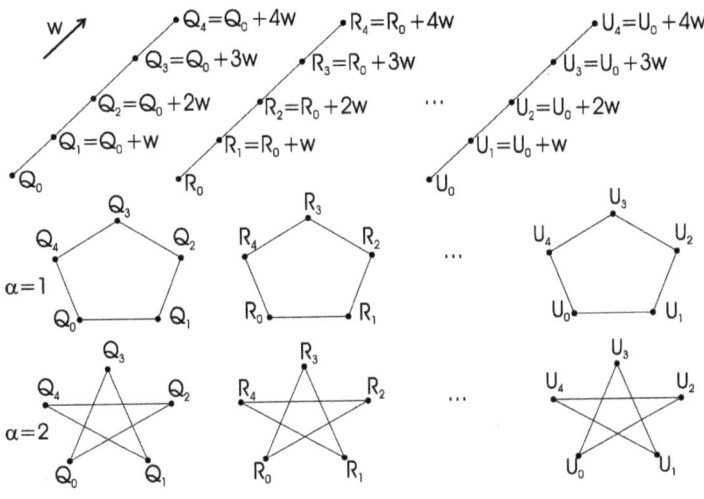

Figure 1: The construction of 2-factors for $p = 5$

2-factor is the union of the p-cycles Γ_α constructed above, one from each line of the parallel class. The procedure just described is illustrated in Figure 1 for $p = 5$.

What if we replace the assumption that the group action is doubly transitive on vertices by the weaker assumption of a primitive action?

In the case of 1-factorizations, A. Bonisoli and S. Bonvicini proved in [9] that there is a wealth of examples arising from simple geometric considerations. Our construction starts from the affine line parallelism of $AG(r, 2)$. The 1-factors in this 1-factorization are the classes of parallel lines. Since we are dealing with the binary field, lines are sets of cardinality 2. The idea is to consider two complementary hyperplanes W and $\overline{W}$ in $AG(r, 2)$. We take two distinct classes of parallel lines in $AG(r, 2)$, both of which are parallel to W. We alter the two original 1-factors in $AG(r, 2)$, by pasting the lines of one direction in W together with the lines of the other direction in $\overline{W}$ and similarly by taking the lines of the second direction in W together with the lines of the first direction in $\overline{W}$. We replace the original 1-factors with the new ones.

The replacement is repeated a number of times so as to obtain a 1-factorization consisting of some "standard" 1-factors arising from whole parallel classes of lines and of some pairs of 1-factors of "mixed" type. If the replacement is done in a suitable manner, the modified 1-factorization still retains a group G of affine transformations acting primitively on the points of the underlying geometry and hence on the vertices of the complete graph. Such a 1-factorization cannot be doubly transitive, since by the result of Cameron and Korchmáros it should coincide with the affine line parallelism, which has clearly been destroyed.

The group G can be described as follows. The r-dimensional vector space V over $GF(2)$ is identified with $GF(2^r)$. Let G consist of all mappings $g : V \to V$ of the form $g(\mathbf{x}) = b \cdot \mathbf{x} + \mathbf{c}$, as the vector $\mathbf{c}$ varies in V and the element b varies in a subgroup B of the multiplicative group $GF(2^r)^*$. If $|B|$ is a 2-primitive divisor of $2^r - 1$, that is a divisor of $2^r - 1$ not dividing $2^m - 1$ for $m < r$, then G acts on V as a primitive permutation group. For $r \neq 6$ the existence of a 2-primitive divisor is assured by Zsigmondy's Lemma, see for instance [38, Theorem 6.2].

Theorem 2.3 [9, Proposition 9] *There exist primitive 1-factorizations of K_{2^r} which are not doubly transitive.*

The idea of "mixed" factors can be modified to yield 2-factorizations of K_{p^r} where p is an odd prime. G. Mazzuoccolo proved the following result.

Theorem 2.4 [39, Proposition 1] *If $p^r - 1$ admits a p-primitive divisor d with $d \leq \lfloor (p^r - 1)/p(p-1) \rfloor$, then there exist primitive 2-factorizations of K_{p^r} which are not doubly transitive.*

On the other hand he proved in Lemma 1 of [39] that if $\mathcal{F}$ is a primitive hamiltonian 2-factorization of K_{p^r}, then $r = 1$ and $\mathcal{F}$ is the unique doubly transitive hamiltonian 2-factorization of K_p described after Theorem 2.1.

Exhaustive analysis of the primitive permutation groups on a specified number v of elements may establish the non-existence of primitive 1-factorizations or 2-factorizations of K_v, but I am aware of no further general result in this direction.

3 Decompositions and factorizations with an automorphism group acting sharply transitively on vertices

Let G be a finite group and let v be an even positive integer. Does there exist a 1-factorization of the complete graph K_v admitting G as an automorphism group acting sharply transitively on the vertex-set $V(K_v)$? *Sharply transitive* means that given any two (not necessarily distinct) vertices x, y there exists exactly one automorphism g in the given group G mapping x to y.

A slightly more general version of this question appears as Problem 4 in the list of [48], namely the word "sharply" does not appear there. The two versions are equivalent for abelian groups since every transitive abelian permutation group is sharply transitive. In the theory of finite permutation groups the term "regular" is generally used instead of "sharply transitive", see [31]: since the word "regular" is highly reserved in graph theory I prefer to avoid its use in this context.

A sharply transitive permutation group is one which is transitive and has trivial one-point-stabilizers. The order of such a group is therefore equal to the cardinality of the unique orbit on the underlying set. In particular if the action of the group is assumed to be sharply transitive on the vertices of the underlying graph, we have that the order of the group is equal to the order of the graph. That is the first necessary condition coming into play.

The question can be formulated for arbitrary decompositions. Let $\mathcal{F}$ be a Γ-decomposition of the complete graph K_v. Does there exist an automorphism group G of $\mathcal{F}$ acting sharply transitively on the vertex-set $V(K_v)$? What if the isomorphism type of G is also specified, say cyclic, elementary abelian, dihedral and the like?

In his paper [47] of 1938 J. Singer observed that a finite projective space $PG(r, q)$ admits a cyclic automorphism group acting sharply transitively on points, a so called

"Singer cycle". The desarguesian projective plane $PG(2,q)$ is thus an example of a "cyclic" plane and, more generally, of a "cyclic" block design. It is thus probably within this context that we can trace the habit of labelling the Γ-decomposition under consideration with the adjective identifying the isomorphism type of the group. A cyclic 1-factorization of K_v is thus one admitting a cyclic automorphism group acting sharply transitively on $V(K_v)$, and we can analogously define, more generally, a cyclic Γ-decomposition of K_v.

3.1 Existence vs. non-existence: a restricted spectrum problem

For a given graph Γ, the set of all positive integers v for which a Γ-decomposition of K_v exists is generally called the "spectrum" for Γ. Solving the spectrum problem for Γ, that is determining the spectrum for Γ, is generally the first issue which is approached for any specific graph Γ which is of interest.

This process typically involves two steps:

i) find a reasonable set of necessary conditions;

ii) figure out that the necessary conditions in this set are also sufficient.

The first obvious necessary condition under i) states that the size (number of edges) of the graph Γ must be a divisor of the size of K_v:

$$|E(\Gamma)| \text{ divides } |E(K_v)| = v(v-1)/2.$$

If Γ is a regular graph of degree r, then the second obvious necessary condition under i) states that the degree of a vertex in Γ must be a divisor of the degree of the same vertex in K_v:

$$r \mid (v-1).$$

Step ii) is usually tackled by direct constructions, generally involving some sort of recursive procedure and the "ad hoc" treatment of some low values of v.

As an explicit example I will mention just one possible choice of Γ: the Petersen graph $P(5,2)$. The spectrum problem for the Petersen graph was solved by P. Adams and D. E. Bryant in [1], yielding

$$v \equiv 1 \text{ or } 10 \mod 15, \quad v > 10.$$

The spectrum problem was solved in [6] for graphs with at most four vertices and in [5] for graphs with five vertices. A detailed illustration of the spectrum problem for m-cycles is contained in [16, Section 2].

Clearly, for each fixed choice of the graph Γ and of the sharply transitive group G, the values of v for which a G-invariant Γ-decomposition of K_v exists must be in the spectrum for Γ.

In other words, for each choice of the isomorphism type of the group G (for example, cyclic, dihedral, elementary abelian) we can define the *restricted* spectrum for Γ as the set of all positive integers v for which a G-invariant Γ-decomposition of K_v exists.

It may well happen that a specific value of v is in the spectrum for Γ but is not in the restricted spectrum for Γ with respect to the current choice of G, in other words a Γ-decomposition of K_v exists, but there is no Γ-decomposition of K_v admitting G as a sharply transitive automorphism group. That is trivially the case if no finite group of order v of the required isomorphism type exists. For example, although $v = 40$ is in the spectrum for the Petersen graph $P(5,2)$, it does not even make sense to look for an elementary abelian $P(5,2)$-decomposition of K_{40} because 40 is not the order of an elementary abelian group, which has to be a prime power.

It may also happen that v does lie in the spectrum for the graph Γ, a finite group G of order v of the desired isomorphism type does exist and yet a Γ-decomposition of K_v admitting G as a sharply transitive automorphism group does not exist.

Probably the most famous example for this situation is the non-existence theorem [33] of A. Hartman and A. Rosa for cyclic 1-factorizations of K_v when v is a power of 2, $v \geq 8$. The proof is achieved by direct methods, but it does require a fairly tricky counting argument. It is thus no wonder that there are not very many such non-existence statements around.

Speaking of 1-factorizations, the papers [36] and [37] prove non-existence in some cases under the additional assumption that the group G fixes at least a 1-factor. It is proved in [11] that if $p > 3$ is a prime, then there cannot exist a 1-factorization of K_{4p} admitting a sharply transitive automorphism group fixing $2p+1$ 1-factors, the largest number of fixed 1-factors which is allowed for by general bounds.

3.1.1 k-matching decompositions
A matching consisting of k edges will be called a k-*matching*. A k-matching decomposition of K_v is a Γ-decomposition of K_v in which Γ is assumed to be a k-matching for some fixed k. The necessary condition under i) above for the existence of a k-matching decomposition of K_v is that k must divide $v(v-1)/2$. A k-matching decomposition of K_v is called a $(0,1)$-*factorization* of K_v by R. Rees in [42]. If v is even and $k = v/2$ then the k-matching is a perfect matching and the k-matching decomposition is a 1-factorization. Apart from that, we have $k < v/2$ and so each k-matching of the decomposition misses some vertex. For example, if v is odd and $k = (v-1)/2$, we are speaking of a near-1-factorization of K_v. The result of Hartman and Rosa motivated R. Rees in studying cyclic k-matching decompositions of K_v: the situation differs substantially from the case of cyclic 1-factorizations, since for each v and for each admissible value of k strictly less than $v/2$ the complete graph K_v admits a cyclic k-matching decomposition.

In the paper [10] A. Bonisoli and S. Bonvicini studied k-matching decompositions of K_v admitting a sharply transitive automorphism group which is either non-cyclic abelian or dihedral. As I will mention in Subsection 3.2 these objects exist when $k = v/2$, that is when the decompositions in question are 1-factorizations. It was much to our surprise to discover that we could prove non-existence for $k < v/2$ under some numerical constraints. As an example I mention the following result.

Proposition 3.1 [10] *Assume $v = 2^m$ with $m > 1$. Let $k = 2^r \cdot d$, with $0 \leq r < m-2$ and $d \equiv 1 \pmod 2$. If $d \geq 2^{r+1}$ then there cannot exist an elementary abelian k-matching decomposition of K_v.*

For instance, in the case $v = 1024$ Proposition 3.1 excludes the admissible values $k = 3, 11, 22, 31, 33, 44, 62, 66, 93, 124, 132, 186, 248, 264, 341, 372$.

We can summarize the phenomenon as follows. When v is a power of 2 and $k = v/2$ the required k-matching decomposition does not exist when G is cyclic [33], while it does exist when G is either non-cyclic abelian [19] or dihedral [11]. Moving from $v/2$ to smaller admissible values of k may yield precisely the opposite behavior: we have existence if G is cyclic [42] and, in certain cases, non-existence if G is either non-cyclic abelian or dihedral [10].

3.2 Constructions: starter like methods

The notion of a *starter* in connection with 1-factorizations was developed first for groups of odd order [49, Chapter 10], [30]. That has to do with the 1-rotational situation, see Section 4. If the group action is assumed to be sharply transitive on the whole set of vertices, then one has to deal with groups of even order. The notion of a starter in a group of even order which is appropriate to this situation was developed by M. Buratti in [19] in connection with abelian groups.

Let G be a finite group of even order v, which we denote additively even though it may well be non-abelian. We identify the vertices of the complete graph K_v with the group elements of G and we shall occasionaly write K_G rather than K_v. We identify each group element $g \in G$ with the permutation $V(K_v) \to V(K_v)$, $x \mapsto x+g$. In this manner G is represented as a sharply transitive permutation group on $V(K_v)$. This representation is classically known as the right regular permutation representation of G, see [31, Section 1.3], where the word "right" is to remind us that we are adding the group element g to the right, which is obviously relevant only for non-abelian groups.

This action of G on $V(K_v)$ induces actions on the subsets of $V(K_v)$ and on sets of such subsets. Hence if $g \in G$ is an arbitrary group-element and S is any subset of $V(K_v)$ then we write $S + g = \{x + g : x \in S\}$. In particular, if $S = [x, y]$ is an edge, then $[x, y] + g = [x + g, y + g]$. Furthermore, if B is a collection of subsets of $V(K_v)$, then we write $B + g = \{S + g : S \in B\}$. In particular, if B is a collection of edges of K_v then $B + g = \{[x + g, y + g] : [x, y] \in B\}$.

The G-orbit of an edge $[x, y]$ has either length v or $v/2$ and we speak of a *long* orbit or a *short* orbit, respectively, and we call $[x, y]$ a *long* edge or a *short* edge, respectively. The latter case occurs if and only if the edge $[x, y]$ is fixed by an involution in the group.

If U is a subgroup of G then a system of distinct representatives for the left cosets of U in G will be called a *left transversal* for U in G. If $[x, y]$ is an edge in K_G we define

$$\partial([x,y]) = \begin{cases} \{x-y, y-x\} & \text{if } [x,y] \text{ is long} \\ \{x-y\} & \text{if } [x,y] \text{ is short} \end{cases}$$

$$\phi([x,y]) = \begin{cases} \{x,y\} & \text{if } [x,y] \text{ is long} \\ \{x\} & \text{if } [x,y] \text{ is short} \end{cases}$$

If S is a set of edges of K_G we define

$$\partial(S) = \bigcup_{e \in S} \partial(e) \qquad \phi(S) = \bigcup_{e \in S} \phi(e)$$

where in either case the union may contain repeated elements and so, in general, will return a multiset.

Definition 3.2 [19, Definition 2.1] A *starter* in a group G of even order is a set $\Sigma = \{S_1, \ldots, S_k\}$ of subsets of $E(K_G)$ satisfying the following conditions.

- $\partial(S_1) \cup \partial(S_2) \cup \cdots \cup \partial(S_k) = G \setminus \{0\}$.

- for $i = 1, 2, \ldots, k$ the set $\phi(S_i)$ is left transversal for a suitable subgroup H_i of G containing all the involutions fixing the short edges in S_i.

The existence of a starter in a finite group G of even order v is equivalent to the existence of a 1-factorization of the complete graph K_v admitting G as an automorphism group acting sharply transitively on vertices [19]. For each index i we form a 1-factor as $\cup_{e \in S_i} Orb_{H_i}(e)$, whose stabilizer in G is the subgroup H_i; the G-orbit of this 1-factor, which has length $|G : H_i|$ is then included in the 1-factorization.

According to the definition of a starter, the existence of a 1-factorization of K_v admitting a given group G as an automorphism group acting sharply transitively on vertices can thus be completely tested within G. It should be remarked, however, that the notion of a starter, although entirely formulated within the group G, cannot avoid referring to the action of G on edges. That is why some constructions are better understood through a direct description of how edge-orbits contribute to the formation of 1-factors.

As a matter of fact, it was precisely in this manner that A. Hartman and A. Rosa in [33] gave direct constructions for cyclic 1-factorizations of K_v in all cases in which v is an even positive integer other than a power of 2.

If G is a finite abelian group of even order v, other than a cyclic group of 2-power order, then M. Buratti showed in [19] that G always admits a starter. A. Bonisoli and D. Labbate proved in [11] that the same is true if G is a dihedral group, and it is also true if G is an arbitrary group of even order v with $v/2$ odd, regardless of the isomorphism type of G.

The existence of a starter is also guaranteed in each non-cyclic group of 2-power order admitting a cyclic subgroup of index 2, in particular in a generalized quaternion group. This was established in [12] by A. Bonisoli and G. Rinaldi, who also proved the existence of a starter in a dicyclic group of order v when $v/2$ is an even integer greater than 2.

G. Rinaldi showed in [43] that a further class of groups admitting a starter is that of nilpotent groups in which the Sylow 2-subgroup is either abelian or admits a cyclic subgroup of index 2.

The whole of combinatorics is full of "doubling" constructions, and so it is no surprise that methods of this nature were devised for starters as well. S. Bonvicini proved a sufficient condition for a group G to have a starter, once a subgroup H of index 2 is known to have one.

Proposition 3.3 [13, Proposition 3.1] *Let H be a group of even order admitting a starter $\Sigma_H = \{S_1, S_2, \ldots, S_r\}$ and assume that H_1, H_2, ..., H_r are the subgroups of H such that $\phi(S_i)$ is a left transversal of H_i in H for $i = 1, 2, \ldots, r$. Let G be a group admitting H as a subgroup of index 2 and assume that there exists $b \in G - H$ such that $b + H_i + b = H_i$ holds for every $i = 1, 2, \ldots, r$. Then there exists a starter Σ_G in G containing Σ_H.*

The original starter in H is thus canonically contained in the larger starter in G, whence justifying the "doubling" terminology. For example, an application of the method shows that if H is an abelian group of even order admitting a starter, then so does the so called *generalized dihedral* group $Dih(H)$, see [45].

A sufficient condition for a group of 2-power order to admit a starter was given by S. Bonvicini in [14]: if the Frattini subgroup, which is the intersection of all maximal subgroups, is elementary abelian then the starter exists. The condition may seem exotic in the sense that the Frattini subgroup is an invariant which usually occurs only in group theoretical considerations.

Question 3.4 *Are cyclic groups of 2-power order the only groups of even order not admitting a starter? Are cyclic groups of 2-power order the only groups of 2-power order not admitting a starter?*

Despite the results illustrated above and some exhaustive investigation for specific orders (S. Bonvicini considered $v = 64$ and $v = 128$), the questions remain open.

Sharply transitive 2-factorizations of K_v can also be approached through the appropriate notion of a starter. In fact M. Buratti and G. Rinaldi showed in [22, Theorem 2.4] that a finite group G of odd order v can be realized as an automorphism group of a 2-factorization of K_v acting sharply transitively on vertices if and only if G admits a *2-starter*. It was then a simple matter for them to show that a group of odd order always admits a 2-starter [22, Theorem 2.6]. The existence problem is thus settled in this situation and the attention shifts towards different questions (such as requesting that the cycles within the 2-factors have prescribed lengths).

3.3 Elementary abelian Petersen decompositions

Denote by $P(n,k)$ the *generalized Petersen* graph. That is the graph with vertex-set $V(P(n,k)) = \{x_0, x_1, \ldots, x_{n-1}, y_0, y_1, \ldots, y_{n-1}\}$ and edge-set $E(P(n,k)) = \bigcup_{i=0}^{n-1} \{[x_i, x_{i+1}], [y_i, y_{i+k}], [x_i, y_i]\}$, where subscripts are meant modulo n. In order to avoid trivial cases it is generally assumed $n \geq 3$ and $1 \leq k \leq n-2$. In particular, $P(5,2)$ is the *standard Petersen graph* and its spectrum is known as I mentioned in Subsection 3.1. The following result partially solves the restricted spectrum problem for generalized Petersen graphs when the group is assumed to be elementary abelian.

Proposition 3.5 [8, Theorem 3.2] *Let p be an odd prime and assume $p^r \equiv 1$ mod 6. If $2 \leq k \leq p-2$ then there exists an elementary abelian $P(p,k)$-decomposition of K_{p^r}.*

Our motivation for choosing the elementary abelian group was the circumstance that the standard Petersen graph is known to "live" inside certain finite desarguesian planes. From an affine point of view such a plane arises from a finite vector space, whose additive group is elementary abelian. It therefore seemed reasonable to be able to find a decomposition into Petersen graphs inside such a group.

4 Different assumptions on the group action

4.1 1-rotational factorizations

A 1-factorization of K_v, v even, is said to be 1-*rotational* if it admits a cyclic automorphism group fixing one vertex and acting sharply transitively on the remaining ones [3]. Many constructions for factorizations of complete graphs are 1-rotational in nature. The idea comes from a drawing of the vertices of the complete graph as the vertices of a regular euclidean polygon with one special vertex being the center of the polygon. The basic step consists in finding one of the factors through some more or less geometric pattern; once that is achieved successfully, the remaining factors are obtained by rotating the "initial" factor through the center of the polygon. The textbook proof that a complete graph K_v of even order is of class I with respect to edge colorings amounts to showing that the 1-factorization GK_v is 1-rotational. Another famous example is Walecki's construction for hamiltonian cycle decompositions of K_v, v odd [2].

Although the connection with plane rotations may well become meaningless, the set of assumptions for the 1-rotational situation makes perfect sense even if we change the isomorphism type of the involved automorphism group. We can thus define a Γ-decomposition of K_v to be 1-rotational of a given isomorphism type if it admits an automorphism group of the specified type fixing one vertex and acting sharply transitively on the remaining ones. Again we may want to identify the isomorphism type of the group in question by placing the corresponding adjective before the term 1-rotational.

As a matter of fact it was precisely the study of 1-rotational 1-factorizations of K_v for an arbitrary group G of order $v-1$, not just a cyclic one, that led to the notion of a starter in a group of odd order, [49, Chapter 10], [30]: every group G of odd order admits a starter and can therefore be realized as a 1-rotational group for some 1-factorization.

M. Buratti and G. Rinaldi in [23] generalized this notion and introduced the definition of a k-*starter* in a finite group G of order divisible by $k > 1$. The complete graph is realized on the vertex set $G \cup \{\infty\}$, where $\infty \notin G$, and it is agreed that each group element in G fixes ∞. A k-factor F of the complete graph is said to be a k-starter if its stabilizer under the action of G has order k and the multiset ΔF, consisting of all differences $x - y$ and $y - x$ as the edge $[x, y]$ runs over the edges of F not through ∞, covers all non-trivial elements of G.

It turns out that a k-factorization of a complete graph which is 1-rotational for the group G exists if and only if G admits a k-starter [23, Theorem 2.3]. Furthermore, the group G acts transitively on the k-factors in such a k-factorization [23, Theorem 2.1] and so, in particular, these k-factors are pairwise isomorphic. That is not surprising, but it is a circumstance that remains unnoticed when dealing with 1-factorizations, as any two 1-factors of a given graph are isomorphic.

It should be noticed that there are currently two different definitions of a 2-starter in the context of 2-factorizations. For a group G of odd order, it amounts to a 2-factorization admitting G as an automorphism group acting sharply transitively on vertices [22]. For a group G of even order, it amounts to a 2-factorization admitting G as a 1-rotational group [23]. There is fortunately no great danger of misunderstanding.

… The existence of a 1-rotational 2-factorization of a complete graph for a given group G of even order does force some structural properties. For example, the involutions in G must form a single conjugacy class [23, Theorem 3.5]. If G is a dihedral group that implies $|G| \equiv 2 \pmod{4}$. Consequently, since the order of G is $v-1$ in the 1-rotational case, we have $v \equiv 3 \pmod 4$ as a necessary condition for the existence of a dihedral 2-factorization of K_v. On the other hand, it can be shown that for each such value of v the dihedral group of order $v-1$ admits a 2-starter. As a matter of fact [23, Theorem 5.1] completely characterizes the 2-starters of the dihedral group. Again it might be easier to visualize the corresponding 2-factorization through the usual polygon representation of the dihedral group, as I do in the next statement.

Proposition 4.1 *For $v \equiv 3 \pmod{4}$ there exists a dihedral 1-rotational 2-factorization of K_v.*

Proof Set $V(K_v) = \mathbf{Z}_{v-1} \cup \{\infty\}$. Let G be the permutation group on $V(K_v)$ generated by the permutations

$$\alpha : x \mapsto x+2 \qquad \beta : x \mapsto -x+1$$

both of which fix ∞. Define the *starter* 2-factor as the union of the following cycles (see Figure 2 for $v = 15$):

$$(\infty, 0, 1)$$
$$(\tfrac{v+1}{4} + \tfrac{v-1}{2}, \tfrac{v-1}{2}, \tfrac{v+1}{2}, \tfrac{v+1}{4})$$
$$(2, \tfrac{v-1}{2} - 1, \tfrac{v+1}{2} + 1, v-2)$$
$$(3, \tfrac{v-1}{2} - 2, \tfrac{v+1}{2} + 2, v-3)$$
$$\ldots$$

The remaining 2-factors are obtained as the images of the starter under the powers of α, that is under rotations through even multiples of $2\pi/(v-1)$. It is clear that this set of 2-factors forms a 2-factorization with the required properties. □

The notion of a 1-rotational decomposition can be generalized even further by considering decompositions admitting an automorphism group fixing any given number of vertices, say k, and acting sharply transitively on the remaining ones. G. Mazzuoccolo and G. Rinaldi in [41] considered the case of 1-factorizations with this property and called them *k-pyramidal 1-factorizations*. If k is assumed to be at least 2, they prove that k, as well as the order of the group, must be even. Furthermore, G must contain precisely $k-1$ involutions. They also present several constructions, in particular they show that an abelian group with $k-1$ involutions can always be realized as a k-rotational group for an appropriate 1-factorization of some complete graph.

4.2 2-factorizations which are doubly transitive on factors

There is a unique 2-factorization of K_5 up to isomorphism, and it is hamiltonian. Its automorphism group acts doubly transitively on vertices, whence transitively on

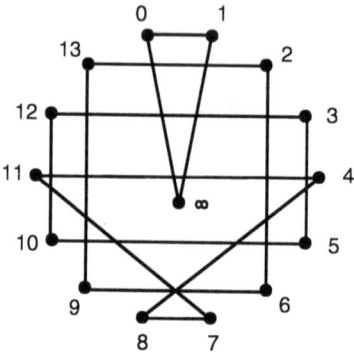

Figure 2: A 2-factor from a dihedral 2-factorization of K_{15}

edges. Consequently the automorphism group acts transitively on factors and so, since there are only two factors in this case, it also acts doubly transitively on factors.

Are there any other 2-factorizations of K_v admitting an automorphism group acting doubly transitively on factors? G. Mazzuoccolo considered this problem in [40] and found that the answer is negative if the 2-factorization is assumed to be hamiltonian. On the other hand he produced infinite families of examples in the non-hamiltonian case. Note that the action on 2-factors may well be non-faithful in this situation.

5 Concluding remarks

Are doubly transitive Γ-designs classified? The answer is affirmative if Γ is a complete graph K_k: W. M. Kantor classified doubly transitive 2-$(v, k, 1)$ designs in [34]. Let us formulate the question in the language of graph decompositions.

Let $\mathcal{F}$ be a Γ-decomposition of K_v admitting an automorphism group G acting doubly transitively on the vertex-set $V(K_v)$. What can we say in general about Γ and G?

Let Γ' be one of the subgraphs in $\mathcal{F}$ and let H be the stabilizer of Γ' in G. Let $[x, y]$ and $[x', y']$ be arbitrary edges in Γ'. By the assumption of double transitivity there exists g in G mapping x to x' and y to y', respectively. Since g is an auomorphism of $\mathcal{F}$, we have that g maps the unique subgraph of the decomposition containing $[x, y]$ to the unique such subgraph containing $[x', y']$. This subgraph is Γ' in either case and consequently g fixes Γ', that is $g \in H$. We can clearly repeat the argument and find $f \in H$ mapping x to y' and y to x'. Consequently not only is H transitive on the edges of Γ', but it is even *arc-transitive*, see [50, D18 p. 491].

A necessary condition for the existence of a doubly transitive Γ-decomposition of K_v is that Γ be an arc-transitive graph. Are there possible choices for Γ other than complete graphs? For example, are doubly transitive k-cycle decompositions classified? Are there examples when Γ is one of the graphs $K_{3,3}$, Q_8 or $P(5,2)$? Even if not, it would still be interesting to figure out how much symmetry we can impose on such decompositions and still have examples (and, possibly, a classification in each case).

The latter question makes sense even in the 1-rotational case: can we assume an action on the non-fixed vertices that is "more" symmetric than just sharply transitive? An inspection of some known examples might give some indication in this respect.

The number of isomorphism classes of groups of a given order is unbounded as the order increases. Settling existence for the decomposition under consideration is thus a never-ending task in the sharply transitive, 1-rotational or k-pyramidal case. The choice of the isomorphism types of groups to be investigated may thus be viewed primarily as a matter of taste. On the other hand it is a fact that a 1-factorization of a complete graph has a trivial automorphism group with probability approaching 1 as the number of vertices tends to infinity. Even a little "quantity" of symmetry can therefore make the factorization or decomposition somehow remarkable.

Acknowledgements

The author is a member of G.N.S.A.G.A. of the Italian I.N.d.A.M. and was supported financially by the Italian Ministry M.I.U.R. through the PRIN 2005 project *Strutture geometriche, combinatoria e loro applicazioni*. He is indebted to the referee for a careful reading of the manuscript and many useful suggestions.

References

[1] P. Adams & D. E. Bryant, The spectrum problem for the Petersen graph, *J. Graph Theory* **22** (1996), 175–180.

[2] B. Alspach, The wonderful Walecki construction, *Bull. Inst. Combin. Appl.* **52** (2008), 7–20.

[3] L. D. Andersen, Factorizations of graphs, in *Handbook of Combinatorial Designs* (eds. C. J. Colbourn & J. H. Dinitz), Second edition, Chapman & Hall/CRC, Boca Raton (2007), pp. 740–755.

[4] E. Artin, *Geometric Algebra*, Interscience, New York (1957).

[5] J.-C. Bermond, C. Huang, A. Rosa & D. Sotteau, Decomposition of complete graphs into isomorphic subgraphs with five vertices, *Ars Combin.* **10** (1980), 211–254.

[6] J.-C. Bermond & J. Schönheim, G-decomposition of K_n where G has four vertices or less, *Discrete Math.* **19** (1977), 113–120.

[7] A. Bonisoli, M. Buratti & G. Mazzuoccolo, Doubly transitive 2-factorizations, *J. Combin. Des.* **15** (2007), 120–132.

[8] A. Bonisoli, M. Buratti & G. Rinaldi, Sharply transitive decompositions of complete graphs into generalized Petersen graphs, *Innov. Incidence Geom.* **6–7** (2009), 95–109.

[9] A. Bonisoli & S. Bonvicini, Primitive one-factorizations and the geometry of mixed translations, *Discrete Math.* **308** (2008), 726–733.

[10] A. Bonisoli & S. Bonvicini, On sharply transitive k-matching decompositions of the complete graph, submitted.

[11] A. Bonisoli & D. Labbate, One-factorizations of complete graphs with vertex-regular automorphism groups, *J. Combin. Des.* **10** (2002), 1–16.

[12] A. Bonisoli & G. Rinaldi, Quaternionic starters, *Graphs Combin.* **21** (2005), 187–195.

[13] S. Bonvicini, Starters: doubling constructions, *Bull. Inst. Combin. Appl.* **46** (2006), 88–98.

[14] S. Bonvicini, Frattini based starters in 2-groups, *Discrete Math.* **308** (2008), 380–381.

[15] J. Bosák, *Decompositions of graphs*, Kluwer, Dordrecht (1990).

[16] D. E. Bryant, Cycle decompositions of complete graphs, in *Surveys in Combinatorics 2007* (eds. A. Hilton & J. Talbot), *Lond. Math. Soc. Lecture Note Ser.*, 346, Cambridge University Press, Cambridge (2007), pp. 67–97.

[17] D. E. Bryant & S. I. El-Zanati, Graph decompositions, in *Handbook of Combinatorial Designs* (eds. C. J. Colbourn & J. H. Dinitz), Second edition, Chapman & Hall/CRC, Boca Raton (2007), pp. 477–486.

[18] D. E. Bryant & C. A. Rodger, Cycle decompositions, in *Handbook of Combinatorial Designs* (eds. C. J. Colbourn & J. H. Dinitz), Second edition, Chapman & Hall/CRC, Boca Raton (2007), pp. 373–382.

[19] M. Buratti, Abelian 1-factorization of the complete graph, *European J. Combin.* **22** (2001), 291–295.

[20] M. Buratti & A. Del Fra, Existence of cyclic k-cycle systems of the complete graph, *Discrete Math.* **261** (2003), 113–125.

[21] M. Buratti & A. Del Fra, Cyclic hamiltonian cycle systems of the complete graph, *Discrete Math.* **279** (2004), 107–119.

[22] M. Buratti & G. Rinaldi, On sharply vertex transitive 2-factorizations of the complete graph, *J. Combin. Theory Ser. A* **111** (2005), 245–256.

[23] M. Buratti & G. Rinaldi, One-rotational k-factorizations of complete graphs and new solutions to the Oberwolfach problem, *J. Combin. Des.* **16** (2008), 87–100.

[24] M. Buratti & G. Rinaldi, A non-existence result on cyclic cycle-decompositions of the cocktail party graph, *Discrete Math.*, in press.

[25] P. J. Cameron, Minimal edge-colorings of complete graphs, *J. Lond. Math. Soc.* **11** (1975), 337–346.

[26] P. J. Cameron, *Parallelisms of Complete Designs*, Cambridge University Press, Cambridge (1976).

[27] P. J. Cameron & G. Korchmàros, One-factorizations of complete graphs with a doubly transitive automorphism group, *Bull. Lond. Math. Soc.* **25** (1993), 1–6.

[28] P. Dembowski, *Finite Geometries*, Springer, Berlin (1968).

[29] R. Diestel, *Graph Theory*, Springer, Berlin (2005).

[30] J. H. Dinitz, Starters, in *Handbook of Combinatorial Designs* (eds. C. J. Colbourn & J. H. Dinitz), Second edition, Chapman & Hall/CRC, Boca Raton (2007), pp. 622–628.

[31] J. D. Dixon & B. Mortimer, *Permutation Groups*, Springer, New York (1996).

[32] *Handbook of Graph Theory*, (eds. J. L. Gross & J. Yellen), CRC Press, Boca Raton (2004).

[33] A. Hartman & A. Rosa, Cyclic one-factorization of the complete graph, *European J. Combin.* **6** (1985), 45–48.

[34] W. M. Kantor, Homogeneous designs and geometric lattices, *J. Combin. Theory Ser. A* **38** (1985), 66–74.

[35] F. Klein, *Vergleichende Betrachtungen über neuere geometrische Forsuchungen*, Deichert Verlag, Erlangen (1872).

[36] G. Korchmáros, Cyclic one-factorization with an invariant one-factor of the complete graph, *Ars Combin.* **27** (1989), 133–138.

[37] G. Korchmáros, Sharply transitive 1-factorizations of the complete graph with an invariant 1-factor, *J. Combin. Des.* **2** (1994), 185–196.

[38] H. Lüneburg, *Translation Planes*, Springer, Berlin (1980).

[39] G. Mazzuoccolo, Primitive 2-factorizations of the complete graph, *Discrete Math.* **308** (2008), 175–179.

[40] G. Mazzuoccolo, On 2-factorizations whose automorphism group acts doubly transitively on the factors, *Discrete Math.* **308** (2008), 931–939.

[41] G. Mazzuoccolo & G. Rinaldi, k-pyramidal one-factorizations, *Graphs Combin.* **23** (2007), 315–326.

[42] R. Rees, Cyclic $(0,1)$-factorizations of the complete graph, *J. Combin. Math. Combin. Comput.* **4** (1988), 23–28.

[43] G. Rinaldi, Nilpotent 1-factorizations of the complete graph, *J. Combin. Des.* **13** (2005), 393–405.

[44] A. Rosa, Two-factorizations of the complete graph, *Rend. Sem. Mat. Messina Ser. 2* **9** (2003), 201–210.

[45] J. S. Rose, *A Course on Group Theory*, Cambridge University Press, Cambridge (1978).

[46] T. Q. Sibley, On classifying finite edge colored graphs with two transitive automorphism groups, *J. Combin. Theory Ser. B* **90** (2004), 121–138.

[47] J. Singer, A theorem in finite projective geometry and some applications to number theory, *Trans. Amer. Math. Soc.* **43** (1938), 377–385.

[48] W. D. Wallis, One-factorizations of complete graphs, in *Contemporary Design Theory: A Collection of Surveys* (eds. D. H. Stinitz & D. R. Stinson), Wiley, New York (1992), pp. 593–631.

[49] W. D. Wallis, *One-Factorizations*, Kluwer, Dordrecht (1997).

[50] M. E. Watkins, Automorphisms, in *Handbook of Graph Theory* (eds. J. L. Gross & J. Yellen), CRC Press, Boca Raton (2004), pp. 485–504.

Università di Modena e Reggio Emilia,
Dipartimento di Scienze e Metodi dell'Ingegneria,
via Amendola 2 – padiglione Morselli,
42100 Reggio Emilia,
Italy
arrigo.bonisoli@unimore.it

Combinatorics of optimal designs

R. A. Bailey and Peter J. Cameron

Abstract

To a combinatorialist, a design is usually a 2-design or balanced incomplete-block design. However, 2-designs do not necessarily exist in all cases where a statistician might wish to use one to design an experiment. As a result, statisticians need to consider structures much more general than the combinatorialist's designs, and to decide which one is "best" in a given situation. This leads to the theory of optimal designs. There are several concepts of optimality, and no general consensus about which one to use in any particular situation.

For block designs with fixed block size k, all these optimality criteria are determined by a graph, the *concurrence graph* of the design, and more specifically, by the eigenvalues of the Laplacian matrix of the graph. It turns out that the optimality criteria most used by statisticians correspond to properties of this graph which are interesting in other contexts: D-optimality involves maximizing the number of spanning trees; A-optimality involves minimizing the sum of resistances between all pairs of terminals (when the graph is regarded as an electrical circuit, with each edge being a one-ohm resistor); and E-optimality involves maximizing the smallest eigenvalue of the Laplacian (the corresponding graphs are likely to have good expansion and random walk properties). If you are familiar with these properties, you may expect that related "nice" properties such as regularity and large girth (or even symmetry) may tend to hold; some of our examples may come as a surprise!

The aim of this paper is to point out that the optimal design point of view unifies various topics in graph theory and design theory, and suggests some interesting open problems to which combinatorialists of all kinds might turn their expertise. We describe in some detail both the statistical background and the mathematics of various topics such as Laplace eigenvalues of graphs.

1 Preliminaries

This first section of the paper sets the scene. We look briefly at the different ways in which a combinatorialist and a statistician view a simple block design such as the Fano plane, and also introduce a running example which would not be recognised as a design under the standard combinatorial definition but is in fact the best design for a statistician in certain circumstances.

Section 2 looks at the way that information about treatment differences is recovered from the results of an experiment conducted using a block design. We look briefly at what makes a good design, and show that all criteria for this can be expressed in terms of the concurrence graph of the design. Having thus focussed our attention on graphs, we give a brief survey of the Laplacian matrix of a graph and its eigenvalues, and mention the interesting question of which graphs can be concurrence graphs of block designs with given block size.

Section 3 covers the three most important kinds of optimality, described by the letters A (average), D (determinant) and E (extreme), and how they look when expressed in terms of the concurrence matrix.

In Section 4 we look at sparse graphs. For trees and unicyclic graphs, we identify the A-, D- and E-optimal graphs, observing that for unicyclic graphs the different criteria give very different answers. We also give extensions to more general classes of graphs. These results have applications in the design of microarray experiments in genetics.

In Section 5 we look at other kinds of design and optimality criterion, including the case of block designs which do not have constant block size. In the final section we look briefly at computational issues and describe how the GAP share package DESIGN can be used for some of these computations.

We note that lack of communication between mathematicians and statisticians has led to some duplication of effort in this area. For example, Kelmans and Chelnokov [62] showed in 1974 that the graph on n vertices with $\lambda n(n-1)/2$ edges with the largest number of spanning trees is the λ-fold complete graph. Cheng [32] pointed out that this follows from an early result on optimal designs by Kiefer in 1958 [65]. Wild [105] has attempted to bring the two communities closer together; this paper represents another attempt.

1.1 Introduction

Combinatorialists and statisticians understand different things by the phrase "block design". To a combinatorialist, the archetypal block design is the Fano plane shown in Figure 1.

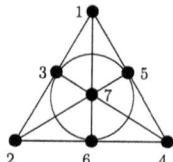

Figure 1: The Fano plane

The design has seven points and seven blocks, each block being a set of three points; any two distinct points are contained in a unique block.

A statistician might introduce this design with an example like the following. Seven different makes of fertilizer are to be tested in an experiment. Twenty-one plots of land are available for the experiment, three plots on each of seven farms in different parts of the country. We can apply one fertilizer to each plot, then grow a crop on all the plots and measure the yield. How should we allocate fertilizers to the plots? Clearly the worst thing we could do would be to put fertilizer 1 on the three plots on Farm A, fertilizer 2 on the plots on Farm B, etc. If we did this, then any difference in yield between these plots could be the result of the fertilizer, but could equally be the result of differences in fertility, soil structure or climate between the farms. It is not hard to believe that the best way to assign the fertilizers to the farms is using a scheme like the one in Table 1.

This is the Fano plane in disguise: if we identify the seven types of fertilizer with the points of the plane, and the seven farms with the lines, then a fertilizer is used on a farm precisely when the corresponding point and line are incident.

There is another step in the design of the experiment, namely *randomization*: we could number the fertilizers and farms in random order, and assign the three

	Farm A	Farm B	Farm C	Farm D	Farm E	Farm F	Farm G
Plot 1	1	1	1	2	2	3	3
Plot 2	2	4	6	4	5	4	5
Plot 3	3	5	7	6	7	7	6

Table 1: The Fano plane

fertilizers randomly to the three plots on each farm. We will not discuss further the techniques and reasons for randomization here.

The main thing to notice is that the most important things in the experimental design, the individual plots or experimental units on which the measurements are made, are not immediately visible in the mathematician's picture of the Fano plane. A little thought shows that plots correspond to the *flags* (the incident point-line pairs) in the plane.

The combinatorialist and statistician agree that in this case the best thing to do is to use the Fano plane. But what if there are only six fertilizers and six farms? There is no 2-design with six points and six blocks, having three points in each block; but the statistician still has the job of designing the experiment so as to extract as much information as possible about differences between the fertilizers. Even worse, what should we do when there are different numbers of plots available on each farm?

1.2 Combinatorial design

A combinatorialist usually defines a *block design* in one of two ways:

- a set of *points*, with a collection of subsets called *blocks*, satisfying some further conditions;

- two sets (a set of *points* and a set of *blocks*) with a binary relation of *incidence* between them.

The two views are almost equivalent. Given the first definition, we define the incidence relation between points and blocks to be the membership relation: point p and block B are incident if $p \in B$. Conversely, given two sets and an incidence relation, we can identify any block with the set of points incident with it; the difference is that we obtain a multiset of subsets of the point set, that is, we allow *repeated blocks*. However, we will freely mix the language associated with the two viewpoints.

Typical of the additional conditions imposed are those of being a t-(v, k, λ) design, where $1 \leq t \leq k \leq v$ and $\lambda > 0$: this requires that there are v points altogether, that any block contains k points, and that any set of t points is contained in exactly λ blocks. Thus the Fano plane is a 2-$(7, 3, 1)$ design.

The difference between the two viewpoints is important for the existence question. If repeated blocks are allowed, it is easy to show that t-(v, k, λ) designs in which not every k-set is a block exist for every t, k, v satisfying $t < k < v - t$ (for some λ). However, the existence of t-designs without repeated blocks is a much more difficult question: none with $t > 5$ were known until the 1980s, and their existence for all t was shown by Teirlinck [96] in 1987. A different proof was found by Ajoodani-Namini [1] in 1996. Ajoodani-Namini's proof is very much simpler, but

Teirlinck's gives designs where the value of λ (though large) is a fixed function of t, and does not grow with v. We refer to the paper by Khosrovshahi and Tayfeh-Reziae in this volume [64] for more information on t-designs.

In the case of 2-designs, there is a powerful existence theory due to Wilson [106]. There are two easy necessary conditions on the parameters for the existence of such a design:

$$\begin{aligned} k-1 &\text{ divides } (v-1)\lambda, \\ k &\text{ divides } v(v-1)\lambda. \end{aligned} \quad (1.1)$$

For the number of blocks containing a point is equal to $(v-1)\lambda/(k-1)$, while the total number of blocks is $v(v-1)\lambda/k(k-1)$. These conditions, together with *Fisher's inequality* $b \geq v$, are known to be sufficient if $k \leq 4$ [51]. Wilson showed that, for any given k and λ, these conditions are sufficient for the existence of a 2-(v, k, λ) design if v is large enough (in terms of k and λ).

We are far from having a comparable theory for t-designs with $t > 2$, however.

It is quite common in the combinatorial literature for the unadorned word "design" to denote a 2-design. This will not be our viewpoint; we now explain why statisticians require a much more general concept.

1.3 Statistical design

In a fairly simple kind of comparative experiment, a number v of different treatments will be applied to experimental units and the responses measured and analysed. The design is an allocation of treatments to units; because of randomization, it is better to think of the design as a partition of the set of units into v parts, so that one treatment can be allocated to each part. We use v for the number of treatments to avoid conflict with the t used in the theory of t-designs; think of v for "varieties" – in an agricultural experiment the treatments applied to the fields might be different varieties of wheat.

If the experimental units are all alike, then it seems reasonable to partition them into parts as nearly equal as possible; any such partition is as good as any other, and there is nothing interesting from a combinatorial point of view. But things are not always so simple. The next level of complication is that the units come already partitioned into "blocks", so that the units within a block are alike, but differ systematically from those in a different block (such as the twenty-one fields on seven different farms in our example). In this case, the relationship between the two partitions gives the design a non-trivial combinatorial structure.

Accordingly, we define a *block design* to be a set of "units" or "plots" carrying two partitions: a treatment partition with v parts, and a block partition with b parts.

This looks very different from a combinatorialist's block design, but we will see that it can be recast into a very similar (but rather more general) form. To do this, we follow the combinatorialist and take the set of v treatments as the basic object. Each block now consists of a multiset of treatments, namely, the treatments applied to the plots in that block. The block design is called *binary* if no treatment occurs more than once in each block. In a binary design, a block is now identified with a set of treatments; so a binary block design can be thought of as a multiset of subsets of the set of treatments (it is a multiset to allow the possibility of repeated blocks). In general, however, we have a multiset of multisets of treatments.

There may be good scientific reasons for allowing a treatment to occur more than once in a block; it may have a self-interaction which is important but would not be detected if this were not done. In the simple models we consider in this paper, such interactions will not be relevant. But this does not mean that the best designs are always binary.

In the discussion of Tocher's paper [99], David Cox said

> I suspect that ... balanced ternary designs are of no practical value.

(Here a design is *ternary* if the maximum number of occurrences of a fixed treatment in a block is two.) Contrary to this suspicion, it may well happen that optimal designs are non-binary; we will see examples.

However, since we will assume that we are not interested in effects involving repetition of a single treatment and simply want to compare the treatments, then a block in which only one treatment occurs (an arbitrary number of times) is simply wasted effort, since it can give no information about treatment differences. So we will make one restriction on the designs we consider: *no block can contain just a single treatment*.

Another way to represent a block design is by its *Levi graph*, defined as follows. The graph is bipartite, with vertices in one class of the bipartition being the treatments and those in the other being the blocks; for each experimental unit, the graph has an edge connecting the treatment on that unit (i.e. the part of the treatment partition containing it) with its block (the part of the block partition containing it). This graph may have multiple edges; indeed, it is simple if and only if the design is binary. Now we can obtain the "multiset of multisets" representation by taking the treatments as points, and the neighbourhoods of the blocks as the multisets. (If n_{ij} edges join point i to block j, then point i occurs n_{ij} times in the multiset corresponding to j.)

These numbers give us another simple way of representing the design. We always consider designs for v treatments in b blocks. Let N be the $v \times b$ *incidence matrix*: the entry n_{ij} is the number of times that treatment i occurs in block j. Thus the column sums of N are the block sizes, which we shall usually assume are all equal to k, while the sum r_i of the i-th row is the number of times which treatment i occurs overall, which is the *replication* of i. Thus the design is binary if $n_{ij} \in \{0, 1\}$ for all i and j.

Put $\Lambda = NN^\top$, which is called the *concurrence matrix*. If the design is binary then λ_{ij} is just the number of blocks in which treatments i and j occur; in particular, $\lambda_{ii} = r_i$.

For more information on design of comparative experiments, we refer to [8].

1.4 An example

Example 1.1 Here is an example of a block design which we will use as a running example in this paper. This design was first given by Tocher [99] in the paper cited above. There are five treatments, labelled $1, 2, 3, 4, 5$, and 21 units, divided into seven blocks of three, corresponding to the columns of Table 2. Its Levi graph is in Figure 2.

1	1	1	1	2	2	2
1	3	3	4	3	3	4
2	4	5	5	4	5	5

Table 2: A block design

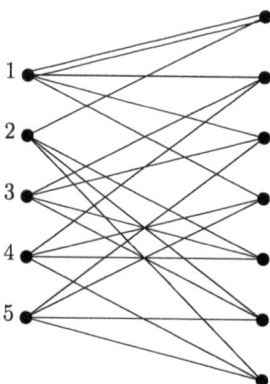

Figure 2: The Levi graph of the design in Table 2

Every block has size three, though of course the first block is a multiset. The point 1 occurs five times, including twice in the first block; the remaining points occur four times each. Every pair of points has concurrence precisely 2: the two concurrences of 1 and 2 are accounted for by the first block. So all off-diagonal elements in the concurrence matrix are equal to 2. On the other hand, the diagonal entries are $(7, 4, 4, 4, 4)$. (The two occurrences of treatment 1 in the first block contribute 4 to the first entry of $NN^\top$.)

1.5 Graphs

As hinted above, our graphs can have multiple edges but not loops. (In Section 5.4, we briefly have a more general concept, edge-weighted graphs.) A *simple graph* is one with no multiple edges; a *regular graph* is a graph where all vertices have the same valency. Graphs will almost always be connected. The *adjacency matrix* of a graph has rows and columns indexed by vertices, with (i, j) entry equal to the number of edges connecting i and j.

A *distance-regular graph* is a simple connected regular graph of diameter d for which there exist parameters c_m, a_m, b_m (for $0 \le m \le d$, but c_0, a_0 and b_d are undefined) such that, if i and j are vertices at distance m, then the numbers of vertices adjacent to j and at distance $m-1, m, m+1$ respectively from i are c_m, a_m, b_m respectively.

There is a rich theory of distance-regular graphs, for which we refer to [16, 23, 48]. In particular, note that the eigenvalues of the adjacency matrix of a distance-regular graph are the same as those of a tridiagonal matrix of order $d+1$ having c_m, a_m, b_m in row $m+1$; their multiplicities can also be calculated from these parameters. A *strongly regular graph* is a distance-regular graph with diameter 2. (Some authors allow "strongly regular graphs" to be disconnected; such graphs are disjoint unions of complete graphs of the same size.)

1.6 Partial balance

An *association scheme* of rank $s+1$ on the set of v treatments is a set of symmetric $v \times v$ matrices $A_0, \ldots, A_s$ with entries 0 and 1 such that $A_0 = I$, $A_0 + \cdots + A_s = J$, and the product $A_i A_j$ is a linear combination of $A_0, \ldots, A_s$ for all i and j. Here J denotes the $v \times v$ all-1 matrix. Thus if $s = 1$ then the scheme is trivial with $A_1 = J - I$. A pair of distinct treatments are said to be *i-th associates* if the corresponding entry of A_i is 1.

An association scheme of rank 3 is either a complementary pair of strongly regular graphs, or is *group-divisible* (one associate class being a uniform partition with the diagonal removed)—the second alternative is unnecessary if the more general definition of "strongly regular graph" is adopted. More generally, the distance classes in a distance-regular graph of diameter s define an association scheme of rank $s+1$.

The set of all real linear combinations of $A_0, \ldots, A_s$ is the *Bose–Mesner algebra* of the association scheme. This algebra is commutative, and so has a basis of minimal idempotents as well as the basis formed by the adjacency matrices A_i. Many techniques for calculation in association schemes depend on transforming between these two bases: see [5].

A binary block design is defined to be *partially balanced* with respect to a given association scheme if its concurrence matrix is in the Bose–Mesner algebra of that scheme. If the scheme has rank $s+1$, the design is said to have s associate classes. Partially balanced block designs with two associate classes were introduced by Bose and Nair in [21] when statisticians found that they needed block designs for some parameter sets for which no 2-design exists. Allowing two different concurrences seemed to be the next level of complexity; and restricting to partially balanced designs gave a simple method of calculating matrix inverses (see Section 2.1) in the days before computers were widely available.

2 Optimality

We begin this section with the statistical background: the information matrix of a block design, its generalized inverse, and the role of the latter in estimating treatment differences from experimental data. We want these estimates to be as accurate as possible: this leads to various notions of optimality. For designs with constant block size, optimality criteria depend on the concurrence graph of the design, specifically on the eigenvalues of its Laplacian matrix. Kiefer's Theorem (Theorem 2.3) asserts that 2-designs, when they exist, are optimal in any reasonable sense. We conclude with a discussion of variance-balanced designs (with constant block size), a generalization of 2-designs.

2.1 The information matrix and its generalized inverse

Put
$$C = R - \frac{1}{k}\Lambda,$$
where $R = \mathrm{diag}(r_1, \ldots, r_v)$. This is called the *information matrix* of the design, for reasons that will become clear in Section 2.2. It is a $v \times v$ symmetric matrix with zero row and column sums.

The trace of C is equal to $\left(\sum_{i \neq j} \lambda_{ij}\right)/k$. We will need to know when this is maximized (for given v, b, k). If the point i has m_i occurrences in a block, then the contribution of that block to the trace is
$$\sum_{i \neq j} m_i m_j = k^2 - \sum_i m_i^2;$$
since $\sum_i m_i = k$, the contribution is maximized when the m_i are as nearly equal as possible. We conclude:

Proposition 2.1 *A design with v points and b blocks of size k maximizes the trace of C (among all such designs) if and only if each point occurs $\lfloor k/v \rfloor$ or $\lceil k/v \rceil$ times in each block. In particular, if $k < v$, the trace of C is maximized by the binary designs; the maximum value is $b(k-1)$.*

In general, designs satisfying the conditions of this result are called *generalized binary*; such a design is obtained by adding a fixed number of occurrences of each point to each block of a binary design.

Let u be the all-1 vector in $\mathbb{R}^v$. (We adopt the statisticians' convention that vectors are column vectors.) Then u is an eigenvector of C with eigenvalue 0, which we call the *trivial eigenvalue*. In general, the multiplicity of the zero eigenvalue is equal to the number of connected components in the Levi graph (this follows from Proposition 2.9). This is why we say "non-trivial" rather than "non-zero" to indicate the eigenvalues on vectors orthogonal to u. Connectedness ensures that this eigenvalue has multiplicity 1. In this case, the design too is said to be *connected*. From now on, we shall assume that we are considering only connected designs.

Since it is symmetric, C can be written in spectral form as
$$C = \sum \theta_i P_i,$$
where θ_i are the distinct eigenvalues and P_i are the eigenprojectors. It is convenient to number the eigenvalues so that $\theta_0 = 0$; then, for a connected design, P_0 is the projector onto the space $\langle u \rangle$ spanned by u, so $P_0 = v^{-1} J$, where J is the $v \times v$ all-1 matrix.

The matrix C is singular, but we need some kind of inverse. The matrix B is said to be a *generalized inverse* of C if it satisfies
$$BCB = B \quad \text{and} \quad CBC = C. \tag{2.1}$$

Since C is invertible on the space $\langle u \rangle^\perp$ and zero on the space $\langle u \rangle$, in some ways the most natural generalized inverse is the one defined by
$$BC = CB = I - P_0 \quad \text{and} \quad BP_0 = P_0 B = 0.$$

This one is called the *Moore–Penrose generalized inverse*, written C^-. It is given by

$$C^- = \sum_{i \neq 0} \frac{1}{\theta_i} P_i. \tag{2.2}$$

It is also a symmetric matrix with zero row and column sums. If B is any other generalized inverse, then $C^- = (I - P_0)B(I - P_0)$.

Statisticians who evaluate block designs have developed several methods for calculating the Moore–Penrose generalized inverse of the information matrix C. Here are some of their methods.

(a) Express C in spectral form and use Equation (2.2). This method is good when it is easy to find the eigenspaces. Then it can work for whole families of matrices of different sizes. For example, in an association scheme the eigenprojectors are just (sums of some of) the minimal idempotents of the scheme.

(b) Use a computer (algebra) package to compute

$$C^- = (C + P_0)^{-1} - P_0,$$

where $P_0 = v^{-1}J$. The trick is that $C + P_0$ is invertible and is the identity on $\langle u \rangle$, so its inverse is also the identity on $\langle u \rangle$.

(c) (This is the method favoured by electrical engineers: see later.) Delete the last row and column of C, to leave an invertible matrix $\tilde{C}$. Put $\tilde{C}^{-1}$ in the top left hand corner, and zeros in the last row and column, then pre- and post-multiply by $I - P_0$.

(d) More generally, to calculate the Moore–Penrose generalized inverse of a symmetric partitioned matrix, statisticians often use the formula in Lemma A.3 of [89], which uses a result of [88]. The formula in Exercise 2.7 of [84, page 33] is similar.

(e) Start with a symmetric matrix B with zero row sums. If there is any simultaneous permutation of the rows and columns of C that leaves it unaltered, reduce the number of unknowns in B so that it is also invariant under this permutation. Then use brute (hand) force to solve Equations (2.1).

(f) Find powers of C until they, together with $I - P_0$, are linearly dependent. This gives an equation of the form

$$\text{polynomial in } C \text{ with no constant term} = I - P_0.$$

The degree of the left-hand side is equal to the number of distinct non-trivial eigenvalues of C. Factorize the left-hand side as CB. Pre- and post-multiply B by $I - P_0$ to get C^-. Since B has constant row and column sums, in practice this means subtracting c/v from every entry in B, where c is the row sum. If there are not many distinct eigenvalues, this method can be quicker than (a), because the eigenprojectors do not need to be calculated.

A square matrix is *completely symmetric* if its diagonal elements are constant and its off-diagonal elements are constant; that is, if it is a linear combination of I and J. If an information matrix is completely symmetric, it has the form $C = c(I - P_0)$, and so its Moore–Penrose inverse $C^- = c^{-1}(I - P_0)$ is also completely symmetric.

2.2 Estimation and variance

In the experiment, the scientist measures a response on each experimental unit. Suppose that experimental unit ω is in block j and is allocated treatment i. We assume that there are numbers $\tau_1, \ldots, \tau_v$ and $\beta_1, \ldots, \beta_b$ such that the expected response on unit ω is $\tau_i + \beta_j$. Notice that this is the real world here, not algebra or arithmetic: if we measure the same thing on the same treatment in the same block on two different occasions we will not usually get the same answer. This is partly because of errors of observation, and partly because of inherent variability in the underlying experimental material. For a good discussion of this, see [81].

Thus, writing Y_ω for the response on unit ω, we have to assume that Y_ω is a random variable whose expectation is $\tau_i + \beta_j$. That is,

$$\mathrm{E}(Y_\omega) = \tau_i + \beta_j \qquad \text{if } \omega \text{ is in block } j \text{ and has treatment } i. \tag{2.3}$$

We are not interested in the β parameters, but do want to estimate the τ parameters.

What is an *estimator*? It is a function of the data, but not of the unknown values which we are trying to estimate. Because the responses are random variables, each estimator is also a random variable. The most important property of an estimator is that it should be *unbiased*, which means that the expected value of the estimator is equal to the unknown parameter which we are trying to estimate.

In Equations (2.3), we can obviously add a constant to all the τ parameters and subtract it from all the β parameters without changing the model, so we cannot estimate the τ_i. In fact, we can adjust constants in this way separately in each connected component of the Levi graph. However, this is the only source of ambiguity, as we shall show. More precisely, we shall show that if the vector x is in the image of C (equivalently, if x is orthogonal to the null space of C), then there is a linear combination of the responses whose expected value is equal to $x^\top \tau$. When the design is connected, the null space of C is just $\langle u \rangle$, so x is orthogonal to the null space if and only if $x^\top u = 0$; that is, the coefficients in the vector x sum to zero. Such a vector is called a *contrast vector*, and the corresponding linear combination $x^\top \tau$ is called a *contrast* in the treatment parameters $\tau_1, \ldots, \tau_v$.

Statisticians usually rewrite Equations (2.3) in the following vector form:

$$\mathrm{E}(Y) = X\tau + Z\beta. \tag{2.4}$$

Here Y, τ and β are vectors in $\mathbb{R}^{bk}$, $\mathbb{R}^v$ and $\mathbb{R}^b$ respectively. The matrix X has bk rows and v columns: the entry in row ω and column i is equal to 1 if experimental unit ω gets treatment i; otherwise it is zero. The $bk \times b$ matrix Z is defined similarly, using blocks instead of treatments.

The matrix $k^{-1}ZZ^\top$ is the orthogonal projector onto the subspace $\mathcal{B}$ of $\mathbb{R}^{bk}$ consisting of vectors which are constant on each block. (If the blocks have different sizes $k_1, \ldots, k_b$, then the matrix is $ZK^{-1}Z^\top$, where $K = \mathrm{diag}(k_1, \ldots, k_b)$: we shall return to this form in Section 5.4.) Put $Q = I - k^{-1}ZZ^\top$. To remove the β parameters from Equation (2.4), we premultiply by Q. The expectation operator commutes with linear transformations, so we obtain

$$\mathrm{E}(QY) = QX\tau. \tag{2.5}$$

That is, the expected value of the transformed data QY lies in the column space $\mathcal{T}$ of QX. It is fairly natural to use the *least-squares principle* to say that our estimate of $QX\tau$ should be the vector in $\mathcal{T}$ which is closest to the transformed data vector QY in the sense that the sum of the squares of the differences in coordinates is minimized. By Pythagoras' Theorem, we can obtain this by projecting the vector QY orthogonally onto $\mathcal{T}$.

Note that $X^\top X = R$ and $X^\top Z = N$. Hence $X^\top QX = C$, where C is the information matrix defined in the previous section. Consider the symmetric matrix $S = QXC^- X^\top Q$. Any vector in $\mathcal{T}$ has the form $QX\tau$, and

$$SQX\tau = QXC^- X^\top QQX\tau = QXC^- X^\top QX\tau = QXC^- C\tau = QX(I-P_0)\tau = QX\tau$$

because the columns of XP_0 are constant vectors, hence in $\mathcal{B}$. On the other hand, if w is a vector in $\mathbb{R}^{bk}$ which is orthogonal to $\mathcal{T}$, then $X^\top Qw = 0$ and so $Sw = 0$. It follows that S is the orthogonal projector onto $\mathcal{T}$.

Thus the least-squares principle gives the vector of *fitted values* as SQY. That is, our estimator of $QX\tau$ is SQY, so our estimator of $X^\top QX\tau$ is $X^\top SQY$. We have already shown that $SQX = QX$, and $QS = S$, so $X^\top SQY = X^\top QSQY = X^\top QY$. Moreover, $X^\top QX = C$, so our estimator of $C\tau$ is $X^\top QY$. If the vector x in $\mathbb{R}^v$ is orthogonal to the null space of C, there is a vector z in $\mathbb{R}^v$ such that $Cz = x$. Then $x^\top \tau = z^\top C\tau$, whose estimator is $z^\top X^\top QY$.

We have used the least-squares principle to obtain an estimator for $x^\top \tau$ for any vector x orthogonal to the null space of the information matrix C. In particular, if the design is connected then this gives an estimator for every contrast. The estimator is a linear combination of the data.

What are the properties of this estimator? If $Cz = x$ then

$$\mathrm{E}(z^\top X^\top QY) = z^\top X^\top Q\,\mathrm{E}(Y) = z^\top X^\top QX\tau = z^\top C\tau = x^\top \tau,$$

so the estimator is unbiased.

The expectation, or mean, of a random variable is its first moment; the second moment (about the mean) is its variance. Now we consider the variance of the estimator. The smaller the variance, the more likely it is that our estimate will be close to the true value. For example, if the estimator of the difference $\tau_1 - \tau_2$ has a normal distribution with variance V_{12}, then the probability that the modulus of the difference between the true value and the estimate is less than $\sqrt{V_{12}}$, $2\sqrt{V_{12}}$, $3\sqrt{V_{12}}$ is 0.683, 0.954, 0.997 respectively.

We assume that the variance of each response is an unknown constant σ^2, and that different responses are independent. This information can be summarized in the variance-covariance matrix for the vector Y, as $\mathrm{Cov}(Y) = \sigma^2 I$. Now, the variance of the estimator $z^\top X^\top QY$ of $x^\top \tau$ is given by $z^\top X^\top Q(\mathrm{Cov}(Y))QXz$. Under our simple assumptions about $\mathrm{Cov}(Y)$, this reduces to $(z^\top Cz)\sigma^2$, which is $(x^\top C^- x)\sigma^2$.

Any other linear estimator of $x^\top \tau$ must have the form $(z^\top X^\top + w^\top)QY$, for some w in $\mathbb{R}^{bk}$. For this estimator to be unbiased, we need $0 = \mathrm{E}(w^\top QY) = w^\top Q\,\mathrm{E}(Y) = w^\top QX\tau$ for all τ, and so $w^\top QX = 0$. Now, the variance of this estimator is

$$\begin{aligned}
(z^\top X^\top + w^\top)Q(\mathrm{Cov}(Y))Q(w + Xz) &= (z^\top X^\top + w^\top)Q(w + Xz)\sigma^2 \\
&= (z^\top Cz + w^\top Qw + 2w^\top QXz)\sigma^2 \\
&= (z^\top Cz + w^\top Qw)\sigma^2.
\end{aligned}$$

As Q is positive semi-definite, $w^\top Q w$ cannot be negative, so this variance cannot be smaller than the variance of the least-squares estimator.

In summary, we have proved the following classical theorem.

Theorem 2.2 *Suppose that* $\mathrm{Cov}(Y) = I\sigma^2$ *and that Equation (2.4) holds. If the design is connected and x is any contrast vector in $\mathbb{R}^v$, then the least-squares estimator of $x^\top \tau$ is $x^\top C^- X^\top QY$, where $Q = I - k^{-1} ZZ^\top$. This estimator is unbiased, and has minimum variance among all unbiased linear estimators of $x^\top \tau$. The variance is equal to $(x^\top C^- x)\sigma^2$.*

In particular, to estimate the difference $\tau_i - \tau_j$ we use the contrast x with $x_i = 1$, $x_j = -1$ and $x_l = 0$ for $l \notin \{i,j\}$. To calculate the variance V_{ij} of this estimator, which is called a *pairwise variance*, we look at the 2×2 submatrix of C^- in the ith and j-th row and column:

$$V_{ij} = (C^-_{ii} + C^-_{jj} - C^-_{ij} - C^-_{ji})\sigma^2. \quad (2.6)$$

If C is completely symmetric, then these pairwise variances are constant. Such a design is called *variance-balanced*. Note that 2-designs are variance-balanced.

2.3 Optimality criteria

A statistician designing a comparative experiment knows the number v of treatments to be compared and the number of experimental units available. In our case, the experimental units are partitioned into b blocks of size k, where b and k are known. Let $\mathcal{D}(v, b, k)$ denote the class of all (not necessarily binary) block designs with v treatments, and b blocks of size k. Which design in this class is best?

One obviously desirable feature is to minimize the average value of the pairwise variances V_{ij}. Since C^- has zero row and column sums, Equation (2.6) shows that for each fixed i,

$$\sum_{j \neq i} V_{ij} = [(v-1)C^-_{ii} + (\mathrm{Tr}(C^-) - C^-_{ii}) + 2C^-_{ii}]\sigma^2 = (\mathrm{Tr}(C^-) + vC^-_{ii})\sigma^2,$$

where $\mathrm{Tr}(C)$ is the trace of C. Hence the average value $\bar{V}$ of the pairwise variances is equal to $2\sigma^2 \mathrm{Tr}(C^-)/(v-1)$.

Let $\theta_1, \ldots, \theta_{v-1}$ be the non-trivial eigenvalues of C, now listed with multiplicities and in non-decreasing order. Then

$$\mathrm{Tr}(C^-) = \frac{1}{\theta_1} + \cdots + \frac{1}{\theta_{v-1}}.$$

Therefore

$$\bar{V} = \frac{2\sigma^2}{v-1}\left(\frac{1}{\theta_1} + \cdots + \frac{1}{\theta_{v-1}}\right) = 2\sigma^2 \times \frac{1}{\text{harmonic mean of } \theta_1, \ldots, \theta_{v-1}}.$$

Thus a design is said to be *A-optimal* (in some class) if it minimizes the average pairwise variance in that class; equivalently, if it maximizes the harmonic mean of $\theta_1, \ldots, \theta_{v-1}$ in that class. Here 'A' stands for 'average'.

This is not the only kind of optimality. When statisticians present an estimate for the difference $\tau_i - \tau_j$, they usually also present a confidence interval for the difference. If they always present a 95% confidence interval, the true value should lie within the confidence interval in 19 cases out of 20. The smaller the confidence interval the better. Under the assumption that the responses are multivariate normal, the length of this confidence interval is proportional to $\sqrt{V_{ij}}$, and therefore minimizing the variance is the same as minimizing the length of the confidence interval. However, the volume of the confidence ellipsoid for the estimate of $(\tau_1, \ldots, \tau_v)$ (in the $(v-1)$-dimensional space orthogonal to u) is proportional to the square root of the reciprocal of $\Pi_{i=1}^{v-1} \theta_i$ (see [4, page 49]). A design is *D-optimal* if it minimizes this volume; equivalently, if it maximizes the geometric mean of $\theta_1, \ldots, \theta_{v-1}$. Here 'D' stands for 'determinant'.

In some experiments, the treatments are combinations of different quantities of inputs that can be measured on continuous scales. A multivariate polynomial model may be fitted rather than the model in Equation (2.4). Since the scales for the different variables may not be comparable, it is important to use an optimality criterion that is invariant to reparametrization of the model. The D-criterion is the only one of the popular criteria that is so invariant. However, for block designs with v qualitative treatments, the A-criterion seems more natural.

Some people are concerned about worst cases. A design is said to be *E-optimal* if it maximizes the minimum θ_1 of the non-trivial eigenvalues of C; equivalently, it minimizes the maximum variance of the estimator of $x^\top \tau$ over all contrast vectors x with $x^\top x = 1$. Here 'E' stands for 'extreme'. Somewhat similarly, a design is said to be *MV-optimal* if it minimizes the maximum of the V_{ij}. This last criterion is not a function of the eigenvalues.

The eigenvalue optimality criteria all fall under the umbrella of Φ_p-optimality. For p in $(0, \infty)$, a design is Φ_p-optimal if it minimizes

$$\left(\frac{\sum_{i=1}^{v-1} \theta_i^{-p}}{v-1} \right)^{\frac{1}{p}}.$$

A-optimality corresponds to $p = 1$; the limit as $p \to 0$ gives D-optimality; the limit as $p \to \infty$ gives E-optimality.

All these criteria depend on the design only through its information matrix, which is a $v \times v$ real symmetric matrix with row and column sums zero. Let $\mathcal{X}_v$ denote the set of all matrices satisfying this condition. An *optimality criterion* is a function $\Phi : \mathcal{X}_v \to \mathbb{R}$. Now a design in some class $\mathcal{D}$ of designs is said to be Φ-optimal if it minimizes the value of $\Phi(C)$ over all information matrices C of designs in $\mathcal{D}$.

The optimality criteria Φ that we consider all satisfy the following conditions:

(a) Φ is convex;

(b) for every C in $\mathcal{X}_v$, the function $\alpha \mapsto \Phi(\alpha C)$ is monotonic non-increasing for non-negative α;

(c) Φ is invariant under any simultaneous permutation of rows and columns by the same permutation—that is, re-labelling the treatments does not affect Φ.

(Any function of the eigenvalues satisfies condition (c).)

A design is said to be *universally optimal* in some class if it minimizes every Φ satisfying the above conditions. Thus universal optimality is stronger than all of the individual criteria, but there may not be any universally optimal design in a given class of designs. Recall that a design in $\mathcal{D}(v,b,k)$ maximizes the trace of C if and only if it is generalized binary. Kiefer [66] proved the following theorem.

Theorem 2.3 *Let $\Phi : \mathcal{X}_v \to \mathbb{R}$ be a function satisfying conditions (a)–(c) above. If there is a design in the class for which C is completely symmetric and has maximum trace, then it is universally optimal. In particular, if $\mathcal{D}(v,b,k)$ contains 2-designs (BIBDs), then the minimum value of $\Phi(C)$ over information matrices of designs in $\mathcal{D}(v,b,k)$ is attained by the 2-designs.*

The A-, D- and E-criteria are all examples of functions satisfying the conditions of Kiefer's theorem. This justifies the assertion that the Fano plane is the best design to choose in our introductory example. In terms of the parameters v,b,k, the necessary conditions (1.1) for the existence of a 2-design or BIBD read

$$v \text{ divides } bk,$$
$$v-1 \text{ divides } bk(k-1).$$

We sketch the proof of Kiefer's Theorem for functions depending on the eigenvalues, in the case $k < v$.

Let $\theta_1, \ldots, \theta_{v-1}$ be the non-trivial eigenvalues of the information matrix of a design in $\mathcal{D}(v,b,k)$. Then $\sum \theta_i = \text{Tr}(C) \leq b(k-1)$, by Proposition 2.1. Under the given hypotheses, $\Phi(C)$ is minimized when the sum attains the upper bound $b(k-1)$ (which means that the design is binary) and all eigenvalues are equal (which means that C is a linear combination of I and J, so that the design is balanced).

We say that the sequence $(x_0, \ldots, x_{v-1})$ is *upper weakly majorized* by the sequence $(y_0, \ldots, y_{v-1})$ if

$$\sum_{i=0}^{n-1} x_i \geq \sum_{i=0}^{n-1} y_i$$

for $n = 1, \ldots, v$. If the sums are equal for $n = v$, we say that $(x_0, \ldots, x_{v-1})$ is *majorized* by $(y_0, \ldots, y_{v-1})$. See Marshall and Olkin [70]. A block design is *Schur-optimal* in a class if the eigenvalues of its information matrix (in non-decreasing order) are upper weakly majorized by those of any competing design. Clearly Schur-optimality is stronger than E-optimality. This seems to make most sense when the sum of the eigenvalues is fixed. Since upper weak majorization is a partial order, there may be no such optimal design within a given class. Giovagnoli and Wynn [47] showed:

Proposition 2.4 *If a design is Schur-optimal within any class then it is also Φ_p-optimal for all p, in particular A-, D- and E-optimal.*

Of course, this is trivial for E-optimality.

An important property of Schur-optimality is given by the following result:

Proposition 2.5 *Suppose that M_1 and M_2 are real symmetric matrices, and that $M_1 - M_2$ is positive semi-definite. Then the eigenvalues of M_1 are upper weakly majorized by those of M_2.*

Proof Let M be any $v \times v$ real symmetric matrix. The smallest eigenvalue μ_0 of M is equal to the minimum of $x^\top M x$ over all unit vectors x. More generally, the nth smallest eigenvalue μ_{n-1} of M is the minimum of $x^\top M x$ over all unit vectors x orthogonal to the eigenvectors associated with the first $n-1$ eigenvalues. It follows immediately that

$$\sum_{i=0}^{n-1} \mu_i = \min_{x_0,\ldots,x_{n-1}} \sum_{i=0}^{n-1} x_i^\top M x_i$$

for $n = 1, \ldots, v$, where the minimum is over all n-tuples of pairwise orthogonal unit vectors.

Now adding a positive semi-definite matrix to M certainly cannot decrease the expression on the right-hand side. □

It follows that, if C_1 and C_2 are the information matrices of two designs, and $C_1 - C_2$ is positive semi-definite, then C_1 beats C_2 on the Schur criterion and hence on all the Φ_p-optimality criteria.

Example 2.6 Let $\mathfrak{D}_1$ be the design in Example 1.1. Since any two distinct points have concurrence 2 in this design, the information matrix is $C_1 = (10I - 2J)/3$, with eigenvalues $0, 10/3, 10/3, 10/3, 10/3$. We will compare $\mathfrak{D}_1$ with the binary design $\mathfrak{D}_2$ obtained by replacing the block $[1, 1, 2]$ by $[1, 2, 3]$. (We use the notation $[a, \ldots]$ to denote a multiset.) A simple computation shows that the information matrix

$$C_2 = \frac{1}{3} \begin{bmatrix} 8 & -1 & -3 & -2 & -2 \\ -1 & 8 & -3 & -2 & -2 \\ -3 & -3 & 10 & -2 & -2 \\ -2 & -2 & -2 & 8 & -2 \\ -2 & -2 & -2 & -2 & 8 \end{bmatrix}$$

of $\mathfrak{D}_2$ has eigenvalues $0, 3, 10/3, 10/3, 13/3$. So $\mathfrak{D}_1$ beats $\mathfrak{D}_2$ on the E-criterion. (We will see later that $\mathfrak{D}_1$ is E-optimal.) However, a further short calculation shows that $\mathfrak{D}_2$ wins on the A- and D-criteria. Indeed, $\mathfrak{D}_1$ beats $\mathfrak{D}_2$ on the Φ_p-criterion if and only if

$$0.9^{-p} + 1.3^{-p} > 2,$$

which holds if and only if $p > 5.32652$ (approximately).

For book-length treatments of optimality, see [4, 82, 91].

2.4 The concurrence graph

We have seen that many optimality criteria of a block design are functions of the non-trivial eigenvalues of its information matrix C. We now give this a graph-theoretic interpretation.

The *concurrence graph* of a block design is defined as follows. The vertices are the points (or treatments); distinct vertices i and j are joined by λ_{ij} edges, where

λ_{ij} is the concurrence of i and j, that is, the number of blocks in which both i and j appear (counted appropriately: that is, λ_{ij} is the (i,j) entry of $NN^\top$, where N is the $v \times b$ incidence matrix of the design—see Section 1.3). Note that the concurrence graph contains no loops, even if the design is not binary.

The *Laplacian matrix* $L(G)$ of a loopless graph G on v vertices is defined as follows: for $i \neq j$, its (i,j) entry is the negative of the number of edges joining i and j; the (i,i) entry is the valency of i (the number of edges incident with i). In other words, $L(G) = D - A(G)$, where D is a diagonal matrix whose diagonal entries are the valencies, and $A(G)$ is the adjacency matrix of G. Note that $L(G)$ is a symmetric matrix with row and column sums zero.

Proposition 2.7 *If a block design has constant block size k (but is not necessarily binary) then its information matrix C is obtained by dividing the Laplacian matrix of its concurrence graph by k.*

Proof We know that $C = R - NN^\top/k$, where R is diagonal. Now $NN^\top$ and $A(G)$ agree at all off-diagonal elements; hence C and $L(G)/k$ agree in all off-diagonal elements. But both C and $L(G)/k$ have row and column sums zero; so $C = L(G)/k$, as required. □

This result, pointed out by Cheng [32] and cited in [91, page 30], has the important consequence that the optimality properties of a design with constant block size k are completely determined by k and its concurrence graph. Given a graph G, there may be many designs which have G as their concurrence graph; so we have potentially made a big simplification to the problem.

Which graphs are concurrence graphs of designs with block size k? The next result answers this question.

Here is a non-standard definition. A *weighted clique* with weights $w_1, \ldots, w_m$ (where $w_1, \ldots, w_m$ are positive integers) has m vertices, numbered $1, \ldots, m$, and has $w_i w_j$ edges joining vertex i to vertex j for $i \neq j$. Its *weight* is the sum of the numbers $w_1, \ldots, w_m$. If all w_i are 1, this is just a complete graph on m vertices, and has weight m.

Proposition 2.8 *Let G be a graph. Then G is the concurrence graph of a block design with block size k if and only if the edge set of G can be partitioned into weighted cliques of weight k. In particular, G is the concurrence graph of a binary block design with block size k if and only if the edge set of G can be partitioned into complete subgraphs of size k.*

Proof From a weighted clique of weight k we can reconstruct a block, in which treatment i occurs w_i times; the concurrences between distinct treatments in this block give precisely the edges of the weighted clique. The converse is clear. □

Remark A given graph may have many such partitions, or none. But even the partition does not uniquely determine the block design. In our running example, two edges joining the vertices 1 and 2 form a weighted clique with weights $(1,2)$ or $(2,1)$, and so can arise from either the block $[1,1,2]$ or the block $[1,2,2]$. Since all optimality properties are determined by the concurrence graph, switching [1, 1, 2] into [1, 2, 2] preserves all such properties.

However, in the special case $k = 2$, then any graph satisfies the conditions: the blocks are the edges of the graph. (Our assumption that there is no block containing only a single treatment guarantees that every block is an edge of the graph rather than a "loop".)

2.5 Laplace eigenvalues

Proposition 2.7 draws our attention to the study of the Laplacian eigenvalues of a graph, and their relationship to graph properties. In this section we give a brief survey of the properties of the Laplace eigenvalues of graphs. Although we speak only of graphs, the results have analogues for edge-weighted graphs (with non-negative weights). For further information and proofs we refer to Mohar's article [72].

The Laplacian matrix is so-called because of its relationship with the Laplacian differential operator $-\nabla^2$ on a manifold. If the manifold is triangulated, then the resulting graph (with its edges weighted by a function of their lengths) is a "discrete approximation" to the manifold, and its Laplacian matrix is an approximation to the Laplacian operator of the manifold. Note that eigenvalues and eigenvectors of $-\nabla^2$ are the solutions of the partial differential equation $\nabla^2 f + \mu f = 0$; the eigenfunctions f are "vibrational modes" of the manifold, and the eigenvalues are related the vibrational frequencies. This led to Mark Kac's famous question, "Can one hear the shape of a drum?" [61], which asks if it is possible for different manifolds or graphs to have the same Laplace eigenvalues. Not surprisingly, the answer, in both the continuous and the discrete cases, is that it is possible.

Let G be a graph with v vertices and e edges; loops are forbidden, but multiple edges are allowed except where we say otherwise. Let $d(i)$ be the valency of vertex i, and let $\delta(G)$ and $\Delta(G)$ be the smallest and largest values of $d(i)$. Recall that the Laplacian matrix $L(G)$ is the $v \times v$ matrix with (i, i) entry $d(i)$ and (i, j) entry minus the number of edges from i to j.

Note that the Laplacian matrix is the sum over edges of matrices of the following form. Each edge between vertices i and j contributes the matrix with submatrix

$$\begin{array}{c} \\ i \\ j \end{array} \begin{array}{cc} i & j \\ \begin{bmatrix} 1 & -1 \\ -1 & 1 \end{bmatrix}, \end{array}$$

all other entries being zero. Thus the matrix for each edge is positive semi-definite. This gives a rather easy proof of the following facts:

Proposition 2.9 *(a) The Laplacian matrix $L(G)$ is positive semi-definite (hence so is any information matrix).*

(b) For any vector x in $\mathbb{R}^v$, we have

$$x^\top L(G) x = \sum_{\text{edges } ij} (x_i - x_j)^2.$$

(c) The nullity of $L(G)$ is equal to the number of connected components of G. In particular, if G is connected, the all-1 vector u spans the null space.

(d) If the graph G_2 is obtained from the graph G_1 by inserting an extra edge, then $L(G_2) - L(G_1)$ is positive semi-definite: so we can never make a design worse on the Schur criterion (and hence any Φ_p criterion) by adding an edge to the graph (Propositions 2.4 and 2.5).

Here is another popular method of proving Proposition 2.9. Choose an arbitrary orientation of the edges of G, and (temporarily) let N be the signed vertex-edge incidence matrix of G, the $v \times e$ matrix whose (i,k) entry is

$$n_{ik} = \begin{cases} +1 & \text{if vertex } i \text{ is the head of the } k\text{th directed edge } f_k, \\ -1 & \text{if vertex } i \text{ is the tail of } f_k, \\ 0 & \text{otherwise.} \end{cases}$$

Then we have

$$NN^\top = L(G),$$

for each edge containing vertex i contributes $+1$ to the (i,i) entry of $NN^\top$, while each edge joining vertices i and j contributes -1 to the (i,j) entry. This shows part (a) of Proposition 2.9. Part (b), giving the quadratic form associated with $L(G)$, is easily proved using the fact that

$$x^\top L(G) x = (N^\top x)^\top (N^\top x).$$

From now on we assume that G is connected. Let $\mu_0, \ldots, \mu_{v-1}$ be the eigenvalues of $L(G)$, arranged in non-decreasing order, so that $\mu_0 = 0$ and $\mu_1 > 0$. (Note that this convention is different from Mohar's [72], who uses $\mu_1, \ldots, \mu_v$.) We often write $\mu_{\max}$ instead of μ_{v-1} for the greatest eigenvalue. Thus we have

$$0 = \mu_0 < \mu_1 \leq \cdots \leq \mu_{\max}.$$

The sum of the Laplace eigenvalues is the trace of $L(G)$, the sum of the valencies of the vertices, which is twice the number of edges; so we have

$$\mu_1 \leq 2e/(v-1) \leq \mu_{\max}.$$

These inequalities can be refined; for example, μ_1 is bounded above by $(d(i)+d(j))/2$, for any two non-adjacent vertices i and j; and

$$\mu_1 \leq v\delta(G)/(v-1) \leq v\Delta(G)/(v-1) \leq \mu_{\max} \leq 2\Delta(G);$$

and $\mu_{\max} = 2\Delta(G)$ if and only if G is bipartite and regular. We will prove the first inequality below; for the rest, see [72].

The Laplace eigenvalues of a simple graph and its complement have a straightforward relation:

Proposition 2.10 Let G be a simple graph on v vertices and $\overline{G}$ its complement. Then $\mu_0(G) = \mu_0(\overline{G}) = 0$ and

$$\mu_i(\overline{G}) = v - \mu_{v-i}(G)$$

for $i = 1, \ldots, v-1$.

This is because $L(G) + L(\overline{G}) = vI - J$, where J is the all-1 matrix, and the all-1 vector u is the zero eigenvector for both $L(G)$ and $L(\overline{G})$, so that any other common eigenvector is perpendicular to u.

If G is regular of valency d, then $L(G) = dI - A(G)$, where $A(G)$ is the usual adjacency matrix of G, so the Laplace eigenvalues have the form $d - \alpha$, where α runs over the eigenvalues of $A(G)$.

One graph parameter closely connected with the Laplace eigenvalues is the isoperimetric number. A "good" graph (for use as a network, say) will be one without bottlenecks: any set of vertices will have many edges joining it to its complement. So, for a set S of vertices, we let $\partial(S)$ (the *boundary* of S) be the set of edges which have one vertex in S and the other in its complement, and then define the *isoperimetric number* $i(G)$ by

$$i(G) = \min\left\{ \frac{|\partial S|}{|S|} : S \subseteq V(G),\ 0 < |S| \leq v/2 \right\}.$$

(We need only take the size of S up to $v/2$ since a set and its complement have the same boundary.) An infinite class of simple graphs is a *family of expanders* if $i(G)$ is bounded away from zero for graphs in this class.

The following result is due to Alon and Milman [3] and Dodziuk [41].

Theorem 2.11 *Let μ_1 be the second Laplace eigenvalue of G, and $\Delta = \Delta(G)$. Then*

$$\mu_1/2 \leq i(G) \leq \sqrt{(2\Delta - \mu_1)\mu_1}.$$

We will prove the left-hand inequality, demonstrating that a graph with large μ_1 has good expansion properties. In fact we show a slightly stronger result, which will be used a number of times.

Proposition 2.12 *Let G have an edge-cutset of size c whose removal separates the graph into components of sizes m and n. Then $\mu_1 \leq c((1/m) + (1/n))$.*

Proof The second eigenvalue μ_1 is the minimum of $x^\top L x / x^\top x$ over all vectors x perpendicular to the all-1 vector u (which is the eigenvector for the smallest eigenvalue $\mu_0 = 0$).

Now let x be the vector taking the value n on vertices on the side of the cut with size m and the value $-m$ on vertices on the other side. Then x has coordinate sum zero, so is perpendicular to the all-1 vector u. By Proposition 2.9(b), we have $x^\top L x = c(m+n)^2$, whereas $x^\top x = mn^2 + nm^2 = mn(m+n)$. So $\mu_1 \leq c(m+n)/mn = c((1/m) + (1/n))$. □

Now the left-hand inequality in the theorem follows: we can choose the cutset such that $c/m = i(G)$ and $m \leq n$; then $\mu_1 \leq 2c/m = 2i(G)$.

The right-hand inequality is one of a family of results called *inequalities of Cheeger type*, since they are analogues of Cheeger's inequality in the continuous case. The analogue of Cheeger's inequality would be the assertion $i(G) \leq \sqrt{2\Delta\mu_1}$; see [38, 72] for the proof of the stronger result quoted here and references to similar results.

Corollary 2.13 *An infinite class of simple graphs is a family of expanders if and only if μ_1 is bounded away from zero for the graphs in the class.*

The importance of expanders in computer science has resulted in a great deal of attention being paid to the graph parameter μ_1. It is also very closely connected with the rate of convergence of a random walk on the graph.

2.6 Variance-balanced designs

Recall that a block design is variance-balanced (or VB, for short) if its information matrix is completely symmetric. Morgan and Srivastav [74] call such a design a *completely symmetric design*, or CSD; because of conflict with the term "symmetric design", we have avoided this. We denote by $\mathrm{VB}(v, k, \lambda)$ a variance-balanced design with v points, block size k and all pairwise concurrences λ. (Remember that they are not assumed to be binary!)

Note that VB designs are not always optimal:

Example 2.14 Two designs with $v = b = 7$, $k = 6$ can be constructed as follows:

(a) the design whose blocks are all 6-subsets of the set of points;

(b) the design obtained from the Fano plane by doubling each occurrence of a point in a block (so that the first block is the multiset $[1, 1, 2, 2, 3, 3]$).

Both these designs are VB; the first has $\lambda = 5$, the second has $\lambda = 4$. The first, being a 2-design, is better, according to Kiefer's theorem.

Question 2.15 *Given k and λ, for which values of v do $\mathrm{VB}(v, k, \lambda)$ designs exist, and what are the possible numbers of blocks of such designs?*

An asymptotic answer to this problem (that is, conditions which are necessary and sufficient for large enough v) would generalize Wilson's existence theorem for 2-designs [106] described earlier. Two surveys of this class of designs have been given by Billington [18, 19].

Morgan and Srivastav define two new parameters of a VB design, as follows:

$$r = \left\lfloor \frac{bk}{v} \right\rfloor, \qquad p = bk - vr,$$

so that $bk = vr + p$ and $0 \leq p \leq v - 1$. Thus, in a 2-design we have $p = 0$. Note that the use of r does not here imply that the design has constant replication!

Morgan and Srivastav further say that a VB design has *maximum trace* if its parameters satisfy the equation

$$r(k-1) = (v-1)\lambda. \tag{2.7}$$

The reason for the term is as follows. Since $bk < v(r+1)$, some treatment occurs at most r times on the bk plots. Each occurrence contributes at most $k-1$ edges to the concurrence graph, so the valency of this vertex is at most $r(k-1)$. But the concurrence graph of a VB design is regular, with valency $(v-1)\lambda$; so we have

$(v-1)\lambda \leq r(k-1)$, and the trace of the information matrix, which is $v(v-1)\lambda$, is at most $vr(k-1)$; equality for the trace implies that $(v-1)\lambda = r(k-1)$.

The maximum trace condition does imply some nice properties.

The above argument shows that, in a VB design of maximum trace, any point lies in at least r blocks (counted with multiplicity), with equality if and only if the point occurs at most once in each block. Since $bk = vr + p$, it follows that the number of "bad" points (which occur more than once in some block) is at most p. So if $p = 0$, the design is binary, and is a BIBD or 2-design.

In Example 2.14, we have $r = 6$, $p = 0$, and so the first design has maximum trace but the second does not.

Morgan and Srivastav [74] show that a VB design with maximum trace is E-optimal. We refer to their paper for the proof, but give here a complementary result in which maximum trace is not assumed but we require that the effect of non-binary blocks is not too large. A binary block (one in which no treatment is repeated) gives rise to $k(k-1)/2$ edges of the concurrence graph; define the *defect* of a non-binary block to be the difference between $k(k-1)/2$ and the number of edges it contributes.

Proposition 2.16 *If $k < v$, then a VB design on v points is E-optimal if the sum of defects of non-binary blocks is less than $v/2$.*

Proof Let x be the sum of defects. Then the number of edges (which we know to be $\lambda v(v-1)/2$) is $bk(k-1)/2 - x$, so that

$$b = \frac{\lambda v(v-1) + 2x}{k(k-1)}.$$

It is easily verified that the non-trivial Laplacian eigenvalues of the λ-fold complete graph are all equal to λv. So, if our design is not E-optimal, then an E-better design (with the same values of (v, b, k)) has $\mu_1 > \lambda v$. Let δ be the minimal degree of the concurrence graph of such a design. By Proposition 2.12,

$$\lambda v < \mu_1 \leq \delta(1 + 1/(v-1)) = \delta v/(v-1),$$

so that $\delta > \lambda(v-1)$, or $\delta \geq \lambda(v-1) + 1$. Hence the concurrence graph has at least $v(\lambda(v-1) + 1)/2$ edges. Since each block of this design contributes at most $k(k-1)/2$ edges, we have

$$\frac{\lambda v(v-1) + 2x}{k(k-1)} = b \geq \frac{v(\lambda(v-1) + 1)}{k(k-1)},$$

whence $x \geq v/2$. So, if $x < v/2$, then no E-better design can exist. $\square$

For $k = 3$, the defect of a non-binary block is 1, so a design with fewer than $v/2$ non-binary blocks is E-optimal. It follows that the design of Example 1.1 is E-optimal, as claimed earlier.

We now turn to constructions.

If we have two VB designs on the same set of v points with the same block size k, having parameters λ_1 and λ_2, then the multiset union of the block multisets is again VB, with parameter $\lambda_1 + \lambda_2$. The new design is not necessarily of maximum trace; but it is so if one of the VB designs we start with is a 2-design and the other is of maximum trace, or if the sum of their p parameters is less than v.

For example, suppose that $k = 3$. A VB design of maximum trace satisfies $2r = (v-1)\lambda$, so that λ is even or v is odd. Moreover, $\lambda = 1$ is impossible (except for 2-$(v, 3, 1)$ designs), since a non-binary block gives concurrence at least 2. Morgan and Srivastav [74] prove that these necessary conditions are sufficient:

Theorem 2.17 *A* VB$(v, 3, \lambda)$ *design of maximum trace exists whenever* $\lambda(v-1)$ *is even and* $\lambda > 1$.

Proof Since a 2-$(v, 3, 6)$ design exists for all v [51], it is enough to settle the existence question for λ in a complete set of non-zero residues mod 6. Now 2-designs exist in the following cases [51]:

- for $\lambda = 1$ or 5, if $v \equiv 1$ or 3 mod 6;
- for $\lambda = 2$ or 4, if $v \equiv 0$ or 1 mod 3;
- for $\lambda = 3$, if v is odd.

We give a construction of VB designs for $\lambda = 2$ and $v \equiv 2$ mod 3 below; they have $p = 1$, so the union of two copies settles $\lambda = 4$. For $\lambda = 5$ or $\lambda = 7$, with v odd, there is a 2-design unless $v \equiv 5$ mod 6; in that case we can take a 2-design with $\lambda = 3$ and a VB design with $\lambda = 2$ or $\lambda = 4$. □

We now look further at the case $k = 3$, $\lambda = 2$. A block consisting of three points is a triangle in the concurrence graph, while a block consisting of two points (one with multiplicity 2) is a double edge. So we want to partition the edge set of doubled K_v into triangles and double edges. Clearly if a 2-$(v, 3, 2)$ design exists, then this gives a VB design with the smallest number of blocks; this design exists if and only if $v \not\equiv 2$ mod 3. There is a "boring" design obtained by partitioning the edge set into double edges, which has the largest number of blocks.

The smallest case for which no 2-design exists is $v = 5$. There are four possibilities, listed in order of increasing numbers of blocks (and perhaps decreasing interest):

- Six triangles and one double edge: this is realised by the design in our running example (Example 1.1).

- Four triangles and four double edges: take a 2-$(4, 3, 2)$ design (consisting of all the 3-subsets of a 4-set) and join its four points to the fifth point by four double edges.

- Two triangles and seven double edges: take a triangle twice and double the seven uncovered edges.

- Ten double edges: this is the "boring" design.

The values of (r, p) in the four cases are $(4, 1)$, $(4, 4)$, $(5, 2)$ and $(6, 0)$. So the first two have maximum trace but the others do not.

Here is a construction for VB$(v, 3, 2)$ designs having just one non-binary block. In this case, as we have seen, we must have $v \equiv 2 \mod 3$. Our construction uses Steiner triple systems; since there are many of these ([107]), the construction is "prolific", producing $v^{(1+o(1))v^2/3}$ non-isomorphic designs.

Suppose first that $v \equiv 2 \mod 6$. In this case, there exist Steiner triple systems of orders $v \pm 1$. Take two such systems, on the point sets $\{1, \ldots, v+1\}$ and $\{1, \ldots, v-1\}$ respectively; let the sets of blocks be $\mathcal{B}_1$ and $\mathcal{B}_2$. Without loss of generality, suppose that the third point of the block B of $\mathcal{B}_1$ containing v and $v+1$ is $v-1$.

Now we take the point set of the new design to be $\{1, \ldots, v\}$. For the blocks, we first remove the block B from $\mathcal{B}_1$; then we replace each occurrence of $v+1$ in any other block in $\mathcal{B}_1 \cup \mathcal{B}_2$ with v; the resulting blocks together with $[v-1, v-1, v]$ make up the design.

We have to check that $\{v-1, v\}$ lies only in $[v-1, v-1, v]$, while every other pair $\{i, j\}$ lies in two blocks. For the first, note that the only other candidate, namely B, has been removed. For the second, there are two cases:

- $j = v, i \neq v-1$: in $\mathcal{B}_1$, there is one block containing i and v, and one containing i and $v+1$ (in which $v+1$ is replaced by v). No block of $\mathcal{B}_2$ can occur.

- $v \notin \{i, j\}$: one block of $\mathcal{B}_1$ and one of $\mathcal{B}_2$ contain $\{i, j\}$, and these two points are unchanged in these blocks.

There is a similar construction when $v \equiv 5 \mod 6$. In this case, both $v-2$ and $v+2$ are orders of Steiner triple systems; and, if $v \geq 17$, we may assume that the larger system contains a subsystem of order 7. Choose the numbering so that the larger system uses the points $a, \ldots, g$, with blocks $abc, ade, afg, bdg, bef, cdf, ceg$ forming the subsystem of order 7, and abc is a block of the smaller system, while d, e, f, g are not in the smaller system. Now:

- Delete these eight blocks from the two systems.

- Any block remaining contains at most one of d, e, f, g. Replace the point e by d and the point g by f everywhere.

- Add new blocks $bdf, cdf, abd, acd, abf, acf$ and bbc.

We leave it to the diligent reader to check that this construction works: only bbc contains b and c, while every other pair lies in two blocks. Note that it gives $(v+2)(v+1)/6 + (v-2)(v-3)/6 - 1 = (v^2 - v + 1)/3$ blocks, the correct number.

For $v = 11$, we can ask the computer to produce a design; there are plenty to choose from (see Section 6.1).

Question 2.18 *For which v, λ does a VB$(v, 3, \lambda)$ design without repeated blocks exist?*

This question is of no statistical significance; but, as explained earlier, it may be taken as a challenge by combinatorialists. More generally one could ask for the possible numbers of distinct blocks in such a design.

3 Optimal designs

In this section we first state some general results on optimality, and then look more closely at the three most important optimality criteria: D, A and E. We will see that they are related to important graph-theoretic notions for the concurrence graph: number of spanning trees, resistance as an electrical network, and expander properties.

3.1 General results

There are a few general results, showing that certain kinds of designs (if they exist) are optimal for a wide variety of criteria. The first such result (Theorem 2.3) was due to Kiefer [66], and shows that 2-designs are optimal (if they exist).

If a 2-design with the given parameters does not exist, then this theorem gives no information. An interesting question is: What happens if the necessary conditions (1.1) are satisfied but there is no design? A special case of this occurs for the values $v = 15$, $b = 21$, $k = 5$. (If a 2-design with these parameters existed, then by the Hall–Connor theorem [50] it would be the block residual of a 2-(22,7,2) design; but such a design does not exist, by the Bruck–Ryser–Chowla theorem [24, 36].) This parameter set has been examined by Reck and Morgan [85, 86], and the A-, D- and E-optimal designs determined.

Some extensions of Kiefer's Theorem are known. In order to state them, we require some more definitions.

A binary equireplicate design is said to be *group-divisible* if the treatments can be divided into "groups", all of the same size, such that two treatments in the same group are contained in λ_1 blocks while two treatments in different groups are contained in λ_2 blocks. Group-divisible designs are partially balanced with respect to the group-divisible association scheme defined earlier.

A *regular-graph design* is a binary equireplicate design in which any pair of vertices is contained in either λ or $\lambda + 1$ blocks, for some λ [58]. The name arises because its concurrence graph consists of the λ-fold complete graph together with a simple regular graph.

Now Chêng [30, 31] proved:

Theorem 3.1 *Let f be any non-negative real function which is strictly decreasing and strictly convex, such that its derivative is strictly concave and $\lim_{x \to 0^+} f(x) = \infty$. For $X \in \mathcal{X}_v$, set $\Phi(X) = \sum f(\mu_i)$, where the μ_i run over the eigenvalues of X having eigenvector orthogonal to the all-1 vector.*

(a) *If $\mathcal{D}(v,b,k)$ contains a group-divisible design with two groups and $\lambda_2 = \lambda_1 + 1$, then the minimum of $\Phi(C)$ over information matrices C of designs in $\mathcal{D}(v,b,k)$ is attained by such a design.*

(b) *If $\mathcal{D}(v,b,k)$ contains a group-divisible design with $\lambda_2 = \lambda_1 + 1$, then the minimum of $\Phi(C)$ over information matrices C of regular-graph designs in $\mathcal{D}(v,b,k)$ is attained by such a design.*

Again the result applies for A-optimality (with $f(x) = 1/x$) and D-optimality (with $f(x) = -\log x$). For E-optimality, we can take $f(x) = x^{-p}$ for $p > 0$ and let $p \to \infty$.

Many optimality results limit the class of competing designs to be binary and equireplicate. For example, Cheng and Bailey [34] proved the following.

Theorem 3.2 *Let f and Φ be as in Theorem 3.1. If there is a regular-graph design in $\mathcal{D}(v,b,k)$ for which (a) the graph in question is strongly regular and (b) the concurrence matrix Λ is singular, then this design minimizes $\Phi(C)$ over all binary equireplicate designs in $\mathcal{D}(v,b,k)$.*

For example, the generalized quadrangle with $v = b = 15$ and $k = 3$ is A-, D- and E-optimal among binary equireplicate designs of this size. The points (treatments) are all unordered pairs from a set of size six, and the blocks are all the partitions of that set into three pairs. A pair is in a given block if it is a part of the corresponding partition.

This result has been extended in [13].

For further general results on optimality we refer to [82, 91].

3.1.1 Equireplicate designs and their duals If design $\mathfrak{D}$ in $\mathcal{D}(v,b,k)$ is equireplicate with replication r and incidence matrix N, then its information matrix C is $rI - k^{-1}NN^\top$, whose eigenvalues lie in $[0,r]$. Statisticians usually compare such a design with the hypothetical one that would be possible in a single large block of vr plots if enough similar plots were available for the experiment. The information matrix for this design is $r(I - v^{-1}J)$, all of whose non-trivial eigenvalues are equal to r. Normalizing by this, the non-trivial eigenvalues of C are divided by r to give what are called the *canonical efficiency factors*, which lie in $(0,1]$ for connected designs. So long as we are considering only equireplicate designs, the eigenvalues of C can be replaced by the canonical efficiency factors in all the standard optimality criteria.

Interchanging the roles of treatments and blocks in $\mathfrak{D}$ gives an equireplicate design $\mathfrak{D}^*$ in $\mathcal{D}(b,v,r)$, which is called the *dual design*. Its incidence matrix is $N^\top$ and its information matrix C^* is $kI - r^{-1}N^\top N$, with eigenvalues in $[0,k]$. The nonzero eigenvalues of $N^\top N$ and $NN^\top$ are the same, including multiplicities, so the canonical efficiency factors of a design and its dual are the same, apart from $|b-v|$ occurrences of the maximum value 1. This observation gives an easy proof of the following result [91, p. 28]:

Theorem 3.3 *On all the standard optimality criteria, a given equireplicate design is optimal among equireplicate designs if and only if its dual design is optimal among equireplicate designs.*

3.1.2 Operations on designs The following elementary lemma, whose proof we omit, allows the non-trivial eigenvalues of various designs obtained from a given one to be calculated.

Proposition 3.4 *Let $\mathfrak{D}_1$ and $\mathfrak{D}_2$ be designs on the same set of v points. Suppose that, for $n = 1, 2$, the design $\mathfrak{D}_n$ has constant block size k_n. Suppose further that the concurrences $\lambda_{ij}^{(n)}$ satisfy*

$$\lambda_{ij}^{(2)} = \alpha + \beta \lambda_{ij}^{(1)}$$

for $i \neq j$. Then the non-trivial eigenvalues $\theta_m^{(n)}$ of the information matrices are related by

$$\theta_m^{(2)} = \begin{cases} \dfrac{\alpha v}{k_2} + \dfrac{\beta k_1}{k_2} \theta_m^{(1)} & \text{if } \beta \geq 0, \\ \dfrac{\alpha v}{k_2} + \dfrac{\beta k_1}{k_2} \theta_{v-m}^{(1)} & \text{if } \beta < 0. \end{cases}$$

In particular, suppose designs $\mathfrak{D}'_1$ and $\mathfrak{D}'_2$ also have v points and block-sizes k_1 and k_2 and their concurrences stand in the same relation, with the same values of α and β. If $\beta < 0$, suppose also that the information matrices of $\mathfrak{D}_1$ and $\mathfrak{D}'_1$ have the same trace. Then $\mathfrak{D}_1$ beats $\mathfrak{D}'_1$ on the Schur criterion if and only if $\mathfrak{D}_2$ beats $\mathfrak{D}'_2$ on the Schur criterion. If $\mathfrak{D}_1$ is E-better than $\mathfrak{D}'_1$, and $\beta > 0$, then $\mathfrak{D}_2$ is E-better than $\mathfrak{D}'_2$.

In general, affine transformations of the form given in this Proposition do not preserve optimality criteria. We look at three special cases.

The complement of the block set Let $\mathfrak{D}_1$ be a simple binary design with constant block size k, and let $\mathfrak{D}_2$ be the design whose blocks are all k-sets which are *not* blocks of $\mathfrak{D}_1$. (If $k = 2$, this is the complement of the graph $\mathfrak{D}_1$.) Then the conditions of Proposition 3.4 hold, with $k_1 = k_2 = k$ and $\alpha = \binom{v-2}{k-2}$, $\beta = -1$.

The complementary design, whose blocks are the complements of the blocks of the original Suppose that $\mathfrak{D}_1$ is binary and equireplicate, with block size k and replication r, and let the blocks of $\mathfrak{D}_2$ be the complements of those of $\mathfrak{D}_1$. A simple inclusion-exclusion argument shows that the conditions of Proposition 3.4 hold, with $k_1 = k$, $k_2 = v - k$, $\alpha = b - 2r$, $\beta = 1$. See also [59].

The design formed by adjoining all points to a block Let $\mathfrak{D}_1$ be as in the preceding paragraph. Form a new design $\mathfrak{D}_2$ with block size $v + k$ by adding one copy of each point to each block. Now the contributions to $\lambda_{ij}^{(2)}$ from a block B are

$$\begin{cases} 4 & \text{if } i, j \in B, \\ 2 & \text{if } i \in B, j \notin B \text{ or } vice\ versa, \\ 1 & \text{if } i, j \notin B. \end{cases}$$

Hence the conditions of Proposition 3.4 hold, with $k_1 = k$, $k_2 = v + k$, $\alpha = b + 2r$, and $\beta = 1$.

John and Williams [59] conjectured that, if $\mathfrak{D}_1$ is A-optimal, then so is $\mathfrak{D}_2$. Example 2.2 of Jacroux and Whittinghill [55] shows that this is not true in general.

Example 3.5 The Petersen graph is A-optimal among equireplicate designs in $\mathcal{D}(10, 15, 2)$ but John and Mitchell [58] found by computer search that its complement (the *triangular graph* $T(5)$, the line graph of K_5) is not A-optimal. The Laplace eigenvalues of $T(5)$ are 5 (four times) and 8 (five times), see Example 3.11; John and Mitchell found that the A-optimal equireplicate design in $\mathcal{D}(10, 30, 2)$ consists of a bipartite graph $K_{4,6}$ into which edges have been inserted to make a 6-cycle on the larger class of the bipartition. This graph has Laplace eigenvalues 5 (twice), 6 (three times), 7 (twice), 8 and 10; so it beats $T(5)$ on the D-criterion as well as the A-criterion, and they are equal on the E-criterion. The two graphs are not Schur comparable.

3.2 D-optimality: spanning trees

D-optimality is closely connected with maximizing the number of spanning trees of the concurrence graph [31, 45]. This is a consequence of Kirchhoff's Matrix-Tree Theorem.

The Matrix-Tree Theorem asserts:

Theorem 3.6 *Let G be a graph with v vertices. Then the following are equal:*

(a) the number of spanning trees of G;

(b) any $(v-1) \times (v-1)$ cofactor of the Laplacian matrix $L(G)$;

(c) the product of the Laplacian eigenvalues $\mu_1, \ldots, \mu_{v-1}$, divided by v.

The fact that if a matrix has row and column sums zero then all its cofactors are equal, and the expression in terms of the non-trivial eigenvalues, are much more general; we give the simple proof at the end of this section.

Hence a block design with block size k on v vertices is D-optimal in some class if and only if its concurrence graph maximizes the number of spanning trees over this class.

For example, suppose that we are considering graphs with v vertices and $\lambda v(v-1)/2$ edges. (This applies if we want a binary block design with v treatments and blocks of size k, where $bk(k-1) = \lambda v(v-1)$, for example.) By Kiefer's theorem, if a 2-design exists, then it is optimal. This is certainly the case if $k = 2$. So, for example, the λ-fold multiple of the complete graph has the largest number of spanning trees of all multigraphs with v vertices and $\lambda v(v-1)/2$ edges, as we noted in the Introduction.

The number of spanning trees of a graph is a topic with a large literature. It was given extra impetus by a conjecture of Merino and Welsh [71], asserting that the number of spanning trees is bounded above by the maximum of the number of acyclic orientations and the number of totally cyclic orientations. A recent investigation [28] suggests that, for sparse graphs, the number of acyclic orientations is larger than the number of totally cyclic orientations, with the reverse being true for dense graphs.

In network theory, a simple graph is called *t-optimal* if it maximizes the number of spanning trees among simple graphs with given number of vertices and edges. Petingi and Rodriguez [80] extend the results in [31], but do not mention [35]. Of course, for D-optimality there is no restriction to simple graphs.

We also mention here that the number of spanning trees of a graph G is an evaluation of the Tutte polynomial of G; in fact, it is $T(G; 1, 1)$. See [76, 94] for more information about the Tutte polynomial of a graph or matroid. The other optimality criteria are not associated with evaluations of the Tutte polynomial. For example, all trees on n vertices have Tutte polynomial x^n; we will see in Section 4 that they differ on the A- and E-criteria.

In the other direction, graphs with the same Laplace eigenvalues may have different Tutte polynomials. For example, the graph $L_2(4)$ (the line graph of the complete bipartite graph $K_{4,4}$) and the Shrikhande graph are strongly regular graphs with the same eigenvalues, but the former has 576 proper 4-colourings (these are just the Latin squares of order 4) while the latter has 240.

3.2.1 Matrices with row and column sums zero

It may be folklore that, if a square matrix has all row and column sums zero, then all its cofactors are equal. (This comes up in the Matrix-Tree Theorem.) Here is a simple proof of a generalization to arbitrary square singular matrices, which works even in non-zero characteristic with some restriction. For any (column) vector x with components $x_1, \ldots, x_v$, set $\|x\| = \sum x_i^2$.

Proposition 3.7 *Let A be a square matrix over a field F. Suppose that there exist vectors x and z satisfying*

- *$\|x\|$ and $\|z\|$ are non-zero (in F);*
- *$x^\top A = 0$ and $Az = 0$.*

Then there is a constant d such that the (i,j) cofactor of A is equal to $x_i d z_j$.

Proof We prove the result for the $(1,1)$ cofactor for convenience; the argument is the same for any cofactor.

Consider the matrix $A' = A + xz^\top$, with (i,j) entry $a_{ij} + x_i z_j$. Perform the following row and column operations on it:

- Let r_i be the ith row of A'. Replace r_1 by $\sum_i x_i r_i$, leaving the other entries unaltered. As a result, the $(1,j)$ entry becomes

$$\sum_i x_i(a_{ij} + x_i z_j) = \|x\| z_j,$$

while elements in rows other than the first are unaltered. The determinant is multiplied by x_1.

- Let c_j be the jth column of this matrix. Replace c_1 by $\sum_j c_j z_j$, leaving the other entries unaltered. As a result, the $(1,1)$ entry becomes

$$\sum_j \|x\| z_j^2 = \|x\| \cdot \|z\|;$$

for $i > 1$, the $(i,1)$ entry becomes

$$\sum_j (a_{ij} + x_i z_j) z_j = x_i \|z\|,$$

and the other elements are unaltered. The determinant is multiplied by z_1.

- Subtract $x_i/\|x\|$ times the first row from the ith row, for all $i > 1$. For $i > 1$, the $(i,1)$ entry becomes 0, while for $i,j > 1$, the (i,j) entry becomes

$$a_{ij} + x_i z_j - (x_i/\|x\|)\|x\| z_j = a_{ij}.$$

The determinant is unaltered by this move.

Since we have a column (the first) in which all but one entry is zero, this entry being $\|x\| \cdot \|z\|$, and the entries in rows and columns different from the first are identical with those of A, we see that

$$x_1 z_1 \det(A') = \|x\| \cdot \|z\| \cdot A_{11},$$

where A_{11} is the $(1,1)$ cofactor. So the Proposition is proved, with

$$d = \det(A')/(\|x\| \cdot \|z\|).$$

□

Now suppose that A is a real symmetric matrix with row and column sums zero, and eigenvalues $\mu_0 = 0, \mu_1, \ldots, \mu_{v-1}$. Taking x and z to be the all-1 vector u, we see that the determinant of $A + uu^\top$ is $v \prod_{i=1}^{v-1} \mu_i$, which gives the result mentioned earlier.

Since $uu^\top = J$, this result is related to method (b) in Section 2.1 for finding a generalized inverse of A.

3.3 A-optimality: electrical networks, random walks, thickets

Given a graph G, we can consider it as an electrical network with a 1-ohm resistance in each edge. If we connect a battery across two of the vertices (or otherwise apply a potential difference to them) then current will flow in the network. Three rules are obeyed.

(a) *Ohm's Law*: In every edge,

$$V = IR. \tag{3.1}$$

Here V denotes the voltage drop across the edge, I denotes the current flowing in the edge, and R is the resistance. Because we consider that $R = 1$ for every edge, we have $V = I$ in every edge. (When we consider designs with non-constant block size in Section 5.4 then the resistances have to be different.)

(b) *Kirchhoff's Voltage Law*: The total voltage drop from one vertex to any other vertex is the same no matter which path we take from one to the other.

(c) *Kirchhoff's Current Law*: At every vertex which is not one of the two connected to the battery, the total current coming in is equal to the total current going out.

Given two vertices i and j, we imagine applying voltage 0 at i and voltage V at j. We use Kirchhoff's Laws [67] to find the total current I from i to j, then use Equation (3.1) to define the *effective resistance* R_{ij} between i and j as V/I.

There are two further laws in electrical networks: the "series" and "parallel" laws, which are easy consequences of the three laws above. If two networks are connected in series at a single vertex then the effective resistance in the combined network is the sum of the two resistances. On the other hand, if they are connected in parallel, joining one "input" vertex to one "output" vertex, then the effective resistance in the combined network is the reciprocal of the sum of the reciprocals of the two resistances.

Tjur [100, 101] and Bailey [7] recognized that these two laws are precisely those that apply to the variance when different estimators are combined. Tjur imagined the Levi graph as an electrical network, Bailey the concurrence graph. In fact, standard textbooks on electrical engineering, such as [14, 39], give the result that, if L is the Laplacian matrix of the graph of the (connected) electrical network, then the effective resistance R_{ij} between i and j is given by

$$R_{ij} = L_{ii}^- + L_{jj}^- - L_{ij}^- - L_{ji}^-. \tag{3.2}$$

Comparing Equations (2.6) and (3.2), and using Proposition 2.7, we see that pairwise variance = effective resistance $\times k\sigma^2$.

This observation is useful as well as interesting. For sparse graphs, or poorly connected graphs, effective resistance can be calculated by hand much faster than the information matrix can be inverted (including data-entry time).

Example 3.8 The poorly connected graph in Figure 3 is taken from [9]. Consider vertices C and D. If current is flowing from C to D then there will be no current in the part of the graph to right of E, and there will be no current in the edge AB, because we can interchange those vertices without changing the graph. This leaves current in the three disjoint paths CAD, CBD and CED. They each have length 2, so the laws permit a current of 1 amp in each of them. Now the total current flowing from C to D is 3, and the voltage drop from C to D is 2, so the effective resistance between C and D is 2/3.

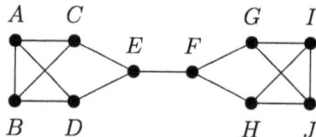

Figure 3: A poorly connected graph

For sparse graphs with a general number of vertices along a long path, such as the one in Figure 4 (taken from [7]), numerical inversion of the information matrix is impossible, but it is perfectly possible to calculate effective resistances in terms of path-lengths.

Other information known to graph theorists and to electrical engineers can now be used to simplify the calculations. For example, [20] gives a formula for the effective resistance between i and j in terms of thickets. A *spanning thicket* for a graph is a subset of its edges which forms a forest with two connected components, with every original vertex being in exactly one component. The formula states that

$$R_{ij} = \frac{\text{number of spanning thickets with } i, j \text{ in different parts}}{\text{number of spanning trees}}. \tag{3.3}$$

This is quite easy to calculate for sparse graphs.

Combinatorics of optimal designs

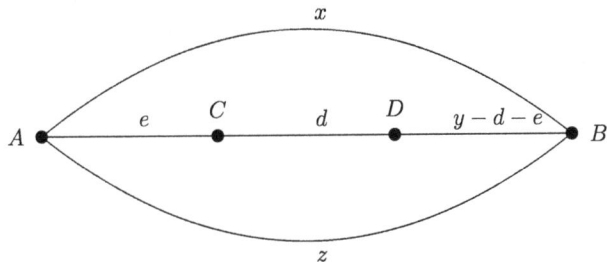

Figure 4: A graph with three parallel paths of lengths x, y and z

Example 3.9 Consider the graph in Figure 4. The only way to make a spanning tree is to remove one edge from two of the parallel paths. To make a spanning thicket, either remove one edge from each of the three parallel paths, or remove two edges from one and one edge from another. For example, to make a spanning thicket separating C and D, remove one edge from the direct path between C and D, and either (i) remove one edge from each of the other two long paths between A and B, or (ii) remove one edge between A and C or one between B and D, and remove one edge from one of the other two long paths. The number of such thickets is $d[xz + (y - d)(x + z)]$. The number of spanning trees is $xy + xz + yz$, so the effective resistance between C and D is $[d(xy + xz + yz) - d^2(x + z)]/(xy + xz + yz)$.

Summing Equations (3.3) over all pairs of distinct vertices gives

$$\sum_{i<j} R_{ij} = \frac{\sum_{\text{thickets } F} |F_1||F_2|}{\text{number of spanning trees}},$$

where the sum is over thickets F consisting a tree F_1 containing i and a tree F_2 containing j: see [92, 102]. This gives a relatively quick method of calculating the A-criterion for sparse graphs.

3.3.1 Foster's formulae and their generalization This section is based on the work of Emil Vaughan [103]. See also [15, 56]. In this section, we consider only simple graphs; the results can be extended to multigraphs but statements are more elaborate.

Let G be a simple connected graph with v vertices. Recall that we are interested in the quantity $\sum_{i,j} R_{ij}$, summed over all pairs of vertices. In 1948, Foster [42] proved the remarkable fact that summing over edges always gives $v - 1$. In this theorem, the notation $i \sim j$ means that we are summing over all unordered pairs $\{i, j\}$ of vertices which are adjacent in the graph.

Theorem 3.10 For any graph G on v vertices,

$$\sum_{i \sim j} R_{ij} = v - 1.$$

Thirteen years later, he proved an analogous formula for the sum of resistances over pairs of vertices at distance at most 2. More precisely, he showed that

$$\sum_{i\sim h\sim j} \frac{R_{ij}}{d(h)} = v - 2.$$

Note that, under the assumption that the graph is regular and that the numbers of common neighbours of two vertices at distance 1 (respectively 2) are constant, this formula gives us the sum of resistances over all pairs of vertices at distance 2. In particular, if the graph is strongly regular, we can compute the sum of resistances over all pairs of vertices.

Example 3.11 *The Petersen graph.* This graph has ten vertices, valency 3, and diameter 2; there are no triangles, and any two non-adjacent vertices have a unique common neighbour. By Foster's first formula, the sum of the resistances between all pairs of adjacent vertices is 9. Also, since there are no triangles, the sum in Foster's second formula is over all pairs of non-adjacent vertices; for each such pair $\{i,j\}$ there is a unique h, and $d(h) = 3$. So the sum of resistances between all non-adjacent pairs is $3 \cdot 8 = 24$, and the total is 33. (Note that, by symmetry, the resistance between two adjacent vertices is 3/5 while that between two non-adjacent vertices is 4/5.)

This can also be found by considering eigenvalues. The eigenvalues of the adjacency matrix are 3, 1, −2 with multiplicities 1, 5, 4 respectively; so the Laplace eigenvalues are 0, 2, 5 with the same multiplicities. Now v times the sum of reciprocals of the non-trivial eigenvalues is $10 \cdot (5/2 + 4/5) = 33$.

New proofs were found by Tetali [97, 98] using the random walk approach, and Palacios [77] extended them to distance 3. Vaughan found a generalization for arbitrary distance, requiring suitable assumptions on the graph; we now discuss this.

Consider the uniform random walk on the graph G: in other words, at each time stage, we move from our present vertex to a neighbouring vertex, all neighbours being equiprobable. Let P_{ij} be the probability of moving from i to j in a single move; thus $P_{ij} = 1/d(i)$ if $i \sim j$, and $P_{ij} = 0$ otherwise. Let $P_{ij}^{(m)}$ be the probability of moving from i to j in exactly m steps. Note that $P_{ii}^{(0)} = 1$ and $P_{ij}^{(0)} = 0$ for $i \neq j$; and $P_{ij}^{(1)} = P_{ij}$. Moreover,

$$P_{ij}^{(m+1)} = \sum_h P_{ih}^{(m)} P_{hj}.$$

If G is a regular graph with valency d, then $P_{ii}^{(m)} = W_{ii}^{(m)}/d^m$, where $W_{ii}^{(m)}$ is the number of walks of length m starting and ending at i.

We say that the graph G is *walk-regular up to length* m if the numbers $W_{ii}, W_{ii}^{(2)}, \ldots, W_{ii}^{(m)}$ are independent of the chosen vertex i. A graph is *walk-regular* if it is walk-regular up to length m for all m. See [48, p. 190]. Note that

- all graphs are walk-regular up to length 1;

- a regular graph is walk-regular up to length 2;
- if a graph is regular, then it is walk-regular up to length m if and only if the probability $P_{ii}^{(p)}$ of returning to vertex i after p steps is independent of i for $p = 1, \ldots, m$;
- a distance-regular graph is walk-regular.

Now Vaughan's Theorem states the following:

Theorem 3.12 *Suppose that G is walk-regular up to length m, for some positive integer m. Then the sum of the resistances between all pairs of vertices at distance at most $m + 1$ is*

$$d^m \left(v \left(1 + \sum_{p=1}^{m} P_{ii}^{(p)} \right) - m - 1 \right).$$

Note that, if $m = 0$, then d is undefined; for $m > 0$, d is the (constant) valency of the graph; for $m = 0, 1$ we recover Foster's two formulae.

Using this theorem, we see that, if the graph G is distance-regular, then the numbers $W_{ii}^{(m)}$ of closed walks, and hence the sum of resistances between all pairs of vertices of G, can be computed in terms of the parameters c_p, a_p, b_p of the graph. Calculation of the eigenvalues is not required to evaluate this.

3.4 E-optimality: spectral gap and root systems

A design is E-optimal if the smallest non-trivial eigenvalue of its information matrix is as large as possible; equivalently, if the smallest non-trivial eigenvalue μ_1 of the Laplacian of its concurrence graph is as large as possible.

We have noted the importance of μ_1 in the theory of expanders, random walks, etc. We see that maximizing μ_1, for graphs with v vertices and e edges, is equivalent to finding the E-optimal design with block size 2 having v points and e blocks. For larger block size, we have to take graphs which are concurrence graphs, as described in Section 2.4.

There is a lot of interest in graphs of fixed valency d having large second Laplace eigenvalue. Alon and Boppana (see [2]) showed (see also [49, 90]):

Theorem 3.13 *In any infinite family of connected regular graphs of valency d, the smallest non-trivial Laplace eigenvalue μ_1 satisfies*

$$\limsup \mu_1 \leq d - 2\sqrt{d-1}.$$

A graph G of valency d is said to be a *Ramanujan graph* if $\mu_1(G) \geq d - 2\sqrt{d-1}$. Explicit Ramanujan graphs were constructed by Margulis [69] and by Lubotzky, Phillips and Sarnak [68]; an accessible account with all of the background material explained appears in [38]. These graphs have good expansion properties, and also often have small diameter and large girth.

However, more often we will not be dealing with graphs with fixed valency; and, of course, if the block size of the designs is greater than 2, then the concurrence graphs will not have the large girth typical of Ramanujan graphs. Indeed, the graphs may have repeated edges!

3.4.1 Root systems

As we have seen, E-optimality involves maximizing the least non-trivial Laplace eigenvalue of the concurrence graph G. If G is simple, then this is equivalent to minimizing the greatest Laplace eigenvalue of its complement $\overline{G}$. Moreover, if $\overline{G}$ is regular, then this is equivalent to maximizing the smallest eigenvalue of its adjacency matrix.

There is another situation in which similar considerations apply, suggested to the authors by Ching-Shui Cheng. Suppose that a design has constant block size k, constant replication r, and has the property that all pairwise concurrences lie in the set $\{\lambda - 1, \lambda, \lambda + 1\}$ for some number λ. Then the concurrence matrix has the form $aI + bJ - M$, where M is a symmetric matrix with entries 0 and ± 1 having zero diagonal and constant row sums. Once again, maximizing the smallest Laplace eigenvalue is equivalent to maximizing the smallest eigenvalue of M.

Now, there is a structure theorem for such matrices with smallest eigenvalue -2 or greater, which we now describe. This approach follows [27] but is a bit different: that paper considered only $\{0, 1\}$ matrices and did not impose the constant row-sum condition. The approach is based on the classical theory of root systems.

A *root system* is a finite non-empty set S of vectors in Euclidean space $\mathbb{R}^d$ satisfying the following conditions:

- if $x \in S$ and $\alpha \in \mathbb{R}$, then $\alpha x \in S$ if and only if $\alpha \in \{\pm 1\}$;
- for all $x, y \in S$, we have $2(x \cdot y)/(x \cdot x) \in \mathbb{Z}$;
- for all $x \in S$, the reflection ρ_x of $\mathbb{R}^d$ in the hyperplane perpendicular to x maps S into itself.

We can assume (and shall always do so) that S spans $\mathbb{R}^d$. The second condition is called the *crystallographic restriction*. The significance of the quantity referred to is that the reflection ρ_x is given by

$$\rho_x : y \mapsto y - \frac{2(x \cdot y)}{x \cdot x} x.$$

A root system is *indecomposable* if it is not contained in the union of two perpendicular subspaces of $\mathbb{R}^d$. There is no loss in assuming this, since an arbitrary root system is a direct sum of indecomposable ones.

The indecomposable root systems were classified by Cartan and Killing in the early twentieth century in connection with the classification of simple Lie algebras over the complex numbers. See [22, 53] for an account. An alternative proof in the spherical case appears in [27] and [48, Chapter 12].

A root system S is called *spherical* if all vectors in S have the same length. If S is spherical, we may re-scale so that $x \cdot x = 2$ for all $x \in S$; then the crystallographic condition asserts that $x \cdot y$ is an integer, and Cauchy's inequality says that $|x \cdot y| < 2$ for $y \neq \pm x$, so we have $x \cdot y \in \{0, \pm 1\}$ for $y \neq \pm x$. This says that any two such vectors lie at an angle of $60°$, $90°$ or $120°$. Moreover, if two vectors x, y make an angle of $120°$, then the vector $x + y$ in their plane making an angle of $60°$ with each is also in S, by the reflection condition.

According to the classification, there are two infinite families $\{A_d : d \geq 1\}$ and $\{D_d : d \geq 4\}$, and three exceptional examples E_6, E_7 and E_8, of indecomposable

Combinatorics of optimal designs

spherical root systems. All of these have very explicit representations. For example, the vectors of D_d can be represented as

$$\{\pm e_i \pm e_j : 1 \leq i < j \leq d\}$$

where $\{e_1, \ldots, e_d\}$ is an orthonormal basis for $\mathbb{R}^d$.

Now the relevance of this to eigenvalues comes from the following two observations.

Extending to a maximal set Let T be any set of vectors in $\mathbb{R}^d$ at angles $60°$, $90°$ and $120°$. Assume that all the vectors in T have length $\sqrt{2}$. Let T' be the set of lines through the origin (1-dimensional subspaces) spanned by these vectors. These lines make angles $60°$ and $90°$ with each other. Extend T' to a maximal set S' of lines at these angles. Then S' is *star-closed*, i.e. if two lines in S' are at an angle of $60°$, then the third line at $60°$ to both is also in S'. (This requires a short calculation, the only calculation in the whole argument.) Now let S consist of the vectors of length $\sqrt{2}$ on the lines of S'. Then S is a spherical root system (the reflection condition follows from the star-closure of S'), is indecomposable if T is, and spans $\mathbb{R}^d$ if T does; and S contains T. So any set T satisfying our conditions is embeddable in a root system.

Representing matrices Now let A be a $v \times v$ symmetric matrix with entries $0, \pm 1$ with zero diagonal, and suppose that the smallest eigenvalue of A is -2 or greater. Then $2I + A$ is positive semi-definite symmetric, and so is the matrix of inner products of a set T of vectors spanning $\mathbb{R}^d$, where d is the rank of $2I + A$ (so that $v - d$ is the multiplicity of the eigenvalue -2 of A, which may be zero.) Since the diagonal entries are 2, all the vectors have length $\sqrt{2}$; the assumption on entries of A shows that the vectors have inner product 0 or ± 1 with one another, so make angles $60°$, $90°$ and $120°$. So T can be embedded into a spherical root system.

This has important consequences for the structure of such matrices.

Graphs Suppose that the entry -1 does not occur in A; then A is the adjacency matrix of a simple graph G with smallest eigenvalue -2 or greater. It is clear that only finitely many such graphs can be "embedded" in the exceptional root systems; moreover, a graph embeddable in A_d or D_d is a *generalized line graph* (see definition below). In particular, if G is regular, then it is one of the following:

- A *line graph*. (If H is a simple graph, then the line graph of H has one vertex for each edge of H; two vertices of the line graph are joined if the corresponding edges of H meet.) We denote the line graph of H by $L(H)$.

- A *cocktail party graph*: this is a complete multipartite graph with parts of size 2; that is, the vertices are paired up, two vertices are joined if and only if they are not in the same pair.

- One of finitely many exceptions (embeddable in the exceptional root systems). These exceptions have all been determined; there are several hundred of them, on at most 28 vertices.

If $L(H)$ is regular, then either H itself is regular, or H is bipartite and *semiregular* (the valency of a vertex depends only on the class of the bipartition containing it) and satisfies some additional strong restrictions (but examples for this case do exist).

Let H be a graph with v vertices, and let $a_1, \ldots, a_v$ be non-negative integers. Then the *generalized line graph* $L(H; a_1, \ldots, a_v)$ consists of the disjoint union of $L(H)$ and cocktail parties C_i on $2a_i$ vertices for $i = 1, \ldots, v$, with an edge from each vertex of C_i to each vertex of $L(H)$ indexed by an edge containing vertex i of H.

Regular signed graphs In general we can regard the matrix A as the adjacency matrix of a *signed graph* (a graph with a sign attached to each edge). Not much can be said in general, but in the case where A has constant row and column sums, see [25] for some results.

3.4.2 Discrepancy We briefly review some results of Morgan [73]. These allow the classification of E-optimal designs for a reasonable range of parameters. This extends the analysis of VB designs given in Section 2.6.

Given v, b, k, we define parameters r, λ, p, q by

$$r = \left\lfloor \frac{bk}{v} \right\rfloor, \qquad p = bk - vr,$$

$$\lambda = \left\lfloor \frac{r(k-1)}{v-1} \right\rfloor, \qquad q = r(k-1) - \lambda(v-1).$$

Thus, the necessary conditions (1.1) for a 2-design are satisfied if and only if $p = q = 0$; in this case, a 2-design is best if it exists. The necessary condition for a 1-design (an equireplicate design) is $p = 0$.

Given a *binary, equireplicate* design in $\mathcal{D}(v, b, k)$, with information matrix C, define its *discrepancy matrix* to be the matrix Δ defined by

$$C = \frac{v\lambda + q}{k} I - \frac{\lambda}{k} J - \frac{1}{k} \Delta.$$

Then Δ has diagonal entries zero and, for $i \neq j$, has off-diagonal entries $\lambda_{ij} - \lambda$, where λ_{ij} is the concurrence of i and j.

We see that the design is a 2-design if and only if its discrepancy matrix is zero. (This is the reason for the name: the matrix measures the "imbalance" of the design.) Also, it is a regular-graph design if and only if all entries of Δ are 0 or 1. We see that E-optimality is equivalent to maximizing the largest eigenvalue of Δ.

Using this approach, the E-optimal designs with $v \leq 15$ have been found. Not all are regular-graph designs (that is, the discrepancy matrix of the optimal design may have entries which are negative or greater than 1). The results are all available at designtheory.org.

One striking observation to emerge from the computation is that, while the E-optimal designs are not necessarily A-optimal, they are very close to optimality in terms of the A-criterion.

Morgan's extensive catalogue should form the basis for further work on E-optimality.

4 Graphs with few edges

In this section, after looking briefly at trees, we determine the optimal unicyclic graphs (connected graphs with equally many vertices and edges), and then turn to sparse graphs (with average valency at most 5/2). We will see that different optimality criteria select very different graphs.

4.1 Trees

As a warm-up, let us consider trees. Any tree has a unique spanning tree, so the D-criterion makes no distinction between one tree and another. In a tree, the resistance between two terminals is just the length of the path joining them; so A-optimality requires that the sum of distances between pairs of vertices is minimized. It is clear that the star $K_{1,v-1}$ does this, since the distance between any two non-adjacent vertices is 2 in the star, and no other tree can do as well or better.

Now we turn to E-optimality. A short calculation shows that the non-trivial Laplace eigenvalues of the star are 1 (multiplicity $v-2$) and v (multiplicity 1). The fact that any other tree has $\mu_1 < 1$ follows from Proposition 2.12. A tree with more than four vertices which is not a star possesses an edge whose removal leaves at least two points in one component and at least three in the other; so the Proposition gives

$$\mu_1 \leq (1/2 + 1/3) = 5/6.$$

The only other tree is the path of length 3, for which a short calculation shows that $\mu_1 = 2 - \sqrt{2}$.

So the best choice of a design in this case (v treatments, $v-1$ blocks of size 2) is to compare all the other treatments with a fixed "reference" treatment.

4.2 Unicyclic graphs

What happens for unicyclic graphs?

Such a graph has a single cycle, of length s say, with trees attached to some vertices (possibly none) in the cycle. The only way to make a spanning tree is to remove one of the edges in the cycle, so the number of spanning trees is maximized when $s = v$; that is, when the graph is just a cycle. This proves:

Theorem 4.1 *Let $v \geq 3$. Among unicyclic graphs with v vertices, the D-optimal graph is the cycle.*

For $k = 2$, such designs are called *loop designs* in the microarray literature.

Now we turn to E-optimality, where the results are rather different. We have seen that the minimum eigenvalue for the star is 1, and Proposition 2.9(d) shows that this is not reduced if an edge is added.

The smallest eigenvalue of the Laplacian matrix of the cycle of size v is equal to $2(1 - \cos(2\pi/v))$, which is greater than 1 when $v \leq 5$, equal to 1 when $v = 6$ and less than 1 when $v \geq 7$.

Each effective resistance is an evaluation of $x^\top L^- x$ for a contrast vector x with $x^\top x = 2$. Consequently, $\mu_1 \leq 2/R_{ij}$ for all pairs of distinct vertices i and j. If vertices i and j are in trees attached to vertices m and n in the cycle (where m

and n may be the same), then the effective resistance R_{ij} is the sum of the distance from i to m, the distance from j to n, and R_{mn}, which is strictly bigger than 2 (and so $\mu_1 < 1$) unless $m = n$ and $\{i, m\}$ and $\{j, n\}$ are edges. Thus a unicyclic graph cannot be E-optimal unless all the trees are leaves and they are attached to the same vertex of the cycle.

If there are at least two leaves attached to some vertex, then there is a Laplacian eigenvalue equal to 1: its eigenvector is the contrast between those two leaves. For a single leaf attached to a cycle of size 2 or 3, the least eigenvalue is $3 - \sqrt{3}$ or 1, respectively.

For $s \geq 3$, the effective resistance between two vertices in the cycle at distance 2 is $2(s-2)/s$, which is bigger than 1 when $s \geq 5$. If there is a leaf attached to a vertex in this cycle, then there is an effective resistance bigger than 2, so the graph cannot be E-optimal.

For $v \geq 5$, consider the graph consisting of a 4-cycle with $v - 4$ leaves attached to one vertex i. The Laplacian has an eigenvalue 1 with multiplicity $v - 5$: the eigenvectors are the pairwise contrasts among the leaves. The contrast between the other two neighbours of i is an eigenvector with eigenvalue 2. The remaining three non-trivial eigenvalues must take one constant value on all the leaves, and a (possibly different) constant value on the other two neighbours of i. Then a short calculation shows that the remaining three non-trivial eigenvalues are the zeros of the polynomial

$$f(\mu) = \mu^3 - (v+3)\mu^2 + (4v-2)\mu - 2v.$$

Then $f(0) < 0$ and $f(1) = v - 4 > 0$, so one of the eigenvalues lies in $(0, 1)$.

Putting all these observations together shows:

Theorem 4.2 *Let $v \geq 3$. Among unicyclic graphs with v vertices, the E-optimal graph is*

- *the cycle, if $v \leq 5$;*

- *a triangle with $v-3$ leaves attached to a vertex, or a star with one edge doubled, if $v \geq 7$;*

- *either of the above if $v = 6$.*

Thus the E-optimal designs are very different from the D-optimal designs when $v \geq 7$.

What about A-optimality? Again, suppose that trees are attached to two or more vertices in the cycle. If the trees are all moved to the same vertex then the sum of the pairwise effective resistances between tree vertices and cycle vertices is unchanged, as are the within-tree resistances, but the resistances between vertices in different trees is reduced. Further, suppose that a tree with more than one vertex is attached to vertex i of the cycle, in such a way that j is the only vertex of the tree joined to i. If the edge $\{i, j\}$ is contracted and vertex j is reattached to i as a leaf, then the resistances between tree vertices and i are interchanged with the resistances between tree vertices and j, resistances between tree vertices and the rest of the cycle decrease, and all other resistances are unchanged. This shows that, just as for E-optimality, the only candidates for A-optimal graphs are those consisting of a cycle of length s with $v - s$ leaves attached to one vertex, for some s.

Combinatorics of optimal designs

A short calculation shows that the sum of the pairwise effective resistances in such a graph is

$$[-s^3 + 2vs^2 + s(13 - 12v) + 12v^2 - 14v]/12.$$

For $v \leq 7$, this polynomial in s is monotonic decreasing on $[0, v]$. For $v = 8$ it attains its minimum on $[0, 8]$ at 8; for higher values of v it attains its minimum on $[0, v]$ between 3 and 4, with the integer achieving the least value being 4 if $9 \leq v \leq 12$ and 3 if $v \geq 12$. This proves the following.

Theorem 4.3 *Let $v \geq 3$. Among unicyclic graphs with v vertices, the A-optimal graph is*

- *the cycle, if $v \leq 8$;*
- *a 4-cycle with $v - 4$ leaves attached to one vertex, if $9 \leq v \leq 11$;*
- *a triangle with $v - 3$ leaves attached to a vertex, if $v \geq 13$;*
- *either of the last two if $v = 12$.*

This result was obtained for $v \leq 12$ by computational search in [60, 63, 108], and by the above analysis in [7, 100].

Together, the last three results show that, for $k = 2$ and $v = b$, not only are the D-optimal designs very different from the A- and E-optimal designs for $v \geq 9$ but also the ranking on the D-criterion is very different from those on the other two criteria. This was generalized in [7] to show that, for any fixed value of $b - v$, if $k = 2$ then there is a threshold v_0 such that if $v \geq v_0$ then the D-optimal design has no leaves but the A-optimal design does have leaves. The next two results have a similar flavour.

4.3 Sparse graphs

Theorem 4.4 *Let $\mathcal{G}$ be the set of connected graphs with v vertices and e edges, where $e > v$. If the graph G in $\mathcal{G}$ has a vertex l which is either a leaf or, more generally, has all its edges joining it to the same other vertex, then G is not D-optimal.*

Proof First suppose that l is a leaf. Since $e > v$, G is not a tree. Therefore it contains at least one edge ij which is not an isthmus. So some spanning trees of G omit the edge ij. Note that l cannot be equal to i or j. All spanning trees for G contain the unique edge through l. Now form the graph G' by removing the edge through l and the edge ij, and inserting new edges il and jl. Each spanning tree for G which contains ij becomes a spanning tree for G' by replacing ij and the edge through l by il and jl. Each spanning tree for G which omits ij gives two spanning trees in G': each omits the old edge through l and includes one of il, jl. Therefore G' has more spanning trees than G, and so G is not D-optimal.

Secondly suppose that l is joined to vertex m by s edges, and to no other vertices of G, where $s \geq 2$. Let i be any other vertex of G. Form the graph G'' by removing one edge between l and m and inserting an edge between l and i. Each tree of G contains one of the edges between l and m: these all give a tree in G'', using the replaced edge if necessary. However, the moved edge creates extra cycles, so there are some trees in G'' that contain two edges through l and do not correspond to any tree in G. Therefore G'' has more spanning trees than G, and so G is not D-optimal. □

Theorem 4.5 *Let $\mathcal{G}$ be the set of connected graphs with v vertices and e edges. If $20 \leq v \leq e < 5v/4$ then the E-optimal graphs in $\mathcal{G}$ have leaves.*

Proof We know that the star $K_{1,v-1}$ has least non-trivial eigenvalue 1 and that adding edges to the star cannot decrease this. Our technique is repeated use of Proposition 2.12 to ban portions of a graph that would force a non-trivial eigenvalue smaller than 1.

Let G be an E-optimal graph in $\mathcal{G}$ and suppose that G has no leaves. For $d \geq 2$, let n_d be the number of vertices of valency d. Call the vertices of G with valency 2 *dots*, the remaining vertices *nodes*. Call a path from one node to another a *side*: this may go through dots, and the two end nodes may be the same.

Suppose that there are three or more dots in one side. The edges at the end of the side make a cutset of size two, with parts of size s and $v - s$, for some $s \geq 3$. Hence $\mu_1 \leq 2(1/s + 1/(v-s))$, which is less than 1 if $s \geq 3$ and $v \geq 20$. So this cannot happen.

Consider a node i of valency 3. If there is a side making a loop at i, it must contain at least one dot, because G is loopless. The remaining edge at i now makes a cutset of size 1 with parts of size s and $v - s$ for $2 \leq s \leq 3$: by Proposition 2.12, this cannot happen. Therefore, the three sides at i are distinct. The edges at the far ends of those sides make a cutset of size three. Since $3(1/4 + 1/16) < 1$, the size of the part containing i is at most three, so there can be no more than two dots in total in these three sides.

Now consider a node j of valency 4. Similar arguments show that it may have one side forming a loop with precisely one dot, in which case there are no dots on the remaining two sides, and that otherwise the total number of dots in the four sides at j is at most four, because $4(1/6 + 1/14) < 1$.

Counting the dots according to nodes at the end of their sides counts them twice, so the preceding arguments show that

$$2n_2 \leq 2n_3 + 4n_4 + \sum_{d \geq 5} 2dn_d.$$

Moreover,

$$v = n_2 + n_3 + n_4 + \sum_{d \geq 5} n_d$$

and

$$2e = 2n_2 + 3n_3 + 4n_4 + \sum_{d \geq 5} dn_d.$$

Hence

$$5v - 4e = n_2 - n_3 - 3n_4 + \sum_{d \geq 5}(5 - 2d)n_d \leq -n_4 + \sum_{d \geq 5}(5 - d)n_d \leq 0.$$

This contradiction shows that G must have leaves. □

5 Further directions

In this section, after looking first at statisticians' view of the theory already presented, allowing for additional practical considerations we have not touched on, we give a number of open problems, and indicate some alternative optimality criteria and what to do if the block size is not constant. Remarkably, the case of non-constant block size leads us to consider the Laplacians of weighted graphs, for which the mathematical theory has already been developed.

5.1 Statistical and practical considerations

When there are no blocks, the information matrix is just the diagonal matrix of replications R. Then the optimal design on all the criteria is as equally replicate as possible. People with even a little statistical training are used to this, and so tend to favour equireplicate designs. One argument in favour of equal replication is that, if the model in Equation (2.4) is fitted and it is found that the block parameters differ little from each other, then the blocks can be ignored and the data re-analysed, at which stage an equireplicate design is best. However, this argument is controversial, and not accepted by all statisticians.

Many blocked experiments have so-called *complete blocks*: $k = v$ and every treatment occurs once in each block. In some areas of science, both practical experimenters and regulatory authorities expect experiments to be like this.

When blocks are smaller, it may happen that blocks are naturally grouped into larger "superblocks". For the design in Table 1, it may be that the farms can be grouped by soil-type; in another experiment, different superblocks may be dealt with at different times. If each superblock contains v experimental units then it is often advocated that a *resolved* block design should be used; that is, one in which each treatment occurs once in each superblock. Since any differences between superblocks are already accounted for in Equation (2.4), the arguments in favour of resolved designs are that (i) if a whole superblock is lost then the remaining design is equireplicate, which begs the question; (ii) if the first analysis suggests that blocks do not differ within each superblock, then the data can be reanalysed with parameters for superblocks but not blocks, which is as controversial as the previous re-analysis strategy. Nevertheless, resolved designs are requested sufficiently often in practice that some work has been done on optimal designs within this class: see [12].

Because of this background, statisticians tend to be horrified at the A- and E-optimal designs found in Section 4, where most treatments have replication 1. On the other hand, some scientists find them completely natural, having absorbed the dictum that all treatments should be compared with a reference, or control, treatment. Even then, there is concern that if the reference treatment is very different from the others then Equation (2.4) may not hold.

Robustness is a relevant issue here. If it is possible that one or two observations may be lost from the experiment, and if it is important to compare all treatments, then a design is robust if all comparisons can be made no matter which two observations are lost, so it must have every treatment replicated at least three times. On the other hand, in an experiment to compare 300 mutants of yeast to the "wild-type", losing two observations from a star-like design does not compromise the information about the remaining 298 mutants, while losing two observations from a 301-cycle makes many comparisons impossible and gives others very large variances.

Sometimes experimenters may be forced to use given, unequal replication because some treatments are more easily available than others, or because safer ones must be tested before riskier ones. Usually it is still true that all comparisons are interesting, but the search for an optimal design must be restricted to one whose replications are feasible.

In many situations, an optimal design for the given restrictions is not known. However, there may be little practical difference between designs which are close to optimal, so it can be useful to have a realistic bound on the optimality criterion.

One block design may be compared with another. The *efficiency factor* for contrast x in design $\mathfrak{D}_1$ relative to $\mathfrak{D}_2$ is the ratio $x^\top C_2^- x / x^\top C_1^- x$, where C_i is the information matrix of $\mathfrak{D}_i$. To decide which design is best, this ratio needs to be compared with the ratio of the variances of the individual responses for the two designs, which may be different if the block sizes differ.

Often the comparator is taken to be the unblocked design with equal replication $\bar{r}$, defined to be bk/v even if this is not an integer. However, if the replications are forced to be unequal, then the comparator design may be taken to be the unblocked design with the same replications, whose information matrix is just R. Then the efficiency factor for x is $x^\top R^{-1} x / x^\top C^- x$. This leads to the notion of a design being *efficiency balanced* if the non-trivial eigenvalues of $R^{-1/2} C R^{-1/2}$ are all equal, where $R^{1/2}$ is the diagonal matrix whose entries are the positive square roots of the replications. This approach takes no account of the possibility of finding a design that can get the most information *in spite of* being constrained in some way. On the other hand, if the low replications are not forced but correspond to less interesting treatments, then the appropriate approach would seem to be to incorporate weights (on contrasts, or on treatments) into the optimality criterion. This is already done in an extreme way (weights of zero and one) in two situations: (i) one treatment is a control and only comparisons with it are of interest; (ii) the treatments are all combinations of levels of several factors, and higher-order interactions are ignored.

5.2 Conjectures and refutations

It seems so obvious that balanced incomplete-block designs must be optimal on every reasonable criterion that everyone believed Kiefer's theorem long before it was proved. Other conjectures about optimality seem equally obvious, but all too often turn out to be wrong.

Given Kiefer's theorem, surely it must be true that, when there is no 2-design in $\mathcal{D}(v, b, k)$, there must be an optimal design which is binary or generalized binary, which is equireplicate or nearly so, and whose concurrences are as equal as possible? Surely there will be optimal designs with a high degree of symmetry, and surely a design which is optimal on one criterion will not be far from optimal on another? Our running example, and the results in Section 4, show that these plausible conjectures are all wrong.

What about the concurrence graph? Is it true that pairwise variance V_{ij} (equivalently, effective resistance R_{ij}) increases as the concurrence λ_{ij} decreases or as the distance in the graph increases? Example 3.8 shows that this is not true. Vertices C

Combinatorics of optimal designs

and D are at distance 2, with $\lambda_{CD} = 0$ and $R_{CD} = 2/3$, while vertices E and F are at distance 1, with $\lambda_{EF} = 1$ and $R_{EF} = 1$. Nonetheless, the vertices at the furthest distance do have the largest resistance.

Question 5.1 *In a loopless graph, is it true that the maximal value of the effective resistance R_{ij} is achieved for some pair of vertices $\{i,j\}$ whose distance apart in the graph is maximal?*

We have already seen that distance-regular graphs can be interesting. In [9] it is proved that, for a regular-graph design for which the graph in question is distance-regular, pairwise variance is a monotonic increasing function of distance in this graph. Biggs [17] proved the equivalent result for effective resistances in an electrical network.

Some of the false conjectures came from experience with partially balanced incomplete-block designs with two associate classes. These designs have two distinct non-trivial eigenvalues. It is shown in [5] that if a design has two distinct non-trivial eigenvalues then pairwise variance is a decreasing linear function of concurrence.

An association scheme is defined to be *amorphic* if it remains an association scheme after any fusion of its associate classes. If a design is partially balanced with respect to an amorphic association scheme then pairwise variance is again a decreasing function of concurrence [9].

The graph in Figure 3 can be regarded as a design in $\mathcal{D}(10, 15, 2)$. It is a fruitful source of counter-examples, but that may be because it is such a bad design. The Petersen graph in Example 3.5 also gives a design in $\mathcal{D}(10, 15, 2)$: it is much better than the previous design on the A-, D- and E-criteria. In fact, the Petersen design is A-optimal among equireplicate designs. Perhaps the nonintuitive behaviour of variance occurs only in designs that are far from optimal. Tjur [100] gave some heuristic reasons for believing this.

Question 5.2 *Suppose that a design is optimal in $\mathcal{D}(v, b, k)$, or in the equireplicate subclass of that. Is it true that, for that design, if $\lambda_{ij} > \lambda_{mn}$ then $V_{ij} < V_{mn}$? If the design is a regular-graph design, is it true that, if the distance between i and j in the graph in question is less than the distance between m and n then $V_{ij} < V_{mn}$?*

If a design is equireplicate then its information matrix is $r(I - (rk)^{-1}\Lambda)$. Paterson [78] expanded $(I - (rk)^{-1}\Lambda)^{-}$ by the binomial theorem to give an expression for $\mathrm{Tr}(C^-)$ as a convergent sum over $m \geq 2$ of the quantities $\mathcal{W}_m/r^m$, where $\mathcal{W}_m = \sum_i W_{ii}^{(m)}$ is the number of closed walks of length m in the concurrence graph. He proposed ordering designs lexicographically by their sequences $(\mathcal{W}_2, \mathcal{W}_3, \mathcal{W}_4, \ldots)$. He argued heuristically that an A-optimal design should have the earliest sequence. However, he showed that, among non-optimal designs, the ranking on the A-criterion could be the reverse of that on the sequences.

Question 5.3 *Is it true, that among equireplicate designs, a design with the earliest sequence is A-optimal?*

Cheng [33], building on work in [35], gave an asymptotic result, showing that this question has a positive answer when b is sufficiently large. More precisely, he proved the following statement. We call a design *nearly balanced* if its replication numbers r_i differ by at most one and, for fixed i, its concurrences λ_{ij} differ by at most one. Now, given v and k, there is a number $b_0 = b_0(v,k)$ such that, if $b \geq b_0$, then any nearly balanced block design in $\mathcal{D}(v,b,k)$ is better than any design which is not nearly balanced on a wide range of criteria, including all Φ_p criteria.

The earlier result in [35] asserts this for $k=2$ and D-optimality: that is, given v, there is a number $e_0 = e_0(v)$ such that, if $e \geq e_0$, then a graph with v vertices and e edges which is nearly balanced has more spanning trees than one which is not. This extends asymptotically the result of Kelmans and Chelnokov [62] cited in the Introduction.

This approach was used in [79] to obtain lower bounds for the A-criterion in terms of $\mathcal{W}_3$. The results in [6] support the idea that the concurrence graph of a good equireplicate design should have few triangles, but it is possible that this is true only within the range of replications that statisticians typically consider. See also [78, 104].

To achieve the necessary conditions for Kiefer's theorem, the number of blocks must be a quadratic function of the number of treatments. Theorem 4.5 suggests that if b is merely a linear function of v then the optimality criteria may conflict if v is large enough.

Question 5.4 *Let c and p be positive constants. Is it true that there is a threshold v_p such that the Φ_p-optimal designs in $\mathcal{D}(v,cv,k)$ are very different from the D-optimal designs when $v > v_p$? If so,*

(a) *are the rankings on the D-criterion and the Φ_p-criterion very different when $v > v_p$?*

(b) *is it true that the D-optimal designs remain (nearly) equireplicate for all v but the Φ_p-optimal designs have one treatment with much larger replication than the rest when $v > v_p$?*

(c) *is v_p a decreasing function of p?*

5.3 Other kinds of optimality

A large number of other optimality criteria have been proposed for designs where the treatments are combinations of amounts of quantitative variables: see [4]. These are not related to combinatorial design, so we do consider them further.

However, there is a refinement of E-optimality that seems worth pursuing. We saw in Section 4 that about $v/2$ edges need to be added to the star $K_{1,v-1}$ before its least non-trivial eigenvalue changes: as these edges are added, it is the *multiplicity* of the eigenvalue that changes. Among designs with the same least Laplacian eigenvalue, it would be sensible to say that one design is better than another if the multiplicity of that eigenvalue is lower.

Such a refinement is analogous to using *minimum aberration* [44] in factorial designs. These designs are the duals of linear codes. Resolution in the factorial design corresponds to minimum weight in the code; minimum aberration to the smallest number of words that have that weight.

Somewhat analogously, one could modify the Schur criterion so that only the first few partial sums $\sum_{i=0}^{n-1} \theta_i$ of eigenvalues are compared. For example, in the Petersen graph, the partial sums of the Laplace eigenvalues are

$$2, 4, 6, 8, 10, 15, 20, 25, 30,$$

whereas for the graph of Figure 3 they are approximately

$$0.22, 2.22, 5.22, 8.22, 11.51, 15.51, 19.51, 24.51, 30;$$

while neither is Schur-better than the other, the Petersen graph is ahead for the first three steps.

5.4 Non-constant block size

So far we have only considered block designs with constant block size. The theory can in principle be extended to remove this assumption. However, there are situations where it is not reasonable to assume that individual responses in different blocks have the same variance if the blocks have different sizes.

Now we have to replace the matrix Q in Section 2.2 by $I - ZK^{-1}Z^\top$, where K is the diagonal matrix of block sizes. Then the information matrix becomes

$$C = R - NK^{-1}N^\top,$$

where R is the diagonal matrix of replication numbers, and N is the point-block incidence matrix. We define the *weighted concurrence graph* to be the complete graph on the vertex set $\{1, \ldots, v\}$, where the edge $\{i, j\}$ is weighted by the sum of the reciprocals of the sizes of blocks containing both i and j; the *weighted adjacency matrix* of this graph has (i, j) entry equal to this weight for $i \neq j$, and diagonal entries zero. Now an argument identical to that of Proposition 2.7 shows:

Proposition 5.5 *Let A be the weighted adjacency matrix of a block design. Then the information matrix is $D - A$, where D is a diagonal matrix whose entries are chosen so that $D - A$ has row and column sums zero.*

In other words, the information matrix is the Laplacian matrix of the weighted concurrence graph.

The theory of Laplacian eigenvalues extends in a fairly straightforward way to weighted graphs [72]. So in principle the earlier results on optimal designs could be extended to optimality in the class $\mathcal{D}'(v; k_1, \ldots, k_b)$ of designs with v points and blocks of sizes $k_1, \ldots, k_b$. Not much is known in generality about this.

There is a simple connection between pairwise balanced designs and variance-balanced designs (without assuming the designs are binary or have constant block size). Any *pairwise balanced design* (one in which all pairwise concurrences are equal) can be converted to a variance-balanced design by repeating each block a number of times proportional to its size; the inverse operation converts a variance-balanced design into a pairwise balanced design. (This result is due to Hedayat and Stufken [52].) So the existence questions for the two types of design with prescribed number of points and set of block sizes are equivalent.

6 Computing with designs

Some commercial statistical computing packages will find optimal designs. For example, CycDesigN [37] gives A-optimal designs while JMP [57] gives D-optimal designs. The publicly available package R [83] includes some procedures to find A- or D-optimal designs. The designs are usually found by iteration to improve the criterion rather than by exhaustive search.

However, any rational symmetric function of the eigenvalues of a rational matrix (such as the A- and D-criteria) is invariant under the Galois group of its characteristic polynomial, and hence is a rational number. A mathematician will naturally prefer to compute such criteria exactly as rationals. For this, it may be best to turn to a computer algebra package such as GAP [46].

6.1 The DESIGN package

Apart from the fact that it is free, a great advantage of GAP is that it has been extended by many refereed shareware packages, including the DESIGN package [93], for computing with block designs. This package can in principle find all designs satisfying given specifications, up to isomorphism (or perhaps up to a restricted notion of isomorphism where a "structure group" is specified), or it can be set just to return a single design if one exists.

Although it deals directly only with binary block designs, many other types of design can be fitted into this framework. For example, a Latin square of order n can be regarded as a group-divisible design with three "groups" of size n, where each block contains one point from each group, and any two points in different groups belong to a single block.

The DESIGN package is well adapted for finding all binary designs with constant block size k having a given concurrence graph. As input to the program we can give the numbers of points and blocks, the block size, and the number of blocks containing any given pair of points. In particular, if we have computed the $n \times n$ adjacency matrix M of the concurrence graph, then the command

```
BlockDesigns(rec(v:=n, t:=2, blockSizes:=[k],
  tSubsetStructure:=rec(t:=2, lambdas:=l, partition:=p)));
```

will find all such designs, up to isomorphism; here p is the partition of the 2-element subsets $\{i,j\}$ of $\{1,\ldots,n\}$ according to the entry M_{ij} of the matrix M, and l the list of corresponding entries. If we also put isoLevel:=0 in the record passed to the function, then it will return just one such design (if one exists).

Example 6.1 *Variance-balanced designs with block size* 3. We have seen that, if $v \equiv 2 \bmod 3$, the best we can do is to take a block $[1,1,2]$, and to choose the remaining $(v+1)(v-2)/3$ blocks to have all points distinct and to cover the pairs other than $\{1,2\}$ twice. The following DESIGN code produces a list of all such designs (up to isomorphism).

```
designs:=function(pts) # argument must be congruent to 2 mod 3
  local pn;
  pn:=[[[1,2]]];
```

```
      Add(pn, Difference(Combinations([1..pts],2),[[1,2]]));
      return BlockDesigns(rec(v:=pts, blockSizes:=[3],
        tSubsetStructure:=rec(t:=2, lambdas:=[0,2], partition:=pn)));
end;
```

The argument to the function BlockDesigns is a record, of which the component v is the number of points, blockSizes a list of the allowed sizes of blocks, and tSubsetStructure describes how many blocks should contain a given t-subset; the first component t is the value of t, the second lambdas a list of the required concurrence numbers, and the third partition a partition of the set of t-subsets with the same number of parts as the number of components of lambdas; the designs found will have the ith component of lambdas as the number of blocks containing a t-set in the ith part of the partition. So the above code specifies that the blocks should have size 3, that no block should contain $\{1,2\}$, and that any other set of size 2 should be contained in two blocks. We then add the block $[1,1,2]$ to get the required design.

For $v = 5$, there is a unique design, the one we have used as a running example. For $v = 8$, there are 10 designs up to isomorphism. Of these, seven have repeated blocks; the sizes of the automorphism groups of the ten designs are are 12, 6 (three times), 4 (three times), 2 (twice) and 1. (The program returns a generating set for each automorphism group.) The computation takes a couple of seconds on a desktop PC.

There is a reason why knowledge of the automorphism group matters. To the output of the program, we can add either $[1,1,2]$ or $[1,2,2]$ to form a VB design. The resulting designs will be isomorphic if and only if the binary part has an automorphism interchanging 1 and 2. Since just two of the designs on 8 points have such an automorphism, we find altogether 18 isomorphism classes of VB$(8,3,2)$ designs with just one non-binary block. Of these 18 designs, 13 have repeated blocks.

For $v = 11$, the computation is much more substantial, since many designs exist. The DESIGN package finds that, if we prescribe that the design should have an automorphism of order 2 fixing the special points 1 and 2 and three further points, there are 3998 choices for the binary part of the design, leading to 7814 designs altogether. We have not computed the total number for $v = 11$ because of memory limitations.

Some other examples of the use of the package include:

- Construction and analysis of *regular-graph designs* [73]. This class includes many optimal designs in cases where 2-designs do not exist.

- Construction and analysis of 2-designs with repeated blocks [40].

- An implementation of the Jacobson–Matthews Markov chain method [54] for choosing a uniform random Latin square, and an adaptation of it to other kinds of design such as Steiner triple systems [26].

- Analysis of small gerechte designs [10]. These are a particular type of Latin square adapted for dealing with spatial inhomogeneities. Solutions to Sudoku puzzles are the best-known examples.

6.2 The designtheory.org project

The DESIGN package was produced as part of a wider project, available from designtheory.org, and described in the paper [11]. The project was restricted to binary block designs in the first instance, though the above example shows how to work around this restriction in some cases.

Among the aims of the project are:

- An XML specification for the *external representation* of block designs, in which the design and a wide variety of its combinatorial and statistical properties are described in a uniform computer-readable form for exchange between different programs and databases;

- The DESIGN package for GAP, for constructing and analysing block designs, and reading and writing them in the external representation format;

- A growing database of designs, with interfaces to make it easily usable by both combinatorialists and statisticians;

- The *Encyclopaedia of Design Theory*, an on-line resource including descriptions of several classes of design, essays on various topics in design theory and its mathematical and statistical underpinning, and an extensive glossary, as well as links to other resources;

- A collection of preprints on design theory.

Contributions to some or all of these components of the project are welcomed.

Acknowledgements

The authors are especially grateful to Ching-Shui Cheng and J. P. Morgan, whose extensive and detailed comments have greatly improved the accuracy and completeness of this paper. We also thank Emil Vaughan, to whose researches much of the information on the graph-theoretic interpretation of A-optimality are due; Francesca Merola for a number of suggestions; and Angela Dean, Heiko Großmann and other participants at the Isaac Newton Institute programme on Design of Experiments in July–August 2008.

References

[1] S. Ajoodani-Namini, Extending large sets of t-designs, *J. Combin. Theory Ser. A* **76** (1996), 139–144.

[2] N. Alon, Eigenvalues and expanders, *Combinatorica* **6** (1986), 83–96.

[3] N. Alon & V. Milman, λ_1, isoperimetric inequalities for graphs, and superconcentrators, *J. Combin. Theory Ser. B* **38** (1985), 73–88.

[4] A. C. Atkinson & A. N. Donev, *Optimum Experimental Designs*, Oxford University Press, Oxford (1992).

[5] R. A. Bailey, *Association Schemes*, Cambridge Studies in Advanced Mathematics, 84, Cambridge University Press, Cambridge (2004).

[6] R. A. Bailey, Six families of efficient resolvable designs in three replicates, *Metrika* **62** (2005), 161–173.

[7] R. A. Bailey, Designs for two-colour microarray experiments, *J. Roy. Statist. Soc. Ser. C* **56** (2007), 365–394.

[8] R. A. Bailey, *Design of Comparative Experiments*, Cambridge Series in Statistical and Probabilistic Mathematics, 25, Cambridge University Press, Cambridge (2008).

[9] R. A. Bailey, Variance and concurrence in block designs, and distance in the corresponding graphs, *Michigan Math. J.*, in press.

[10] R. A. Bailey, P. J. Cameron & R. Connelly, Sudoku, gerechte designs, resolutions, affine space, spreads, reguli, and Hamming codes, *Amer. Math. Monthly* **115** (2008), 383–404.

[11] R. A. Bailey, P. J. Cameron, P. Dobcsányi, J. P. Morgan & L. H. Soicher, Designs on the Web, *Discrete Math.* **306** (2006), 3014–3027.

[12] R. A. Bailey, H. Monod & J. P. Morgan, Construction and optimality of affine-resolvable designs, *Biometrika* **82** (1995), 187–200.

[13] B. Bagchi & S. Bagchi, Optimality of partial geometric designs, *Ann. Statist.* **29** (2001), 577–594.

[14] N. Balabanian & T. A. Bickart, *Electrical Network Theory*, Wiley, (1969).

[15] E. Bendito, A. Carmona, A. M. Encinas & J. M. Gesto, A formula for the Kirchhoff index, *Internat. J. Quantum Chemistry* **108** (2008), 1200–1206.

[16] N. Biggs, *Algebraic Graph Theory*, Second edition, Cambridge University Press, Cambridge (1993).

[17] N. L. Biggs, Potential theory on distance-regular graphs, *Combin. Probab. Comput.* **2** (1993), 107–119.

[18] E. J. Billington, Balanced n-ary designs: a combinatorial survey and some new results, *Ars Combin.* **17** (1984), 37–72.

[19] E. J. Billington, Designs with repeated elements in blocks: a survey and some recent results, *Congr. Numer.* **68** (1989), 123–146.

[20] B. Bollobás, *Modern Graph Theory*, Graduate Texts in Mathematics, 184, Springer-Verlag, New York (1998).

[21] R. C. Bose & K. R. Nair, Partially balanced incomplete block designs, *Sankhyā* **4** (1939), 337–372.

[22] N. Bourbaki, *Lie Groups and Lie Algebras (transl. A. Pressley)*, Springer, (1998, 2002, 2005).

[23] A. E. Brouwer, A. M. Cohen & A. Neumaier, *Distance-Regular Graphs*, Ergebnisse der Mathematik und ihrer Grenzgebiete (3), 18, Springer-Verlag, Berlin (1989).

[24] R. H. Bruck & H. J. Ryser, The nonexistence of certain finite projective planes, *Canadian J. Math.* **1** (1949), 88–93.

[25] P. J. Cameron, Optimal designs and root systems, *Michigan Math. J.*, in press.

[26] P. J. Cameron, A generalisation of t-designs, *Discrete Math.*, in press.

[27] P. J. Cameron, J.-M. Goethals, J. J. Seidel & E. E. Shult, Line graphs, root systems and elliptic geometry, *J. Algebra* **43** (1976), 305–327.

[28] P. J. Cameron & C. Thomassen, Spanning trees and orientations of graphs, in preparation.

[29] A. K. Chandra, R. Raghavan, W. L. Ruzzo, R. Smolensky & P. Tiwari, The electrical resistance of a graph captures its commute and cover times, *Comput. Complexity* **6** (1996), 312–340.

[30] C.-S. Chêng, Optimality of certain asymmetrical experimental designs, *Ann. Statist.* **6** (1978), 1239–1261.

[31] C.-S. Chêng, Maximizing the total number of spanning trees in a graph: two related problems in graph theory and optimum design theory, *J. Combin. Theory Ser. B.* **31** (1981), 240–248.

[32] C.-S. Cheng, Graph and optimum design theories — some connections and examples, in *Proceedings of the 43rd session of the International Statistical Institute (Buenos Aires, 1981)*, Bull. Inst. Internat. Statist. **49 (1)** (1981), 580–590.

[33] C.-S. Cheng, On the optimality of (M.S)-optimal designs in large systems, *Sankhyā* **54** (1992), 117–125.

[34] C.-S. Cheng & R. A. Bailey, Optimality of some two-associate-class partially balanced incomplete-block designs, *Ann. Statist.* **19** (1991), 1667–1671.

[35] C.-S. Cheng, J. C. Masaro & C. S. Wong, Do nearly balanced multigraphs have more spanning trees?, *J. Graph Theory* **8** (1985), 342–345.

[36] S. Chowla & H. J. Ryser, Combinatorial problems, *Canad. J. Math.* **2** (1950), 93–99.

[37] CycDesigN 3.0, http://cycdesign.co.nz.

[38] G. Davidoff, P. Sarnak & A. Valette, *Elementary Number Theory, Group Theory, and Ramanujan Graphs*, Lond. Math. Soc. Student Texts, 55, Cambridge University Press, Cambridge (2003).

[39] N. Deo, *Graph Theory with Applications to Engineering and Computer Science*, Prentice Hall, New Delhi (1980).

[40] P. Dobcsányi, D. A. Preece & L. H. Soicher, On balanced incomplete-block designs with repeated blocks, *European J. Combin.* **28** (2007), 1955–1970.

[41] J. Dodziuk, Difference equations, isoperimetric inequality and transience of certain random walks, *Trans. Amer. Math. Soc.* **284** (1984), 787–794.

[42] R. M. Foster, The average impedance of an electrical network, in *Reissner Anniversary Volume, Contributions to Applied Mechanics*, J. W. Edwards, Ann Arbor, Michigan (1948), pp. 333–340.

[43] R. M. Foster, An extension of a network theorem, *IRE Trans. Circuit Theory* **8** (1961), 75–76.

[44] A. Fries & W. G. Hunter, Minimum aberration 2^{k-p} designs, *Technometrics* **22** (1980), 601–608.

[45] N. Gaffke, *Optimale Versuchsplanung für linear Zwei-Faktor Modelle*, Ph.D. Thesis, Rheinisch-Westfälische Technische Hochschule, Aachen (1978).

[46] The GAP Group, GAP — Groups, Algorithms, and Programming, Version 4.4.10, 2007, http://www.gap-system.org.

[47] A. Giovagnoli & H. P. Wynn, Optimum continuous block designs, *Proc. Roy. Soc. Lond. Ser. A* **377** (1981), 405–416.

[48] C. Godsil & G. Royle, *Algebraic Graph Theory*, Springer, New York (2001).

[49] R. I. Grigorchuk & A. Zuk, On the asymptotic spectrum of random walks on infinite families of graphs, in *Random Walks and Discrete Potential Theory* (eds. M. Picardello et al.), *Symposia Mathematica*, 39, Cambridge University Press, Cambridge (1999), pp. 188–204.

[50] M. Hall, Jr. & W. S. Connor, An embedding theorem for balanced incomplete block designs, *Canadian J. Math.* **6** (1954), 35–41.

[51] H. Hanani, Balanced incomplete block designs and related designs, *Discrete Math.* **11** (1975), 255–369.

[52] A. Hedayat & J. Stufken, A relation between pairwise balanced and variance balanced block designs, *J. Amer. Statist. Assoc.* **84** (1989), 753–755.

[53] J. E. Humphreys, *Reflection Groups and Coxeter Groups*, Cambridge Studies in Advanced Mathematics, 29, Cambridge University Press, Cambridge (1990).

[54] M. T. Jacobson & P. Matthews, Generating uniformly distributed random Latin squares, *J. Combin. Des.* **4** (1996), 405–437.

[55] M. Jacroux & D. C. Whittinghill III, On the E- and MV-optimality of block designs having $k \geq v$, *Ann. Inst. Statist. Math.* **40** (1988), 407–418.

[56] M. A. Jafarizadeh, R. Sufani & S. Jafarizadeh, Calculating two-point resistances in distance-regular resistor networks, *J. Physics Ser. A (Mathematical and Theoretical)* **40** (2007), 4949–4972.

[57] JMP ®, http://www.jmp.com.

[58] J. A. John & T. J. Mitchell, Optimal incomplete block designs, *J. Roy. Statist. Soc. Ser. B* **39** (1977), 39–43.

[59] J. A. John & E. R. Williams, Conjectures for optimal block designs, *J. Roy. Statist. Soc. Ser. B* **44** (1982), 221–225.

[60] B. Jones & J. A. Eccleston, Exchange and interchange procedures to search for optimal designs, *J. Roy. Statist. Soc. Ser. B* **42** (1980), 238–243.

[61] M. Kac, Can one hear the shape of a drum?, *Amer. Math. Monthly* **73** (1966), 1–23.

[62] A. K. Kelmans & V. M. Chelnokov, A certain polynomial of a graph and graphs with an extremal number of trees, *J. Combin. Theory Ser. B* **16** (1974), 197–214.

[63] M. K. Kerr & G. A. Churchill, Experimental design for gene expression microarrays, *Biostatistics* **2** (2001), 183–201.

[64] G. B. Khosrovshahi & B. Tayfeh-Reziae, Trades and t-designs, in *Surveys in Combinatorics, 2009* (eds. S. Huczynska, J. D. Mitchell & C. M. Roney-Dougal), *Lond. Math. Soc. Lecture Notes Ser.*, 365, Cambridge University Press, Cambridge (2009), pp. 91–112.

[65] J. Kiefer, On the nonrandomized optimality and randomized nonoptimality of symmetrical designs, *Ann. Math. Statist.* **29** (1958), 675–699.

[66] J. Kiefer, Construction and optimality of generalized Youden designs, in *A Survey of Statistical Design and Linear Models* (ed. J. N. Srivastava), North-Holland, Amsterdam (1975), pp. 333–353.

[67] G. Kirchhoff, Über die Auflösung der Gleichenung, auf welche man bei der Untersuchung der linearen Verteilung galvanischer Ströme gefürht wird, *Annals of Physical Chemistry* **72** (1847), 497–508.

[68] A. Lubotzky, R. Phillips & P. Sarnak, Ramanujan graphs, *Combinatorica* **8** (1988), 261–277.

[69] G. A. Margulis, Explicit constructions of graphs without short cycles and low density codes, *Combinatorica* **2** (1982), 71–78.

[70] A. W. Marshall & I. Olkin, *Inequalities: Theory of Majorization and its Applications, Mathematics in Science and Engineering*, 143, Academic Press, Inc., New York–London (1979).

[71] C. Merino & D. J. A. Welsh, Forests, colorings and acyclic orientations of the square lattice, *Ann. Combin.* **3** (1999), 417–429.

[72] B. Mohar, Some applications of Laplace eigenvalues of graphs, in *Graph Symmetry: Algebraic Methods and Applications* (eds. G. Hahn & G. Sabidussi), *NATO ASI Series C*, Kluwer, Dordrecht (1997), pp. 227–276.

[73] J. P. Morgan, Optimal incomplete block designs, *J. Amer. Statist. Assoc.* **102** (2007), 655–663.

[74] J. P. Morgan & S. K. Srivastav, The completely symmetric designs with block-size three, *J. Statist. Plann. Inference* **106** (2002), 21–30.

[75] J. P. Morgan & N. Uddin, Optimal, non-binary, variance balanced designs, *Statistica Sinica* **5** (1995), 535–546.

[76] J. G. Oxley, *Matroid Theory*, Oxford University Press, New York (1992).

[77] J. L. Palacios, Foster's formulas via probability and the Kirchhoff index, *Methodol. Comput. Appl. Probab.* **6** (2004), 381–387.

[78] L. Paterson, Circuits and efficiency in incomplete block designs, *Biometrika* **70** (1983), 215–225.

[79] L. J. Paterson & P. Wild, Triangles and efficiency factors, *Biometrika* **73** (1986), 289–299.

[80] L. Petingi & J. Rodriguez, A new technique for the characterization of graphs with a maximum number of spanning trees, *Discrete Math.* **244** (2002), 351–373.

[81] C. Pritchard, Mistakes concerning a chance encounter between Francis Galton and John Venn, *BSHM Bull.* **23** (2008), 103–108.

[82] F. Pukelsheim, *Optimal Design of Experiments*, Wiley, New York (1993).

[83] The R project for statistical computing, http://www.r-project.org.

[84] C. R. Rao, *Linear Statistical Inference and its Applications*, Second edition, Wiley, New York (1973).

[85] B. Reck & J. P. Morgan, Optimal design in irregular BIBD settings, *J. Statist. Plann. Inference* **129** (2005), 59–84.

[86] B. Reck & J. P. Morgan, E-optimal design in irregular BIBD settings, *J. Statist. Plann. Inference* **137** (2007), 1658–1668.

[87] O. Reingold, S. Vadhan & A. Wigderson, Entropy waves, the zig-zag graph product and constant-degree expanders, *Ann. of Math.* (2), **155** (2002), 157–187.

[88] C. A. Rohde, Generalized inverses of partitioned matrices, *J. Soc. Industrial Appl. Math.* **13** (1965), 1033–1035.

[89] R. Schwabe, *Optimum Designs for Multi-Factor Models*, Lecture Notes in Statistics, 113, Springer-Verlag, New York (1996).

[90] J.-P. Serre, Répartition asymptotique des valeurs propres de l'opérateur de Hecke T_p, *J. Amer. Math. Soc.* **10** (1997), 75–102.

[91] K. R. Shah & B. K. Sinha, *Theory of Optimal Designs*, Lecture Notes in Statistics, 54, Springer-Verlag, New York (1989).

[92] L. W. Shapiro, An electrical lemma, *Math. Magazine* **60** (1987), 36–38.

[93] L. H. Soicher, The DESIGN package for GAP, Version 1.3, 2006, http://designtheory.org/software/gap_design/.

[94] A. D. Sokal, The multivariate Tutte polynomial (alias Potts model) for graphs and matroids, in *Surveys in Combinatorics 2005* (ed. B. S. Webb), *Lond. Math. Soc. Lecture Note Ser.*, 327, Cambridge University Press, Cambridge (2005), pp. 173–226.

[95] S. K. Srivastav & J. P. Morgan, Optimality of designs with generalized group divisible structure, *J. Statist. Plann. Inference* **71** (1998), 313–330.

[96] L. Teirlinck, Nontrivial t-designs without repeated blocks exist for all t, *Discrete Math.* **65** (1987), 301–311.

[97] P. Tetali, Random walks and the effective resistance of networks, *J. Theoret. Probab.* **4** (1991), 101–109.

[98] P. Tetali, An extension of Foster's network theorem, *Combin. Probab. Comput.* **3** (1994), 421–427.

[99] K. D. Tocher, The design and analysis of block experiments, *J. Roy. Statist. Soc. Ser. B* **14** (1952), 45–100.

[100] T. Tjur, Block designs and electrical networks, *Ann. Statist.* **19** (1991), 1010–1027.

[101] T. Tjur, An algorithm for optimization of block designs, *J. Statist. Plann. Inference* **36** (1993), 277–282.

[102] E. Vaughan, personal communication.

[103] E. Vaughan, Foster's formulæ for distance-regular graphs, preprint, (2007).

[104] P. Wild, On circuits and optimality conjectures for block designs, *J. Roy. Statist. Soc. Ser. B* **49** (1987), 90–94.

[105] P. Wild, Statistics, in *Graph Connections*, (ed. L. W. Beineke and R. J. Wilson), *Oxford Lecture Ser. Math. Appl.*, 5, Oxford University Press, New York (1997), pp. 208–226.

[106] R. M. Wilson, An existence theory for pairwise balanced designs: I, Composition theorems and morphisms, *J. Combin. Theory Ser. A* **13** (1972), 220–245; II, The structure of PBD-closed sets and the existence conjectures, *ibid.* **13** (1972), 246–273; III, A proof of the existence conjectures, *ibid.* **18** (1975), 71–79.

[107] R. M. Wilson, Non-isomorphic Steiner triple systems, *Math. Z.* **135** (1974), 303–313.

[108] E. Wit, A. Nobile & R. Khanin, Near-optimal designs for dual channel microarray studies, *J. Roy. Statist. Soc. Ser. C* **54** (2005), 817–830.

School of Mathematical Sciences
Queen Mary, University of London
London E1 4NS, UK
r.a.bailey@qmul.ac.uk,
p.j.cameron@qmul.ac.uk

Regularity and the spectra of graphs

Willem H. Haemers

Abstract

We present some recent results that involve regularity of graphs and their eigenvalues. The following questions are dealt with: when can one deduce from an appropriate spectrum of a graph G that (i) G is regular, (ii) G is distance-regular, (iii) G has a perfect matching?

1 Introduction

Spectral graph theory studies the relationship between the structure of a graph and the eigenvalues of associated matrices. Often one considers the standard $(0, 1)$-adjacency matrix A, but the Laplacian matrix L (defined by $L = D - A$, where D is the diagonal matrix with the degrees) has also received much attention. Other matrices are the signless Laplacian $|L|$ (defined by $|L| = D + A$), the normalized Laplacian $\widehat{L}$ (defined by $\widehat{L} = D^{-\frac{1}{2}} L D^{-\frac{1}{2}}$, provided there are no isolated vertices), and the Seidel matrix S (defined by $S = J - 2A - I$, where J and I are the all-one matrix and the identity matrix, respectively).

A central question is: Given the spectrum (of one of these matrices), what can be said about the structure? Sometimes the eigenvalues uniquely determine the graph. If that is the case we say that the graph is determined by the spectrum (with respect to the matrix under consideration). In recent years this problem has attracted much interest. For many graphs, and with respect to several types of matrices, it has been established whether they are determined by the spectrum or not. See [10] and [11] for a survey.

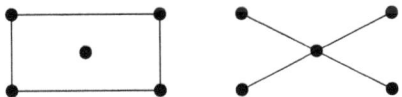

Figure 1: The Saltire: two graphs with cospectral adjacency matrices

Two graphs with the same spectrum for some type of matrix are called *cospectral* with respect to that matrix. A pair of cospectral graphs with respect to the adjacency matrix is given in Figure 1 (we like to call the pair the Saltire, because the superposition of the two gives the Scottish flag). Both graphs have spectrum $\{2, 0^3, -2\}$ (where multiplicities are written as exponents). The Saltire is the smallest example, that is, all graphs on fewer than five vertices are determined by the spectrum of the adjacency matrix. A smallest pair of cospectral graphs with respect to the Laplacian matrix is presented in Figure 2.

If a graph is determined by its spectrum, then clearly all structural properties follow from the spectrum. However, this is not the type of result that this article concentrates on. Here we focus on general structural properties that can be identified from the spectrum. Some famous examples are (see [7]):

Figure 2: Graphs with cospectral Laplacian matrices

- A graph is bipartite if and only if its adjacency spectrum is symmetric around 0.
- The number of components of a graph equals the multiplicity of the Laplacian eigenvalue 0.
- The number of spanning trees of a connected graph on n vertices equals the product of the nonzero Laplacian eigenvalues divided by n.
- The number of edges is half the sum of the Laplacian eigenvalues.
- The number of edges is half the sum of the squares of the adjacency eigenvalues.
- For a connected graph, the diameter is strictly less than the number of distinct eigenvalues.

The last property holds for the adjacency as well as the Laplacian matrix. We also see that the number of edges follows from the adjacency as well as the Laplacian spectrum. However, there is no spectral characterization for connectivity in the case of the adjacency matrix, and no spectral characterization for bipartiteness in the case of the Laplacian matrix. This is illustrated by the examples in Figure 1 and 2.

For the normalized Laplacian $\widehat{L}$, it is known that the graph is connected if and only if $\widehat{L}$ has an eigenvalue 0 of multiplicity 1, and the graph is bipartite if and only if the spectrum is symmetric around 1. A weakness of $\widehat{L}$ is that the number of edges doesn't follow from the spectrum. Indeed, all complete bipartite graphs with n vertices have the same spectrum for the normalized Laplacian.

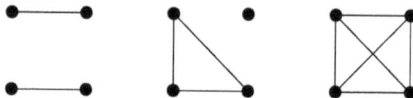

Figure 3: Three graphs, equivalent under Seidel switching

The Seidel matrix $S = J - 2A - I$ has entries -1 (adjacency), 1 (nonadjacency), and 0 (diagonal). If, for a subset of vertices, all corresponding rows and columns are multiplied by -1, we obtain the Seidel matrix S' of another graph, but with the same spectrum (indeed, S and S' are similar by a diagonal matrix $\mathrm{diag}(\pm 1, \ldots, \pm 1)$). The corresponding graphs are called *switching equivalent* and the operation is called *Seidel switching*. For example the three graphs of Figure 3 are switching equivalent; all three have Seidel spectrum $\{1^3, -3\}$. This example shows that in general the Seidel spectrum gives no information on bipartiteness, regularity, the number of edges, or the number of components. Table 1 shows which of the basic graph properties of connectedness, bipartiteness and regularity can be identified from the spectra of

	A	L	$|L|$	$\hat{L}$	S
connected	no	yes	no	yes	no
bipartite	yes	no	no	yes	no
regular	yes	yes	yes	no	no

Table 1: Properties that follow from the spectrum

the aforementioned types of matrices (answers for regular graphs will be explained below).

1.1 Regular graphs

In this article we focus on the relation between eigenvalues and regularity. For a regular graph the spectrum of any type of matrix can be obtained from the spectrum of any other matrix. So for regular graphs it doesn't matter what kind of matrix we take. However, it is not always the case that regularity follows from the spectrum. We note that it is false for the Seidel matrix and the normalized Laplacian. However it is true for the adjacency, the Laplacian, and the signless Laplacian matrix. In Section 2 we will give a precise answer to the question: For which matrices in span$\{A, J, D, I\}$ can regularity be deduced from the spectrum?

It is well known that the property of being strongly regular can be seen from the spectrum (of any type of matrix for which regularity follows from the spectrum). For the more general concept of distance-regular graphs this is no longer true. For many distance-regular graphs there exists a cospectral graph which is not distance-regular. However, in some special cases it can be proved that a graph cospectral with a distance-regular graph is again distance-regular. In Section 3 we will present some of these results.

Section 4 is of a different nature. Here we survey some rather recent results concerning perfect matchings and eigenvalues in regular graphs. König proved that a regular bipartite graph of positive valency has a perfect matching. In general, a connected k-regular graph with an even number of vertices does not need to have a perfect matching. However it is true if the second largest adjacency eigenvalue is at most $k - 1$ (the largest eigenvalue equals k). In fact, it still holds if the second largest eigenvalue is greater than $k - 1$, and the third largest eigenvalue is less than $k - 1 + \varepsilon$, where ε is a positive number depending on k and the number of vertices. We present an explicit formula for the best possible value of ε.

1.2 Tools

In this article we use several theorems and results from linear algebra and graph theory. Most of these are well known and can easily be found in the literature or on the internet. Some general references are [22] for graphs, [20] for linear algebra, and [7] and [16] for both. In this section we present some more specific results related to partitioned matrices. If the proof is not straightforward, it can be found in [16] or [17].

Consider an $n \times n$ matrix

$$N = \begin{bmatrix} N_{1,1} & \cdots & N_{1,m} \\ \vdots & & \vdots \\ N_{m,1} & \cdots & N_{m,m} \end{bmatrix},$$

whose rows and columns are partitioned according to a partitioning $X_1, \ldots, X_m$ of $\{1, \ldots, n\}$. Such a matrix partition is *equitable* if each block $N_{i,j}$ has constant row and column sum. The *quotient matrix* Q is the $m \times m$ matrix whose entries are the average row sums of the blocks $N_{i,j}$ of N.

Lemma 1.1 *Suppose Q is the quotient matrix of an equitable partition of N.*

(i) Every eigenvalue of Q is an eigenvalue of N.

(ii) If N is symmetric, then N has two kinds of eigenvectors: m independent eigenvectors that are constant over each partition class (in other words they are linear combinations of the characteristic vectors of $X_1, \ldots, X_m$), which correspond to eigenvalues of Q, and $n-m$ independent eigenvectors orthogonal to the characteristic vectors of the partition classes.

Note that eigenvectors of the second type, and the corresponding eigenvalues, do not change if some blocks $N_{i,j}$ of N are replaced by $N_{i,j} + cJ$ for some real number c. Next we mention two cases of eigenvalue interlacing. Consider two sequences of real numbers $\lambda_1 \geq \cdots \geq \lambda_n$ and $\mu_1 \geq \cdots \geq \mu_m$ with $m \leq n$. The second sequence is said to *interlace* the first whenever

$$\lambda_i \geq \mu_i \geq \lambda_{n-m+i} \quad (i = 1, \ldots m).$$

Lemma 1.2 *(i) If M is a principal submatrix of a symmetric matrix N, then the eigenvalues of M interlace those of N.*

(ii) If Q is the $m \times m$ quotient matrix of a symmetric partitioned matrix N, then the eigenvalues of Q interlace those of N.

Throughout this article the eigenvalues of a symmetric $n \times n$ matrix N will be denoted by $\lambda_1(N) \geq \cdots \geq \lambda_n(N)$.

2 Regularity from the spectrum

Suppose that G is a k-regular graph on n vertices. Then k is an eigenvalue of the adjacency matrix A with eigenvector $\mathbf{1}$, the all-one vector. By the Perron-Frobenius theorem $k = \lambda_1(A)$ and $|\lambda_i(A)| \leq k$ $(i = 1, \ldots, n)$. We have $nk = \text{trace}(A^2)$, hence

$$\lambda_1(A) = \frac{1}{n} \sum_i \lambda_i^2(A).$$

It is well known that the last equation is also sufficient for regularity. Similarly, it is known that a graph is regular if and only if the eigenvalues of the Laplacian matrix L satisfy

$$n \sum_i \lambda_i(L)(\lambda_i(L) - 1) = (\sum_i \lambda_i(L))^2.$$

Regularity and spectra

Thus regularity can be seen from the spectrum of the adjacency as well as the Laplacian matrix. The following common generalization appeared in [10].

Proposition 2.1 *Let $\alpha \neq 0$. With respect to the matrix $M = \alpha A + \beta J + \gamma D + \delta I$, a regular graph cannot be cospectral with a non-regular one, except possibly when $\gamma = 0$ and $-1 < \beta/\alpha < 0$.*

Proof Without loss of generality we may assume that $\alpha = 1$ and $\delta = 0$. Let d_i, $i = 1, \ldots, n$ be a putative sequence of vertex degrees.

First suppose that $\gamma \neq 0$. Then it follows from trace(M) that $\sum_i d_i$ is determined by the spectrum of M. Since trace$(M^2) = \beta^2 n^2 + (1 + 2\beta + 2\beta\gamma)\sum_i d_i + \gamma^2 \sum_i d_i^2$, it also follows that $\sum_i d_i^2$ can be expressed in terms of the the spectrum of M. Now Cauchy's inequality states that $(\sum_i d_i)^2 \leq n\sum_i d_i^2$ with equality if and only if $d_1 = d_2 = \cdots = d_n$. This shows that regularity of the graph can be seen from the spectrum of M.

Next we consider the case $\gamma = 0$, and $\beta \leq -1$ or $\beta \geq 0$. Since trace$(M^2) = \beta^2 n^2 + (1 + 2\beta)\sum_i d_i$, also here it follows that $\sum_i d_i$ is determined by the spectrum of M (here we only use $\beta \neq -1/2$), and so is $s = \beta n^2 + \sum_i d_i$, which is the sum of the entries of M. We easily have $s = \mathbf{1}^\top M \mathbf{1}$, and Rayleigh's inequalities give

$$\lambda_1(M) \geq \frac{\mathbf{1}^\top M \mathbf{1}}{\mathbf{1}^\top \mathbf{1}} \geq \lambda_n(M).$$

Equality on either side implies that $\mathbf{1}$ is an eigenvector of M, meaning that M has constant row sums (equal to s/n). Hence $s/n = \lambda_1(M)$ or $s/n = \lambda_n(M)$ (which can be deduced from the spectrum of M) implies that the graph is regular. Conversely, if the graph is regular, $\mathbf{1}$ is an eigenvector of M with eigenvalue s/n. The above restrictions for β imply that M or $-M$ is a nonnegative matrix, and the Perron-Frobenius theorem gives that the eigenvector $\mathbf{1}$ corresponds to the spectral radius, which is $\lambda_1(M)$, or $\lambda_1(-M) = -\lambda_n(M)$, respectively. Thus the graph is regular if and only if $\lambda_1(M) = s/n$, or $\lambda_n(M) = s/n$. $\square$

Figure 4: Cospectral graphs with respect to $A - \frac{1}{3}J$

Figure 3 shows that the statement of Proposition 2.1 is false for the Seidel matrix, which corresponds to the case $\beta = -1/2$. In the above proof we saw that for $\beta \neq -1/2$ the number of edges is determined by the spectrum of M. Yet, even if $\beta \neq -1/2$ a regular graph can be cospectral with a nonregular one. Figure 4 gives an example for the case $\beta = -1/3$ (i.e. $M = A - \frac{1}{3}J$). The following result from [5] gives many more such examples.

Theorem 2.2 *For every rational value of $\beta \in (-1, 0)$, there exist a pair of graphs, one regular and one not, that are cospectral with respect to $A + \beta J$.*

Proof Write $\beta = -p/q$, where p and q are positive integers and q is even. We will construct two cospectral matrices M and $\overline{M}$ of the form $qA - pJ$, and order $4q + q^2$. Define

$$M = \begin{bmatrix} K & B \\ B^\top & C \end{bmatrix} \text{ and } \overline{M} = \begin{bmatrix} K & \overline{B} \\ \overline{B}^\top & C \end{bmatrix}.$$

The matrices $K, B, \overline{B}$ and C are built with $q \times q$ blocks having constant row and column sums. The construction is as follows:

$$K = \begin{bmatrix} -pJ & rJ & rJ & -pJ \\ rJ & -pJ & -pJ & -pJ \\ rJ & -pJ & -pJ & -pJ \\ -pJ & -pJ & -pJ & -pJ \end{bmatrix},$$

$$B = \begin{bmatrix} B_{1,1} & B_{1,2} & B_{1,3} & B_{1,4} & \cdots & B_{1,q-1} & B_{1,q} \\ B_{2,1} & \overline{B}_{2,2} & B_{2,3} & \overline{B}_{2,4} & \cdots & B_{2,q-1} & \overline{B}_{2,q} \\ \overline{B}_{3,1} & B_{3,2} & \overline{B}_{3,3} & B_{3,4} & \cdots & \overline{B}_{3,q-1} & B_{3,q} \\ \overline{B}_{4,1} & \overline{B}_{4,2} & \overline{B}_{4,3} & \overline{B}_{4,4} & \cdots & \overline{B}_{4,q-1} & \overline{B}_{4,q} \end{bmatrix},$$

$$\overline{B} = \begin{bmatrix} \overline{B}_{4,1} & \overline{B}_{4,2} & \overline{B}_{4,3} & \overline{B}_{4,4} & \cdots & \overline{B}_{4,q-1} & \overline{B}_{4,q} \\ \overline{B}_{3,1} & B_{3,2} & \overline{B}_{3,3} & B_{3,4} & \cdots & \overline{B}_{3,q-1} & B_{3,q} \\ B_{2,1} & \overline{B}_{2,2} & B_{2,3} & \overline{B}_{2,4} & \cdots & B_{2,q-1} & \overline{B}_{2,q} \\ B_{1,1} & B_{1,2} & B_{1,3} & B_{1,4} & \cdots & B_{1,q-1} & B_{1,q} \end{bmatrix},$$

where $B_{i,j}$ is any $q \times q$ matrix with $p - 1$ entries r and $r + 1$ entries $-p$ in each row and column, and $\overline{B}_{i,j}$ is any $q \times q$ matrix with $p + 1$ entries r and $r - 1$ entries $-p$ in each row and column. So $B_{i,j}$ has row sums $-q$ and $\overline{B}_{i,j}$ has row sums q. Notice that the first $4q$ rows of M all have row sum $q(q - 4p)$, whilst the first $4q$ row sums of $\overline{M}$ take three different values: $q(3q - 4p)$, $q(q - 4p)$ and $q(-q - 4p)$. Also observe that $\overline{B}$ can be obtained from B by reversing the order of the block rows. The matrix

$$C = \begin{bmatrix} C_{1,1} & \cdots & C_{1,q} \\ \vdots & & \vdots \\ C_{q,1} & \cdots & C_{q,q} \end{bmatrix}$$

should be chosen so that C is symmetric with diagonal entries $-p$, and all row and column sums equal to $q(q - 4p)$ (which makes all row sums of M equal). All blocks $C_{i,j}$ must have constant row and column sums. There are many ways to achieve this. For instance, take $C_{1,1} = C_{1,2} = C_{1,q} = -pJ$, $C_{1,q/2+1} = rJ$ and for the remaining values of i take for $C_{1,i}$ any $q \times q$ matrix with p times r and r times $-p$ in each row and column. Then put $C = \text{circulant}(C_{1,1}, \ldots, C_{1,q})$.

It is clear that M represents a regular graph and $\overline{M}$ represents a nonregular graph. What remains to be proved is that M and $\overline{M}$ are cospectral. First observe that the given partition of M (and $\overline{M}$) into $q + 4$ blocks of size $q \times q$ is an equitable partition. The corresponding quotient matrices are cospectral, because one can be obtained from the the other by multiplying the first four rows and columns by -1. By Lemma 1.1, the eigenvalues of the quotient matrices are eigenvalues of M and of $\overline{M}$ with eigenvectors in the space $\mathcal{V}$ spanned by the characteristic vectors of the partition. The other eigenvalues have eigenvectors in $\mathcal{V}^\perp$. These eigenvalues are not changed if any block $M_{i,j}$ of M is replaced by $M_{i,j} + cJ$ for some real number c. Define

$$M' = \begin{bmatrix} O & B \\ B^\top & C \end{bmatrix} \text{ and } \overline{M}' = \begin{bmatrix} O & \overline{B} \\ \overline{B}^\top & C \end{bmatrix}.$$

Then for the eigenvectors in $\mathcal{V}$, the matrices M' and M have the same eigenvalues, and so do $\overline{M}'$ and $\overline{M}$. But since $\overline{B}$ can be obtained from B by a row permutation, M' and $\overline{M}'$ are cospectral. The conclusion is that M and $\overline{M}$ have the same eigenvalues for the eigenvectors in $\mathcal{V}$ and for the eigenvectors in $\mathcal{V}^\perp$. Therefore M and $\overline{M}$ have the same spectrum. □

The only β for which we have not seen whether regularity can be deduced from the spectrum of M are the irrational β between -1 and 0. Surprisingly, for these β the answer is affirmative. This is a direct consequence of the following result.

Theorem 2.3 *Consider two graphs G and G', with adjacency matrices A and A', respectively. The following are equivalent:*

(i) $A + \beta J$ and $A' + \beta J$ are cospectral for all $\beta \in \mathbf{R}$,

(ii) $A + \beta J$ and $A' + \beta J$ are cospectral for two distinct values of β,

(iii) $A + \beta J$ and $A' + \beta J$ are cospectral for an irrational value of β.

Proof For a graph G with adjacency A we define the generalized characteristic polynomial: $p_G(x,y) = \det(xI + yJ - A)$. Thus $p_G(x,y)$ can be interpreted as the characteristic polynomial of $A - yJ$, and $p_G(x,0)$ is the characteristic polynomial of A. Moreover, $p_G(x,y)$ has integral coefficients. It follows that the degree in y of $p_G(x,y)$ is 1. Indeed, for an arbitrary square matrix M it is known that $\det(M + yJ) = \det M + y\Sigma \operatorname{adj} M$, where $\Sigma \operatorname{adj} M$ denotes the sum of the entries of the adjugate (adjoint) of M. It is also easily derived from the fact that by Gaussian elimination in $xI + A - yJ$ one can eliminate all y's, except for those in the first row. So we may write

$$p_G(x,y) = \sum_{i=0}^{n} (a_i + b_i y)x^i.$$

It is clear that $p_G(x,y) \equiv p_{G'}(x,y)$ if and only if $A - yJ$ and $A' - yJ$ are cospectral for all $y \in \mathbf{R}$. Suppose that $A - yJ$ and $A' - yJ$ are cospectral for some $y = \hat{y}$, but not all, values of y. Then $p_G(x,\hat{y}) = p_{G'}(x,\hat{y})$ for all $x \in \mathbf{R}$, whilst $p_G(x,y) \not\equiv p_{G'}(x,y)$. If this is the case, then $a_i + \hat{y}b_i = a'_i + \hat{y}b'_i$ with $b_i \neq b'_i$ for some i. This implies $\hat{y} = -(a_i - a'_i)/(b_i - b'_i)$. Thus we have proved that $\hat{y}$ is rational, and that there is only one possible value of $\hat{y}$. □

By this theorem, cospectrality for an irrational β implies cospectrality for the adjacency matrix, and hence G is regular if and only if G' is. Therefore:

Corollary 2.4 *There exist a pair of cospectral graphs with respect to the matrix $M = \alpha A + \beta J + \gamma D + \delta I$, where one graph is regular and the other one is not, if and only if $\gamma = 0$ and β/α is a rational number strictly between -1 and 0.*

A more extended version of Theorem 2.3 appeared in [12]; the equivalence of (i) and (ii) is due to Johnson and Newman [21].

3 Distance-regularity

Here we deal with the question: When can we deduce from the spectrum of a graph that it is distance-regular? Since distance-regular graphs are regular (of degree k, say), it doesn't really matter which type of matrix we take, as long as regularity can be deduced from the spectrum (see Corollary 2.4). We choose the adjacency matrix and write $k = \lambda_1 \geq \cdots \geq \lambda_n$ for the spectrum.

In terms of the adjacency matrix A, distance-regularity can be defined as follows. A graph is *distance-regular* whenever it is connected and for every vertex the distance partition of A is equitable with the same quotient matrix. (The *distance partition* of a connected graph with respect to a vertex x partitions the vertex set into $X_0, \ldots, X_d$, where X_i is the set of vertices at distance i from x.) The quotient matrix Q of a distance partition is a tri-diagonal matrix, which can be written as:

$$Q = \begin{bmatrix} a_0 & b_0 & & & \\ c_1 & a_1 & b_1 & & \\ & \ddots & \ddots & \ddots & \\ & & c_{d-1} & a_{d-1} & b_{d-1} \\ & & & c_d & a_d \end{bmatrix}.$$

In the case of a distance-regular graph the entries of Q are integers, and it follows easily that

$$a_0 = 0, \ b_0 = k, \ c_1 = 1 \text{ and } a_i + b_i + c_i = k \text{ for } i = 0, \ldots, d \text{ (take } c_0 = b_d = 0\text{)}.$$

If k_i is the number of vertices at distance i from any given vertex, then

$$k_0 = 1, \ k_1 = k, \text{ and } k_i b_i = k_{i+1} c_{i+1} \text{ for } i = 0, \ldots, d-1.$$

So the array $(b_0, \ldots, b_{d-1}; c_1, \ldots, c_d)$ determines all other parameters a_i and k_i. It is called the *intersection array*, and the parameters a_i, b_i, c_i are usually called the *intersection numbers*. We will call Q the *intersection matrix*.

Some trivial examples of distance-regular graphs are the complete graphs (case $d = 1$), and the polygons (case $k = 2$). In both of these cases the graphs are characterized by their spectra. A distance-regular graph with diameter 2 is the same as a connected strongly regular graph. The following result for strongly regular graphs is well known.

Theorem 3.1 *A connected graph is strongly regular if and only if it is regular with exactly three distinct adjacency eigenvalues.*

We will prove a more general result below. Theorem 3.1 implies that the property of being strongly regular can be seen from its spectrum. Many other distance-regular graphs are known to exist; see [2] for an extensive treatment. We mention two famous families.

The *Hamming graphs* $H(q,d)$ have vertex set $\{1, \ldots, q\}^d$ (i.e. the words of length d over an alphabet with q symbols); two vertices are adjacent if the two words differ in exactly one place. A Hamming graph is distance-regular of diameter d with intersection numbers $b_j = (d-j)(q-1)$, $c_j = j$ $(0 \leq j \leq d)$.

Regularity and spectra

The *Johnson graphs* $J(m,d)$ have as vertices the d-subsets (i.e. the subsets of size d) of an m-set ($m \geq 2d$); two vertices are adjacent whenever the subsets intersect in $d-1$ points. A Johnson graph is distance-regular with intersection numbers $b_j = (d-j)(m-d-j)$, $c_j = j^2$ ($0 \leq j \leq d$).

The (adjacency) eigenvalues of a distance-regular graph are precisely the eigenvalues of the intersection matrix Q. The multiplicities can also be deduced from the intersection matrix Q. So, the intersection numbers determine the spectrum. Conversely, if we have the eigenvalues of a distance regular G we can compute the intersection numbers. This implies that cospectral distance-regular graphs must have the same intersection array. However, it is not necessarily the case that a graph G has to be distance-regular if it has the spectrum of a distance-regular graph. An important tool for the constructions of such cospectral graphs is the following result of Godsil and McKay [15].

Lemma 3.2 *Let G be a graph and let $\{D, C_1, \ldots, C_t\}$ be a partition of the vertex set of G. Suppose that for each vertex $x \in D$ and every $i \in \{1, \ldots, t\}$, x has 0, $\frac{1}{2}|C_i|$ or $|C_i|$ neighbors in C_i. Moreover, suppose that $\{C_1, \ldots, C_t\}$ is an equitable partition of $G \setminus D$. Make a new graph G' as follows. For each $x \in D$ and $i \in \{1, \ldots, t\}$ such that x has $\frac{1}{2}|C_i|$ neighbors in C_i, delete the corresponding $\frac{1}{2}|C_i|$ edges and join x instead to the $\frac{1}{2}|C_i|$ other vertices in C_i. Then G and G' have the same adjacency spectrum.*

Theorem 3.3 *For the Johnson graph $J(m,d)$ there exists a graph cospectral with $J(m,d)$, which is not distance-regular, if and only if $d \geq 3$.*

Proof Johnson graphs with $d=2$ (also known as *triangular graphs*) are strongly regular, a property which can be deduced from the spectrum. For $d \geq 3$, let V be the vertex set of $J(m,d)$, that is, V is the set of all d-subsets of the set $X = \{1, \ldots, m\}$. Let C be the set of d-subsets of X with precisely three elements in $Y = \{1, 2, 3, 4\}$, and define $D = V \setminus C$. Partition C into $t = \binom{m-4}{d-3}$ subsets $C_1, \ldots, C_t$, each consisting of the four d-subsets that have the same intersection with $X \setminus Y = \{5, \ldots, m\}$. Now $\{C_1, \ldots, C_t\}$ is an equitable partition of C. Moreover, for each vertex w in D, and each C_i, we have that if w intersects Y in at most one element, then w has no neighbors in C_i; if w intersects Y in two elements, then w has either two or zero neighbors in C_i; and if w intersects Y in four elements, then w has either zero or four neighbors in C_i. We thus have a switching partition as required in Lemma 3.2, and by switching we obtain a graph that is cospectral with $J(m,d)$. We now claim that this cospectral graph is not distance-regular. Indeed, take $x = \{2, \ldots, d+1\}$ and $y = \{4, \ldots, d+3\}$ (note that $m \geq 2d \geq d+3$). Then, after switching, x and y are nonadjacent vertices with two common neighbors (namely $\{1, 4, \ldots, d+2\}$ and $\{1, 4, \ldots, d+1, d+3\}$). Since the Johnson graph has intersection number $c_2 = 4$ (not 2), the cospectral graph is not distance-regular. □

The above construction appeared in [13]. That paper contains many more examples of distance-regular graphs for which the switching method of Lemma 3.2 applies. For the Hamming graphs, the above method doesn't seem to work. For these graphs another method for the construction of cospectral graphs can be used (see [19]).

Theorem 3.4 *For the Hamming graph $H(d,q)$ with $d = q \geq 3$ there exists a graph cospectral with $H(d,q)$ which is not a distance-regular graph.*

Proof Consider the following incidence structure with points and lines. Points are the vertices of $H(d,q)$, and a line is a set of vertices consisting of the words that are identical in all but one coordinate. So there are q^d points and dq^{d-1} lines. The lines are cliques of size q in $H(d,q)$. Let N be the incidence matrix of this structure (rows correspond to points and columns to lines). Then $NN^\top = dI + A$, where A is the adjacency matrix of $H(d,q)$, and $N^\top N = qI + B$, where B is the adjacency matrix of a graph $H'(d,q)$ defined on the lines, with two vertices adjacent if the lines intersect. If $d = q$ then $NN^\top$ and $N^\top N$ have the same spectrum, and so have A and B. Therefore $H(d,d)$ and $H'(d,d)$ are cospectral. We claim that if $d \geq 3$ then $H'(d,d)$ is not distance-regular. Indeed, there are two types of pairs of lines at distance two (meaning that they don't intersect, but there exists a line that intersects both). If they are parallel, there are q lines that intersect both, and otherwise there is just one line that intersects both. So in $H'(d,d)$ two nonadjacent vertices can have 1 or q common neighbors, and therefore $H'(d,d)$ is not distance-regular. □

The above statement is also true if $d > q \geq 3$ (by taking graph products with $H'(q,q)$). But it is false if $d \leq 2$ (the strongly regular case), if $d = 3$, $q = 2$ (the cube), and if $d = 3$ and $q \geq 36$ (see [1] for this and the latest developments on spectral characterizations of the Hamming graphs).

An important tool in proving distance-regularity from the spectrum is the following theorem due to Fiol and Garriga [14].

Theorem 3.5 *Let G be a graph whose spectrum is feasible for a distance-regular graph of diameter d (this means that the spectrum corresponds to an intersection matrix, hence k_d can be computed from the spectrum). Suppose that for each vertex of G, the number of vertices at distance d equals k_d (that is, the correct number in case of distance-regularity). Then G is distance-regular.*

The approach of Fiol and Garriga is more general, and the proof is rather involved. Recently, Van Dam [8] has given a more elementary proof, of a less general result, which still implies the above theorem. By use of Theorem 3.5 we can easily establish some spectral characterizations of distance-regularity.

Theorem 3.6 *If G is a graph whose spectrum is feasible for a distance-regular graph with diameter d and girth $g \geq 2d - 1$, then G is distance-regular.*

Proof The girth of a regular graph can be deduced from the adjacency spectrum. (Indeed, since $\mathrm{trace}(A^j) = \sum_{i=1}^n \lambda_i^j(A)$, the spectrum determines the number of closed walks of length j, and, since the graph is regular, also the number of closed walks of length j that contain a cycle.) Thus we know that G is regular of girth g. We consider the distance partition of G and define the numbers a_i, b_i, c_i in the same way as we did for the distance-regular case. Similarly we define k_i for $i = 1, \ldots, d$. Since there are no cycles of length less than $2d - 1$, it follows that $c_1 = \cdots = c_{d-1} = 1$ and $a_1 = \cdots = a_{d-2} = 0$. Hence $b_i = k - 1$ and $k_i = k(k-1)^i$, for $i = 1, \ldots, d-1$. Therefore $k_d = n - \sum_{i=0}^{d-1} k_i$ has the value it should have in the case of distance-regularity, and so by Theorem 3.5, G is distance-regular. □

Because every graph has girth at least 3, the condition of Theorem 3.6 is fulfilled if $d = 2$, which proves Theorem 3.1. An example of a graph with larger diameter satisfying this condition is the Coxeter graph (with diameter 4 and girth 7). We refer to [3], [9], or [18] for direct proofs of Theorem 3.6.

Theorem 3.7 *If G is a graph whose spectrum is feasible for a bipartite distance-regular graph with diameter d and girth $g \geq 2d - 2$, then G is distance-regular.*

Proof As in the above proof, we again consider the distance partition of G and define the numbers a_i, b_i, c_i and k_i ($i = 0, \ldots, d$). (At this point the diameter d' of G, may be smaller than d; then we take $k_i = a_i = b_i = 0$ for $d' < i \leq d$.) We know that G is regular and bipartite of girth $g \geq 2d - 2$; therefore $c_1 = \cdots = c_{d-2} = 1$ and $a_1 = \cdots = a_d = 0$. Hence $b_i = k - 1$ and $k_i = k(k-1)^i$, for $i = 1, \ldots, d-2$. Furthermore $c_d = c_{d-1} + b_{d-1} = k$, $b_{d-1}k_{d-1} = c_d k_d$, $b_{d-2}k_{d-2} = k_{d-1}c_{d-1}$, and $k_{d-1} + k_d = n - \sum_{i=0}^{d-2} k_i$. This system of five equations with five unknowns (b_{d-1}, c_{d-1}, c_d, k_{d-1} and k_d) has a unique solution for k_d. Now Theorem 3.5 completes the proof. □

So in particular any graph cospectral with a bipartite distance-regular graph of diameter 3, is such a graph. This includes the cube $H(3, 2)$ (see above). Many more such conditions exist, but the proofs become more technical (see [9] and [13]). Let us finish this section by stating (without proof) the characterization result (from [9]) for an important family of distance regular graphs: the collinearity graphs of finite generalized polygons (see [2]).

Theorem 3.8 *If G is a graph whose spectrum is feasible for the collinearity graph of a finite generalized polygon, then G is distance-regular. Moreover, if the diameter is at least 3, then G is the collinearity graph of a generalized polygon.*

4 Perfect matchings

Throughout this section G is a k-regular graph on an even number n of vertices. We deal with the question: When can we see from the spectrum that G has a perfect matching?

It makes sense to assume regularity. It leads to rather strong results, which don't seem to generalize to arbitrary graphs. Moreover, regular graphs have large matchings. There is always a matching of size at least $nk/2(k+1)$ (by Vizing's theorem), and there is a perfect matching if the graph is bipartite with at least one edge (by König's theorem). We saw that the latter condition can be rephrased in terms of the (adjacency) spectrum. In this section we present another sufficient condition in terms of the eigenvalues. As in the previous section, it makes essentially no difference which type of matrix we consider. We take the adjacency matrix and write $k = \lambda_1 \geq \cdots \geq \lambda_n$ for the spectrum.

We know of no example of two cospectral regular graphs where one has a perfect matching and the other one has not. Nevertheless, we do not expect that there is a spectral characterization for the existence of a perfect matching in a regular graph, as we have for bipartiteness, connectivity, regularity, etc. But we will present a criterion that is satisfied for almost all regular graphs. We shall see that G has at least $\lfloor (k - \lambda_2 + 1)/2 \rfloor$ edge-disjoint perfect matchings. For almost all k-regular graphs $\lambda_2 \approx 2\sqrt{k}$. This shows that with high probability G has at least $(k - 2\sqrt{k})/2$ edge-disjoint perfect matchings.

To state the main result we introduce a class of graphs $\mathcal{H}(\ell)$, and a number $\rho(\ell)$. The class $\mathcal{H}(\ell)$ consists of all connected irregular graphs of odd order v with maximum degree ℓ, at least $(\ell v - \ell + 2)/2$ edges, at least four vertices of degree ℓ if ℓ is odd, and at least three vertices of degree ℓ if ℓ is even. The number $\rho(\ell)$ is the infimum (which is actually a minimum) of the largest adjacency eigenvalue taken over all graphs in $\mathcal{H}(\ell)$.

Theorem 4.1 *If G is a connected k-regular graph of even order, and $\lambda_3 < \rho(k)$, then G has a perfect matching.*

Proof Let $V = V(G)$ be the vertex set of G, and suppose that G does not have a perfect matching. By Tutte's perfect matching theorem (see for example [22]), there exists a set $S \subset V$ of size s such that $V \setminus S$ induces a subgraph which has at least $q \geq s + 2$ components $G_1, G_2, \ldots, G_q$ (say), of odd order.

Let t_i denote the number of edges in G between S and G_i. Then clearly $\sum_{i=1}^{q} t_i \leq ks$, $s \geq 1$, and $t_i \geq 1$ (since G is connected). Hence $t_i < k$ and $n_i > 1$ for at least three values of i, say $i = 1, 2$ and 3. Let ρ_i denote the largest adjacency eigenvalue of G_i, and assume $\rho_1 \geq \rho_2 \geq \rho_3$. Then eigenvalue interlacing (Lemma 1.2), applied to the subgraph induced by the union of G_1, G_2 and G_3, gives $\rho_i \leq \lambda_i$ for $i = 1, 2, 3$.

Next we show that $G_3 \in \mathcal{H}(k)$. This will prove the theorem, since then
$$\rho(k) \leq \rho_3 \leq \lambda_3.$$
Suppose that G_3 has e_3 edges; then $2e_3 = kn_3 - t_3 \leq n_3(n_3 - 1)$. We saw that $t_3 < k$ and $n_3 > 1$, hence $2e_3 > kn_3 - k$ and $k < n_3$. More precisely: $2e_3 \geq kn_3 - k + 2$ (because n_3 is odd), $k \leq n_3 - 1$ if k is even, and $k \leq n_3 - 2$ if k is odd. Note that $t_3 < n_3$ shows that G_3 cannot be regular. So it only needs to be shown that G_3 has the required number of vertices of degree k. For the average degree $\bar{d}$ of G_3 we find $2e_3/n_3 = k - t_3/n_3 \geq k - (k-1)/n_3$. Suppose that k is even. Then $\bar{d} \geq k - (n_3 - 2)/n_3 = k - 1 + 2/n_3$, which implies that at least two vertices have degree k. But if two have degree k, then $n_3 - 2$ have degree $k - 1$, which is impossible since n_3 is odd. Similarly if k is odd, then $\bar{d} \geq k - 1 + 3/n_3$ and three vertices of degree k are not possible, so least four vertices have degree k. □

Corollary 4.2 *If $\lambda_2 \leq k - 1$, then G has a perfect matching.*

Proof For any graph the largest adjacency eigenvalue is bounded from below by the average degree (it follows, for example, from Lemma 1.2(ii) with $m = 1$). Therefore $\rho(\ell) \geq (\ell v - \ell + 2)/v > \ell - 1$. If $\lambda_2 \leq k - 1$, then G is connected and $\lambda_3 \leq \lambda_2 \leq k - 1 < \rho(k)$, so G has a perfect matching by Theorem 4.1. □

Corollary 4.3 *G has at least $\lfloor (k - \lambda_2 + 1)/2 \rfloor$ edge disjoint perfect matchings.*

Proof The adjacency matrix of the 1-regular graph (a perfect matching) has smallest eigenvalue -1. Therefore, deleting the edges of a perfect matching in G decreases k by 1, and increases every other adjacency eigenvalue by at most 1 (by Weyl's eigenvalue inequalities). So $k - \lambda_2$ is decreased by at most 2. Hence we can delete a perfect matching at least $\lfloor (k - \lambda_2 + 1)/2 \rfloor$ times, and this proves Corollary 4.3. □

Regularity and spectra

A larger lower bound for $\rho(\ell)$ gives a better sufficient condition for a perfect matching. In fact, it is not hard to slightly improve the lower bound for the average degree of a graph in $\mathcal{H}(\ell)$ and obtain:

$$\rho(\ell) \geq \ell - 1 + \frac{3}{\ell+1} \text{ if } \ell \text{ is even, and } \rho(\ell) \geq \ell - 1 + \frac{4}{\ell+2} \text{ if } \ell \text{ is odd.}$$

The following theorem gives the exact values for $\rho(\ell)$.

Theorem 4.4

$$\rho(\ell) = \begin{cases} \theta & \text{if } \ell = 3, \\ (\ell - 2 + \sqrt{\ell^2 + 12})/2 & \text{if } \ell \geq 4 \text{ and even}, \\ (\ell - 3 + \sqrt{(\ell+1)^2 + 16})/2 & \text{if } \ell \geq 5 \text{ and odd}, \end{cases}$$

where $\theta = 2.85577...$ is the largest root of $x^3 - x^2 - 6x + 2$.

Proof We assume $\ell \geq 4$ and even. The other cases go in a similar (though slightly more complicated) way. Let H be the complete graph $K_{\ell+1}$ from which a matching of size $(\ell-2)/2$ has been deleted. It is easily checked that $H \in \mathcal{H}(\ell)$. Moreover, the adjacency matrix of H has an equitable partition with quotient matrix

$$Q = \begin{bmatrix} 2 & \ell - 2 \\ 3 & \ell - 4 \end{bmatrix}.$$

It follows in a straightforward way that the eigenvalue of H that corresponds to $\lambda_1(Q)$ (see Lemma 1.1) has a positive eigenvector, and therefore must be the largest eigenvalue of H. Therefore $\rho(\ell) \leq \lambda_1(Q) = (\ell - 2 + \sqrt{\ell^2 + 12})/2$.

Next we show that $\rho(\ell) \geq \lambda_1(Q)$. Consider an arbitrary graph H' in $\mathcal{H}(\ell)$ of order m and size e (say). As above we partition the adjacency matrix of H according to the three vertices of degree ℓ and the remaining $m - 3$ vertices. If a denotes the average degree of the subgraph induced by the three vertices of degree ℓ, then H' has quotient matrix

$$Q' = \begin{bmatrix} a & \ell - a \\ 3(\ell-a)/(m-3) & (2e - 6\ell + 3a)/(m-3) \end{bmatrix}.$$

By Lemma 1.2(ii), the largest eigenvalue $\lambda_1(Q')$ is a lower bound for the largest adjacency eigenvalue of H'. Therefore it suffices to show that $\lambda_1(Q') \geq \lambda_1(Q)$. We easily have $\ell - m + 3 \leq a \leq 2$ and $2e \geq \ell m - \ell + a$ (because $e \geq (\ell m - \ell + 2)/2$). If we decrease a diagonal element the largest eigenvalue does not increase; therefore we may without loss of generality take $2e = \ell m - \ell + a$ and

$$Q' = \begin{bmatrix} a & \ell - a \\ 3(\ell-a)/(m-3) & \ell - 4(\ell-a)/(m-3) \end{bmatrix}.$$

Now $2\lambda_1(Q' - \ell I) = (\ell - a)(-4x - 1 + \sqrt{16x^2 + 4x + 1})$ with $x = 1/(m-3)$, $0 < x \leq 1/(\ell - a)$. It is straightforward to show that $\lambda_1(Q' - \ell I)$ is decreasing in x, and therefore $\lambda_1(Q' - \ell I)$ is minimal if $x = 1/(l-a)$ and $a = 2$. This implies that $\lambda_1(Q' - \ell I) \geq \lambda_1(Q - \ell I)$, which proves our claim. □

Finally we shall construct a k-regular graph G without a matching but with $\lambda_3 = \rho(k)$, showing that the defined value of $\rho(\ell)$ is best possible. Again we restrict ourselves to the case of even $k \geq 4$. As in the above proof, H is the complete graph K_{k+1} from which a matching of size $(k-2)/2$ has been deleted. To make G, we take k disjoint copies of H, and add a set S of $k-2$ isolated vertices. Next we insert edges. For every copy of H we insert an edge between each vertex of degree $k-1$ and each vertex in S. It is easily checked that H is k-regular, but has no perfect matching (indeed, deleting S gives $k > |S|$ components of odd order). We only need to prove that $\lambda_3 = \rho(k)$.

Again H admits an equitable partition having quotient matrix

$$Q_H = \begin{bmatrix} 2 & k-2 \\ 3 & k-4 \end{bmatrix}$$

with eigenvalues $(k-2 \pm \sqrt{k^2+12})/2$. This partition of H leads to an equitable partition of G with $2k+1$ parts, and quotient matrix:

$$Q_G = \begin{bmatrix} 2I_k & (k-2)I_k & 0 \\ 3I_k & (k-4)I_k & 1 \\ 0^\top & 1^\top & 0 \end{bmatrix}.$$

Eigenvalue interlacing shows that each of the above eigenvalues of Q_H occurs $k-1$ times as an eigenvalues of Q_G. In turn, Q_G admits an equitable partition with quotient matrix

$$\begin{bmatrix} 2 & k-2 & 0 \\ 3 & k-4 & 1 \\ 0 & k & 0 \end{bmatrix},$$

which gives the three remaining eigenvalues of Q_G: k and $-1\pm\sqrt{3}$. Hence $\lambda_2(Q_G) = \cdots = \lambda_{k-1}(Q_G) = \lambda_1(Q_H)$. The eigenvalues of Q_G are eigenvalues of G (Lemma 1.1), so what remains to be proved is that the other eigenvalues of G are at most $\lambda_1(Q_H)$. By Lemma 1.1(ii), the other eigenvalues do not change if an all-one block of the partitioned matrix A is replaced by an all-zero block. If we do so for every all-one block in A we obtain a nonnegative matrix with maximum row sum $k-3$. So the largest eigenvalue is at most $k-3$, which is less than $\lambda_1(Q_H)$.

For the case k odd, we refer to [6]. In that paper it is proved that the condition $\lambda_3 < \rho(k)$ is also sufficient for existence of matchings of size $(n-1)/2$ in the case that n is odd.

Corollaries 4.2 and 4.3 appeared in [4]. There, it is also proved that the condition $\lambda_1(L) \leq 2\lambda_{n-1}(L)$ is sufficient for a perfect matching in an arbitrary graph with n even and Laplacian matrix L.

References

[1] S. Bang, E. R. van Dam & J. H. Koolen, Spectral characterization of the Hamming graphs, *Linear Algebra Appl.* **429** (2008), 2678–2686.

[2] A. E. Brouwer, A. M. Cohen & A. Neumaier, *Distance-Regular Graphs*, Springer, Heidelberg (1989).

[3] A. E. Brouwer & W. H. Haemers, The Gewirtz graph: An exercise in the theory of graph spectra, *European J. Combin.* **14** (1993), 397–407.

[4] A. E. Brouwer & W. H. Haemers, Perfect matchings and eigenvalues, *Linear Algebra Appl.* **395** (2005), 155–162.

[5] A. A. Chesnokov & W. H. Haemers, Regularity and the generalized adjacency spectra of graphs, *Linear Algebra Appl.* **416** (2006), 1033–1037.

[6] S. M. Cioabă, D. A. Gregory & W. H. Haemers, Matchings in regular graphs from eigenvalues, *J. Combin. Theory Ser. B* **99** (2009), 287–297.

[7] D. Cvetković, P. Rowlinson & S. Simić, *An Introduction to the Theory of Graph Spectra*, Cambridge University Press, to appear.

[8] E. R. van Dam, The spectral excess theorem for distance-regular graphs: a global (over)view, *Electron. J. Combin.* **15(1)** (2008), R129.

[9] E. R. van Dam & W. H. Haemers, Spectral characterizations of some distance-regular graphs, *J. Algebraic Combin.* **15** (2002), 189–202.

[10] E. R. van Dam & W. H. Haemers, Which graphs are determined by their spectrum?, *Linear Algebra Appl.* **373** (2003), 241–272.

[11] E. R. van Dam & W. H. Haemers, Developments on spectral characterizations of graphs, *Discrete Math.* **309** (2009), 576–586.

[12] E. R. van Dam, W. H. Haemers & J. H. Koolen, Cospectral graphs and the generalized adjacency matrix, *Linear Algebra Appl.* **423** (2007), 33–41.

[13] E. R. van Dam, W. H. Haemers, J. H. Koolen & E. Spence, Characterizing distance-regularity of graphs by the spectrum, *J. Combin. Theory Ser. A* **113** (2006), 1805–1820.

[14] M. A. Fiol & E. Garriga, From local adjacency polynomials to locally pseudo-distance-regular graphs, *J. Combin. Theory Ser. B* **71** (1997), 162–183.

[15] C. D. Godsil & B. D. McKay, Constructing cospectral graphs, *Aequationes Math.* **25** (1982), 257–268.

[16] C. D. Godsil & G. Royle, *Algebraic Graph Theory*, Springer-Verlag, New York (2001).

[17] W. H. Haemers, Interlacing eigenvalues and graphs, *Linear Algebra Appl.* **226–228** (1995), 593–616.

[18] W. H. Haemers, Distance-regularity and the spectrum of graphs, *Linear Algebra Appl.* **236** (1996), 265–278.

[19] W. H. Haemers & E. Spence, Graphs cospectral with distance-regular graphs, *Linear and Multilinear Algebra* **39** (1995), 91–107.

[20] R. A. Horne & C. R. Johnson, *Matrix Analysis*, Cambridge University Press, Cambridge (1990).

[21] C. R. Johnson & M. Newman, A note on cospectral graphs, *J. Combin. Theory Ser. B* **28** (1980), 96–103.

[22] D. West, *Introduction to Graph Theory*, Second edition, Prentice Hall, 2001.

Tilburg University,
Tilburg,
The Netherlands
haemers@uvt.nl

Trades and t-designs

G. B. Khosrovshahi and B. Tayfeh-Rezaie

Abstract

Trades, as combinatorial objects, possess interesting combinatorial and algebraic properties and play a considerable role in various areas of combinatorial designs. In this paper we focus on trades within the context of t-designs. A pedagogical review of the applications of trades in constructing halving t-designs is presented. We also consider (N,t)-partitionable sets as a generalization of trades. This generalized notion provides a powerful approach to the construction of large sets of t-designs. We review the main recursive constructions and theorems obtained by this approach. Finally, we discuss the linear algebraic representation of trades and present two applications.

1 Introduction

Let v, k, t and λ be integers such that $v \geq k \geq t \geq 0$ and $\lambda \geq 1$. Let X be a v-set and let $P_i(X)$ denote the set of all i-subsets of X for any i. A t-(v, k, λ) design (briefly t-design) is a pair $D = (X, \mathcal{B})$ in which $\mathcal{B}$ is a collection of elements of $P_k(X)$ such that every $A \in P_t(X)$ appears in exactly λ elements of $\mathcal{B}$. Let N be a natural number greater than 1. A *large set of t-(v, k, λ) designs of size N*, $LS[N](t, k, v)$, is a set of N disjoint t-(v, k, λ) designs $(X, \mathcal{B}_i)$ such that $\{\mathcal{B}_i| 1 \leq i \leq N\}$ is a partition of $P_k(X)$. A $T(t, k, v)$ *trade* is a pair $T = (X, \{T^+, T^-\})$ in which T^+ and T^- are two disjoint collections of elements of $P_k(X)$ such that for every $A \in P_t(X)$, the number of occurrences of A in T^+ is the same as the number of occurrences of A in T^-, i.e. T^+ and T^- are mutually balanced. Note that some elements of $P_t(X)$ may not appear in T^+ and T^-. For simplicity, we write $(X; T^+, T^-)$ instead of $(X, \{T^+, T^-\})$.

The family of t-designs are among the most important and fundamental families of combinatorial designs. They show up in different areas of combinatorics such as coding theory, group theory, finite geometry and so on. They are employed in construction of various combinatorial designs and configurations. In particular, 2-designs, for their statistical optimality properties, are widely used in design of experiments.

Trades are useful combinatorial objects with interesting combinatorial and algebraic properties. They appear in various contexts of combinatorial design theory. Trades were first introduced in the theory of t-designs by some authors under different names in the seventies of the last century (see [17] for details). In recent years Latin trades in connection to Latin squares and graphical trades related to graphical designs have been investigated. For a survey of trades in different contexts we refer the reader to [9]. The reference [22] is a survey specifically devoted to applications of trades in the theory of t-designs.

In this paper, we consider trades within the context of t-designs. Trades are very useful in the study of t-designs with many applications. They are utilized in constructing signed t-designs, nonisomorphic t-designs and t-designs with repeated blocks. They are also employed to determine the spectrum of support sizes of t-designs with repeated blocks, block intersection numbers and defining sets of t-

designs. For a thorough discussion of these applications, see [11, 17, 18, 22] and the references therein.

In the last twenty years trades and their generalization, i.e. (N,t)-partitionable sets, have been successfully used in the study of the existence problem of t-designs and large sets. This study has led to a powerful approach to the construction of large sets of t-designs. The method was developed in the nineties of the last century by Ajoodani-Namini and Khosrovshahi [5] through their work on Hartman's conjecture (halving conjecture) on the existence of halving designs. Since then, many existence results as well as recursive constructions have been obtained using this approach for t-designs in general and for halving designs and large sets of prime sizes in particular. Undoubtedly, the most outstanding result achieved by this method is the proof of halving conjecture for 2-designs by Ajoodani-Namini [1]. We here present an instructive review of the approach of (N,t)-partitionable sets. For simplicity of presentation, our treatment is mainly based on trades, i.e. $(2,t)$-partitionable sets. We demonstrate how one can use trades to find recursive constructions for halving designs or large sets of size 2. The general case of (N,t)-partitionable sets is briefly discussed and it is shown that most of the results for large sets of size 2 can easily be extended to large sets in general. Trades also provide an algebraic setting for the study of t-designs. We describe the linear algebraic representation of trades and present two applications of it.

2 t-Designs and large sets

Let $D = (X, \mathcal{B})$ be a t-(v, k, λ) design. The parameter t is called the *strength* of D by some authors. The elements of X are called *points* and the elements of $\mathcal{B}$ are called *blocks*. When the blocks of D are distinct, the design is *simple*. In this paper, we are mainly interested in simple designs. An easy counting argument shows that if there exists a t-(v, k, λ) design, then

$$\lambda_i = \lambda \frac{\binom{v-i}{t-i}}{\binom{k-i}{t-i}} \quad \text{for} \quad 0 \leq i \leq t,$$

are integers. These are known as the *feasibility conditions*. From these conditions it easily follows that D is an i-(v, k, λ_i) design for any $0 \leq i \leq t$. The pair $(X, P_k(X))$ is a t-$(v, k, \binom{v-t}{k-t})$ design and is called a *complete design*. The following observation is well known.

Lemma 2.1 *Let $0 \leq t \leq k \leq v$. If there exists a t-(v, k, λ) design which is not complete, then $t < k < v - t$.*

The main question concerning t-designs is the existence problem: For given integers v, k, t, λ such that $0 \leq t \leq k \leq v$, $\lambda \geq 1$ and satisfying the feasibility conditions, does there exist a t-(v, k, λ) design? Naturally, the existence problem is in general intractable. However, it has been dealt with and answered in some special cases. For a survey of known results, the interested reader may consult [21, 33] and the references therein.

For some time it was believed that there is no simple t-design for $t \geq 6$. Then, in 1982, Magliveras and Leavitt constructed the first examples of simple 6-designs

[32]. Later, in 1986, Kreher and Radziszowski [29] found the smallest possible 6-design, i.e. a 6-(14,7,4) design. Finally, in 1987, Teirlinck [36] obtained the striking result that simple t-designs exist for all t. In [3], Ajoodani-Namini gave a new proof of this result using a completely different methodology, i.e. the method of (N,t)-partitionable sets which is the subject of this review.

The approach of (N,t)-partitionable sets provides a method to tackle the existence problem of t-designs through the construction of large sets of t-designs. Based on this method, some strong recursive constructions for large sets are obtained. Using these constructions we are able to establish many interesting existence results on t-designs. The method will be explained in detail in the subsequent sections. Here, we review some basic facts on large sets. The following theorems are easy to prove but contain important information.

Theorem 2.2 [3, 27] *If there exists an $LS[N](t,k,v)$, then there exist $LS[N](t-i, k-j, v-l)$ for all $0 \leq j \leq l \leq i \leq t$.*

Theorem 2.3 *There exists an $LS[N](t,k,v)$ if and only if there exists an $LS[N](t, v-k, v)$.*

A set of necessary conditions for the existence of an $LS[N](t,k,v)$ is

$$N \left| \binom{v-i}{k-i} \right. \quad \text{for} \quad 0 \leq i \leq t. \tag{2.1}$$

These conditions are direct consequences of the feasibility conditions for t-designs. They can also be deduced as follows. If the large set $LS[N](t,k,v)$ exists, then for $0 \leq i \leq t$, the set $P_{k-i}(X \setminus \{1,\ldots,i\})$ is partitioned into N equal parts. Note that the necessary conditions (2.1) are not always sufficient. A historic example is the nonexistence of $LS[5](2,3,7)$. It is worth noting that by a celebrated result due to Baranyai, the necessary conditions (2.1) are sufficient for the existence of large sets $LS[N](1,k,v)$ [7, 10, 16].

Let N, t, k and v be integers such that $N > 1$ and $0 \leq t \leq k \leq v$. The quadruple $(N; t, k, v)$ satisfying (2.1) is called a *feasible quadruple*. It is possible to give a better description for feasible quadruples when N is a prime power. Let m and n be positive integers. We denote by $(m)_n$ the remainder obtained when m is divided by n.

Theorem 2.4 [27] *Let p be a prime, α a positive integer and $0 \leq t \leq k \leq v$. The quadruple $(p^\alpha; t, k, v)$ is feasible if and only if there exist distinct positive integers ℓ_i $(1 \leq i \leq \alpha)$ such that $t \leq (v)_{p^{\ell_i}} < (k)_{p^{\ell_i}}$.*

Using Theorem 2.4, we can easily determine all the feasible quadruples when $N = p$ is a prime:

$$(p; t, k, v) = (p; t, mp^z + r, np^z + s), \tag{2.2}$$

where $0 \leq t \leq s < r < p^z$ and $0 \leq m < n$. We can also assume that z is the smallest or the largest number with the properties above to be assured of the uniqueness of the representation (2.2). We present two examples to show the importance of Theorem 2.4.

Example 2.5 By Theorem 2.4, the quadruple $(5; 2, 4, 13)$ is feasible since $2 \leq (13)_5 < (4)_5$ and also the quadruple $(11; 2, 4, 13)$ is feasible since $2 \leq (13)_{11} < (4)_{11}$. Now from (2.1), it is clear that the quadruple $(55; 2, 4, 13)$ is feasible.

Example 2.6 What is the largest value of t for which the parameter set of an LS$[13](t, 9, 18)$ is feasible? By Theorem 2.4, we must have $t \leq (18)_{13^a} < (9)_{13^a}$ and hence $\alpha = 1$ and $t_{\max} = 5$.

3 Trades

We start this section with two examples of trades.

Example 3.1 Let $X = \{1, 2, 3, 4, 5, 6\}$. Let

$$T_1^+ = \{135, 146, 236, 245\},$$
$$T_1^- = \{136, 145, 235, 246\}.$$

Then $T_1 = (X; T_1^+, T_1^-)$ is a T$(2, 3, 6)$ trade. Note that by 123 we mean $\{1, 2, 3\}$, etc.

Example 3.2 Here is another example of a T$(2, 3, 6)$ trade. Let

$$T_2^+ = \{123, 124, 156, 256, 345, 346\},$$
$$T_2^- = \{125, 126, 134, 234, 356, 456\}.$$

Then $T_2 = (X; T_2^+, T_2^-)$ is a T$(2, 3, 6)$ trade.

Lemma 3.3 *A* T(t, k, v) *trade* T *is also a* T(t', k, v) *trade for any* $0 \leq t' < t$.

By letting $t' = 0$ in this lemma, it follows that the number of blocks in T^+ is the same as the number of blocks in T^-, which is called the *volume* of T. By letting $t' = 1$, we also observe that the set of points covered by T^+ is exactly the same as the set of points covered by T^-. This set is called the *foundation* of T. We are now ready to state the following fundamental theorem of trades.

Theorem 3.4 [19] *A nontrivial* T(t, k, v) *trade has foundation size at least* $k + t + 1$ *and volume at least* 2^t.

A trade of volume 0 is called the *trivial trade*. By the above theorem and what follows, there exists a nontrivial T(t, k, v) trade if and only if $t < k < v - t$. A T(t, k, v) trade of foundation size $k + t + 1$ and volume 2^t has a unique structure and is called the *minimal* trade. For example, any minimal T$(2, 3, 6)$ trade is isomorphic to trade T_1 of Example 3.1. We also note that a T$(0, k, v)$ trade $T = (X; T^+, T^-)$ has a simple structure. In fact the only restriction on T to be a T$(0, k, v)$ is that T^+ and T^- contain the same number of blocks. For $t = 1$, it is easy to construct trades. However, for $t \geq 2$ the construction does not seem to be so trivial. The problem becomes harder as t increases. Even for $t = 3$, there are many unsolved problems on the existence of T(t, k, v) trades. In order to construct trades for any t there is a method based on the notion of products of trades which is to be presented in Section 4.

By definition, the blocks of a trade do not have to be distinct. However, trades with distinct blocks (simple trades), are of greater importance. Trades T_1 and T_2 of Examples 3.1 and 3.2, respectively, are examples of simple trades. There is also a trade on $X = \{1, 2, \ldots, 6\}$ which in not simple as the following example illustrates.

Example 3.5 Let

$$T_3^+ = \{123, 125, 134, 145, 246, 246, 356, 356\},$$
$$T_3^- = \{124, 124, 135, 135, 236, 256, 346, 456\}.$$

Then $T_3 = (X; T_3^+, T_3^-)$ is a T(2, 3, 6) trade which is not simple.

In this paper we are only interested in simple trades. Hence, hereafter we assume that all trades are simple.

We need to define the union operation of simple trades. Let $T_i = (X_i; T_i^+, T_i^-)$ be $T(t, k, v_i)$ trades for $i = 1, 2$ with disjoint block sets, i.e. $(T_1^+ \cup T_1^-) \cap (T_2^+ \cup T_2^-) = \emptyset$. Then, the *union* of T_1 and T_2, denoted by $T_1 + T_2$ is a $T(t, k, v_1 + v_2)$ trade defined as $(X_1 \cup X_2; T_1^+ \cup T_2^+, T_1^- \cup T_2^-)$.

Example 3.6 The union of disjoint trades T_1 and T_2 of Examples 3.1 and 3.2, respectively, gives a T(2, 3, 6) trade $T_4 = (X; T_4^+, T_4^-)$, where

$$T_4^+ = \{123, 124, 135, 146, 156, 236, 245, 256, 345, 346\},$$
$$T_4^- = \{125, 126, 134, 136, 145, 234, 235, 246, 356, 456\}.$$

Since the volume of T_4 is equal to $\binom{6}{3}/2 = 10$, T_4 at the same time represents an LS[2](2, 3, 6) and T_4^+ and T_4^- are the block sets of 2-(6,3,2) designs.

Let $T = (X; T^+, T^-)$ be a simple $T(t, k, v)$ trade such that $T^+ \cup T^- = P_k(X)$. Since any t-subset Y of X is contained in $\binom{v-t}{k-t}$ blocks of $P_k(X)$, it follows that Y is contained in $\binom{v-t}{k-t}/2$ blocks of T^+ (T^-). Hence, (X, T^+) and (X, T^-) are t-$(v, k, \binom{v-t}{k-t}/2)$ designs and we also have an LS[2](t, k, v). The converse also holds, of course. Note that the volume of T is $\binom{v}{k}/2$. A simple t-$(v, k, \binom{v-t}{k-t}/2)$ design is called a *halving* design. In our approach to the existence problem of halving designs it is more natural to consider halving t-(v, k, λ) designs as LS[2](t, k, v) or trades of volume $\binom{v}{k}/2$.

4 Product of trades

In this section we present a description of the operation of product of trades using a number of examples. The following definition is a special case of a general notion which was introduced in [5]. The general case is considered later in Section 8. Let X_1 and X_2 be two disjoint sets of cardinality v_1 and v_2, respectively. For $\mathcal{B}_1 \subseteq P_{k_1}(X_1)$ and $\mathcal{B}_2 \subseteq P_{k_2}(X_2)$, we let

$$\mathcal{B}_1 * \mathcal{B}_2 = \{B_1 \cup B_2 | B_1 \in \mathcal{B}_1, \ B_2 \in \mathcal{B}_2\}.$$

Let $T_i = (X_i; T_i^+, T_i^-)$ be a $T(t_i, k_i, v_i)$ trade of volume s_i on X_i for $i = 1, 2$. Then the *product* of T_1 and T_2, denoted by $T_1 * T_2$, is defined as $(X_1 \cup X_2; S^+, S^-)$, where

$$S^+ = (T_1^+ * T_2^+) \cup (T_1^- * T_2^-),$$
$$S^- = (T_1^+ * T_2^-) \cup (T_1^- * T_2^+).$$

The following is an important extension theorem concerning the product of trades. The theorem states that from trades of strength t_1 and t_2, one can produce a trade of strength $t_1 + t_2 + 1$. Since the presence of $+1$ here is quite unexpected, we present the proof to clarify it.

Theorem 4.1 [5] $T_1 * T_2$ *is a* $T(t_1 + t_2 + 1, k_1 + k_2, v_1 + v_2)$ *trade of volume* $2s_1 s_2$.

Proof It is easy to see that S^+ and S^- are disjoint and $|S^+| = |S^-| = 2s_1 s_2$. Therefore, we only need to show that S^+ and S^- are $(t_1 + t_2 + 1)$-balanced. Let B be a $(t_1 + t_2 + 1)$-subset of $X_1 \cup X_2$ and for $1 \leq i \leq 2$, define $B_i = B \cap X_i$ and $r_i = |B_i|$. With no loss of generality, we may assume that $r_1 \leq t_1$ (since if $r_i > t_i$ for $i = 1, 2$, then $|B| \geq r_1 + r_2 + 2$, a contradiction).

Let x be the number of occurrences of B_1 in T_1^+ (T_1^-). Let y and z be the number of occurrences of B_2 in T_2^+ and T_2^-, respectively. Then clearly, $x(y+z)$ is the number of occurrences of B in S^+ and at the same time the number of occurrences of B in S^-. Hence, the assertion holds. □

We also need the following definition. Let $\mathcal{B} \subseteq P_{k_1}(X_1)$ be of cardinality b and let $T = (X_2; T^+, T^-)$ be a $T(t, k_2, v_2)$ trade of volume s. Then, the *product* of $\mathcal{B}$ and T, denoted by $\mathcal{B} * T$, is defined as $(X_1 \cup X_2; S^+, S^-)$, where

$$S^+ = \mathcal{B} * T^+,$$
$$S^- = \mathcal{B} * T^-.$$

Theorem 4.2 [5] $\mathcal{B} * T$ *is a* $T(t, k_1 + k_2, v_1 + v_2)$ *trade of volume* bs.

One may use Theorem 4.1 to construct trades of any strength t starting with trades of strength 0. Theorem 4.1 is the basis of our approach in constructing t-designs and large sets. We illustrate the product of trades and the theorems above by some examples.

Example 4.3 Let T_1 be as given in Example 3.1. Let $\mathcal{B} = \{0\}$. Then $\mathcal{B} * T_1 = (\{0, 1, \ldots, 6\}; S^+, S^-)$ is a $T(2, 4, 7)$ trade by Theorem 4.2, where

$$S^+ = \{0135, 0146, 0236, 0245\},$$
$$S^- = \{0136, 0145, 0235, 0246\}.$$

Example 4.4 Let T_1 be as given in Example 3.1. Let $T_2 = (\{x, y, z, t\}; T_2^+, T_2^-)$ be a $T(1, 2, 4)$ trade where

$$T_2^+ = \{xy, zt\},$$
$$T_2^- = \{xz, yt\}.$$

Then $T_1 * T_2 = (\{1,2,\ldots,6,x,y,z,t\}; S^+, S^-)$ is a T(4,5,10) trade of volume 16, where

$$S^+ = \{135xy, 146xy, 236xy, 245xy, 135zt, 146zt, 236zt, 245zt,$$
$$136xz, 145xz, 235xz, 246xz, 136yt, 145yt, 235yt, 246yt\},$$
$$S^- = \{135xz, 146xz, 236xz, 245xz, 135yt, 146yt, 236yt, 245yt$$
$$136xy, 145xy, 235xy, 246xy, 136zt, 145zt, 235zt, 246zt\}.$$

In the example above, starting with trades of strength 1 and 2, we find a trade of strength 4. Starting with trades of strength 0 and repeating this approach we can find a trade of any arbitrary strength t. This is shown in the next example.

Example 4.5 Minimal trades have a unique structure. They can be constructed through the product of trades. A minimal $T(t, k, k+t+1)$ trade is in fact the product of $t+1$ trades of strength 0. To be more specific, let

$$X = \{x_1, x_2, \ldots, x_{t+1}, y_1, y_2, \ldots, y_{t+1}, z_1, z_2, \ldots, z_{k-t-1}\}$$

be a $(k+t+1)$-set. Let $T_i = (\{x_i, y_i\}; \{x_i\}, \{y_i\})$ be a $T(0,1,2)$ trade of volume 1 for $i = 1, 2, \ldots, t+1$. Then by Theorem 4.1, the product T of T_i $(1 \le i \le t+1)$ is a $T(t, t+1, 2t+2)$ trade. Let $\mathcal{B} = \{z_1, z_2, \ldots, z_{k-t-1}\}$. Then by Theorem 4.2, $\mathcal{B} * T$ is a $T(t, k, k+t+1)$ trade which is minimal. For example trade T_1 of Example 3.1, is the product of trades $(\{1,2\}; \{1\}, \{2\})$, $(\{3,4\}; \{3\}, \{4\})$ and $(\{5,6\}; \{5\}, \{6\})$.

Trade T_2 of Example 3.2 cannot be obtained through the product operation, since then by Theorem 4.1, it should be the product of a trade of volume 1 and a trade of volume 3. But by Theorem 3.4, the trade of volume 1 has strength 0 and it is of the form $(\{x, y\}; \{x\}, \{y\})$ where $x, y \in \{1, 2, \ldots, 6\}$. Therefore every block of T_2 should contain x or y which is not the case.

Combining two operations of union and product of trades we can construct many trades. In [8], this method has been used to construct simple $T(2, 3, v)$ trades for any even foundation size v and any possible volume.

5 Recursive constructions

In this section we make use of the two operations defined in the previous sections, i.e. the product and the union operations of trades, to obtain some recursive constructions for large sets of size 2. We note that although the constructions are given for large sets of size 2, however they can easily be extended to any size N. We will discuss the general case in Section 8.

A description of the approach is as follows: We construct some block disjoint $T(t, k, v)$ trades on a v-set X using the product operation and then take their union. If the resulting trade covers all k-subsets of X, then we have an $LS[2](t, k, v)$. The following lemma provides a formal statement for later use.

Lemma 5.1 *If there exist $T(t, k, v)$ trades $T_i = (X; T_i^+, T_i^-)$ $(1 \le i \le n)$ such that $T_i^+ \cup T_i^-$ partition $P_k(X)$, then there exists an $LS[2](t, k, v)$.*

Proof Let

$$T^+ = \bigcup_{i=1}^n T_i^+$$

$$T^- = \bigcup_{i=1}^n T_i^-.$$

Then $T = (X; T^+, T^-)$ is a T(t,k,v) trade such that $T^+ \cup T^- = P_k(X)$. Therefore, (X, T^+) and (X, T^+) are halving designs and we have an LS[2](t,k,v). □

The recursive constructions we present in this section stem from some binomial identities. We start with the following simple one:

$$\binom{v}{k} = \binom{v}{v-k},$$

which follows from the trivial one-to-one correspondence between k-subsets and $(v-k)$-subsets of a v-set. Translating this to the language of large sets we obtain Theorem 2.3.

The next identity is

$$\binom{v}{k} = \binom{v-1}{k} + \binom{v-1}{k-1},$$

which is obtained by counting the number of k-subsets of a v-set in two ways. The left hand side is straightforward since the number of k-subsets of a v-set is $\binom{v}{k}$. The right hand side follows from counting first k-subsets which do not contain a fixed point x and then counting k-subsets which contain x. The identity and its proof suggests the following for large sets.

Theorem 5.2 *There exists an LS[2](t,k,v) if and only if there exist both an LS[2]$(t,k,v-1)$ and an LS[2]$(t,k-1,v-1)$.*

Proof From LS[2](t,k,v) we obtain an LS[2]$(t,k,v-1)$ and an LS[2]$(t,k-1,v-1)$ using Theorem 2.2.

For the converse, let X be a $(v-1)$-set and $x \notin X$. From the assumption, there is a T(t,k,v) trade $T_1 = (X \cup \{x\}; T_1^+, T_1^-)$ such that $T_1^+ \cup T_1^- = P_k(X)$ and also a T$(t, k-1, v-1)$ trade $T_2 = (X; T_2^+, T_2^-)$ such that $T_2^+ \cup T_2^- = P_{k-1}(X)$. Let $T_3 = \{x\} * T_2$. Then by Theorem 4.2, T_3 is a T(t,k,v) trade. Note that T_i^+ and T_i^- ($i=1,3$) partition $P_k(X \cup \{x\})$. Now from Lemma 5.1 we obtain an LS[2](t,k,v). □

The following theorem is a consequence of Theorem 5.2 and an induction argument.

Theorem 5.3 *If there exist LS[2]$(t, k+i, v)$ for all $0 \leq i \leq l$, then there exist LS[2]$(t, k+i, v+j)$ for all $0 \leq j \leq i \leq l$.*

Trades and t-designs

The following identity is more involved:

$$\binom{v+1}{a+b+1} = \sum_{i=a}^{v-b}\binom{i}{a}\binom{v-i}{b}.$$

It is obtained using the double counting suggested by the following lemma.

Lemma 5.4

$$P_{a+b+1}(\{1,2,\ldots,v+1\}) = \bigcup_{i=a}^{v-b}(P_a(\{1,2,\ldots,i\}) * \{i+1\} * P_b(\{i+2,i+3,\ldots,v+1\})).$$

Proof Sort $(a+b+1)$-subsets of $\{1,2,\ldots,v+1\}$ in lexicographic order and partition them by looking at the elements in the position $a+1$. □

We demonstrate by an example how the lemma can be utilized to construct large sets.

Example 5.5 We construct an LS[2](2,3,10). To do this we first construct an LS[2](2,7,10) and then using Theorem 2.3 we obtain an LS[2](2,3,10). Let $a = b = 3$ and $v = 9$ in Lemma 5.4 and let $X = \{1,2,\ldots,10\}$. We have

$$P_7(X) = \bigcup_{i=3}^{6}\mathcal{B}_i,$$

where

$$\mathcal{B}_3 = P_3(\{1,2,3\}) * \{4\} * P_3(\{5,6,\ldots,10\}),$$
$$\mathcal{B}_4 = P_3(\{1,2,3,4\}) * \{5\} * P_3(\{6,7,\ldots,10\}),$$
$$\mathcal{B}_5 = P_3(\{1,2,\ldots,5\}) * \{6\} * P_3(\{7,8,9,10\}),$$
$$\mathcal{B}_6 = P_3(\{1,2,\ldots,6\}) * \{7\} * P_3(\{8,9,10\}).$$

We show that there are T(2,7,10) trades $T_i = (X; T_i^+, T_i^-)$ for $i = 3,4,5,6$ such that $T_i^+ \cup T_i^- = \mathcal{B}_i$ and hence by Lemma 5.1, there is an LS[2](2,7,10). In Example 3.6, we constructed an LS[2](2,3,6). From this and Theorem 2.2, we obtain an LS[2](1,3,5) and an LS[2](0,3,4). From LS[2](2,3,6) we find a T(2,3,6) trade $T_1 = (Y; T_1^+, T_1^-)$, where $Y = \{5,6,\ldots,10\}$. Now let $T_3 = P_3(\{1,2,3\}) * \{4\} * T_1$ which is, by Theorem 4.2, a T(2,7,10) trade. T_6 is constructed in the same way. From LS[2](0,3,4) we find a T(0,3,4) trade $T_1 = (Y; T_1^+, T_1^-)$, where $Y = \{1,2,3,4\}$ and from LS[2](1,3,5) we find a T(1,3,5) trade $T_2 = (Z; T_2^+, T_2^-)$, where $Z = \{6,7,\ldots,10\}$. Now let $T_4 = T_1 * \{5\} * T_2$ which is, by Theorems 4.1 and 4.2, a T(2,7,10) trade. T_5 is constructed in a similar way.

The partition given by Lemma 5.4, has been used to obtain a recursive construction for large sets of prime sizes in [34].

Here is another example of a useful binomial identity:

$$\binom{u+v+1}{k} = \sum_{i=0}^{k}\binom{u-i}{k-i}\binom{v+i}{i},$$

with a proof given in the next lemma.

Lemma 5.6 *Let $X = \{1, \ldots, u+v+1\}$ and let $X_j = \{1, \ldots, j\}$ and $Y_j = X \setminus X_j$ for $j = 1, \ldots, u+v+1$. Assume that*

$$\mathcal{B}_i = P_{k-i}(X_{u-i}) * P_i(Y_{u-i+1}), \qquad 0 \leq i \leq k.$$

Then the sets $\mathcal{B}_i$ partition $P_k(X)$.

Proof Let $0 \leq j < i \leq k$ and $A \in \mathcal{B}_i$. Then $|A \cap X_{u-i}| = k - i$ and

$$|A \cap X_{u-j}| \leq |A \cap X_{u-i}| + |X_{u-j} \setminus X_{u-i}| - 1$$
$$= k - j - 1.$$

Therefore, $A \notin \mathcal{B}_j$. Hence all $\mathcal{B}_i$ are mutually disjoint.

Now let $A \in P_k(X)$. Let $0 \leq i \leq k$ be the smallest integer such that $|A \cap X_{u-i}| \geq k - i$. Then $|A \cap X_{u-i+1}| \leq k - i + 1$ and therefore,

$$k - i \leq |A \cap X_{u-i}|$$
$$\leq |A \cap X_{u-i+1}|$$
$$\leq k - i.$$

Hence, $|A \cap X_{u-i}| = |A \cap X_{u-i+1}| = k - i$ and $A \in \mathcal{B}_i$. □

An extension of Lemma 5.6 is given in [2] (see also [25]). However, for our purpose, this simplified version is well-suited. Before presenting the construction, we give an example.

Example 5.7 We construct an LS[2](2,3,10) from an LS[2](2,3,6) using the partition given in Lemma 5.6 (compare to the construction given in Example 5.5). Let $u = 6, v = 4$ and $k = 3$ in Lemma 5.6 and let $X = \{1, 2, \ldots, 10\}$. We have

$$P_3(X) = \bigcup_{i=0}^{3} \mathcal{B}_i,$$

where

$$\mathcal{B}_0 = P_3(\{1, 2, \ldots, 6\}),$$
$$\mathcal{B}_1 = P_2(\{1, 2, \ldots, 5\}) * P_1(\{7, 8, 9, 10\}),$$
$$\mathcal{B}_2 = P_1(\{1, 2, 3, 4\}) * P_2(\{6, 7, \ldots, 10\}),$$
$$\mathcal{B}_3 = P_3(\{5, 6, \ldots, 10\}).$$

We show that there are T(2,3,10) trades $T_i = (X; T_i^+, T_i^-)$ for $i = 0, 1, 2, 3$ such that $T_i^+ \cup T_i^- = \mathcal{B}_i$ and hence by Lemma 5.1, there is an LS[2](2,3,10). In Example 3.6, we constructed an LS[2](2,3,6). From this and Theorem 2.2, we obtain an LS[2](1,2,5) and an LS[2](0,1,4). From LS[2](2,3,6) we find a T(2,3,6) trade $T_0 = (Y; T_0^+, T_0^-)$, where $Y = \{1, 2, \ldots, 6\}$. Similarly, we construct a T(2,3,6) trade $T_3 = (Z; T_3^+, T_3^-)$, where $Z = \{5, 6, \ldots, 10\}$. From LS[2](1,2,5) we have a T(1,2,5) trade $T_4 = (\{1, 2, \ldots, 5\}; T_4^+, T_4^-)$ and from LS[2](0,1,4) we find a T(0,1,4) trade $T_5 = (\{7, 8, 9, 10\}; T_5^+, T_5^-)$. Now let $T_1 = T_4 * T_5$ which by Theorem 4.1, is a T(2,3,10) trade. T_2 is constructed in the same way.

Now using the partition provided by Lemma 5.6, we present an important recursive construction.

Theorem 5.8 [5] *If $LS[2](t, i, v+i)$ exist for all $t+1 \le i \le k$ and an $LS[2](t, k, u)$ also exists, then $LS[2](t, k, u + l(v+1))$ exist for all $l \ge 1$.*

Proof It suffices to prove the theorem for $l = 1$. Then the general case easily follows by an induction on l.

Consider the partition given in Lemma 5.6. We show that there exist $T(t, k, u + v + 1)$ trades $T_i = (X; T_i^+, T_i^-)$ for $1 \le i \le k$ such that $T_i^+ \cup T_i^- = \mathcal{B}_i$ and hence by Lemma 5.1, there is an $LS[2](t, k, u + v + 1)$.

Let $0 \le i \le k$. Recall that

$$\mathcal{B}_i = P_{k-i}(X_{u-i}) * P_i(Y_{u-i+1}).$$

First assume that $0 \le i \le t$. By $LS[2](t, k, u)$ which exists by the assumption and Theorem 2.2, we acquire $LS[2](t-i, k-i, u-i)$ and from this we obtain a $T(t-i, k-i, u-i)$ trade $T = (X_i; T^+, T^-)$ such that $T^+ \cup T^- = P_{k-i}(X_{u-i})$. By $LS[2](t, t+1, v+t+1)$ which comes from the assumption and Theorem 2.2, we also obtain an $LS[2](i-1, i, v+i)$. Since $|Y_{u-i+1}| = v+i$, there is a $T(i-1, i, v+i)$ trade $T' = (Y_{u-i+1}; T'^+, T'^-)$ such that $T'^+ \cup T'^- = P_i(Y_{u-i+1})$. Now let $T_i = T*T'$ which is a $T(t, k, u+v+1)$ trade by Theorem 4.1. Next let $i > t$. By the assumption we have an $LS[2](t, i, v+i)$ and so there is a $T(t, i, v+i)$ trade $T = (Y_{u-i+1}; T^+, T^-)$ such that $T^+ \cup T^- = P_i(Y_{u-i+1})$. Now let $T_i = P_{k-i}(X_{u-i}) * T$ which is $T(t, k, u+v+1)$ trade by Theorem 4.2. □

More recursive constructions have been found using this approach. The reader is referred to [2, 3, 25, 27, 34]. In Section 8, we return to these constructions when we mention large sets of any size.

6 Halving designs

In Section 3, we observed that halving t-(v, k, λ) designs, large sets $LS[2](t, k, v)$ and $T(t, k, v)$ trades of volume $\binom{v}{k}/2$ are in principle the same objects. In the sequel, we consider $LS[2](t, k, v)$ as halving designs.

From (2.1), we recall that the parameters of an $LS[2](t, k, v)$ satisfy the following necessary conditions:

$$2 \left| \binom{v-i}{k-i} \right., \qquad 0 \le i \le t. \tag{6.1}$$

A triple (t, k, v) satisfying (6.1) is called *feasible*. From Theorem 2.4, we have the following lemma.

Lemma 6.1 *Let $0 \le t \le k \le v$. Then (t, k, v) is feasible if and only if there is a positive integer z such that $t \le (v)_{2^z} < (k)_{2^z}$.*

The above characterization can be rephrased as follows.

Lemma 6.2 *Let $0 \le t \le k \le v$. Then (t, k, v) is feasible if and only if $k = m2^z + r$ and $v = n2^z + s$ for some integers r, s, z, m, n such that $t \le s < r < 2^z$.*

A long-standing conjecture of Hartman [16] known as the halving conjecture states that the necessary conditions (6.1) are sufficient for the existence of an LS[2](t, k, v). Formally we have the following.

Halving conjecture Let $0 \leq t \leq k \leq v$. Then there exists an LS[2](t, k, v) if and only if

$$2 \left| \binom{v-i}{k-i} \right., \qquad 0 \leq i \leq t.$$

The first author (GBK) proposed the following analogue for large sets of size 3 [4].

Conjecture Let $0 \leq t \leq k \leq v$. Then there exists an LS[3](t, k, v) if and only if

$$3 \left| \binom{v-i}{k-i} \right., \qquad 0 \leq i \leq t.$$

For some results concerning this conjecture, see [26, 34].

In spite of some results concerning the existence of halving designs, the conjecture is still wide open and seems to be far from being resolved in the near future. The following theorem which follows from the recursive constructions given in Section 5 and an induction argument provides a strategy to tackle the halving conjecture. The large sets needed in the following theorem are called *root cases* [27].

Theorem 6.3 [1] *Let t, k and s be positive integers such that $2^s - 1 \leq t < 2^{s+1} - 1$ and $t < k$. Suppose that for every j and n such that $0 \leq j \leq [t/2]$ and $t + 1 \leq 2^n + j \leq k$, there exists an LS[2]$(t, 2^n + j, 2^{n+1} + t)$. Then for any integer $v > k$ such that the triple (t, k, v) is feasible, there exists an LS[2](t, k, v).*

The most outstanding result on halving designs is due to Ajoodani-Namini which establishes the halving conjecture for $t = 2$ in [1]. By the theorem above in order to prove the conjecture for $t = 2$, it suffices to find two infinite families LS[2]$(2, 2^n + 1, 2^{n+1} + 2)$ and LS[2]$(2, 2^n, 2^{n+1} + 2)$. In fact both families exist. The existence of the first family is a result of two well known theorems. By Baranyai's theorem [7], there exists an LS[2]$(1, 2^n, 2^{n+1} + 1)$ and by Alltop's theorem [6] it can be extended to LS[2]$(2, 2^n + 1, 2^{n+1} + 2)$. Ajoodani-Namini constructed the second family using the approach of products of trades [1]. His construction is rather complicated and leaves little hope for an extension to higher values of t. For $t > 2$, there are some partial results. For $t = 3$, the conjecture has been settled for infinitely many values of k [2, 25]. For some other results on small halving designs, see [30].

Now we present a general view, or a road map, on how to attack the halving conjecture. One way is to tackle the problem for any given t. In this scenario, to settle the conjecture for given t, by Theorem 6.3, one has to construct the root cases LS[2]$(t, 2^n + j, 2^{n+1} + t)$, where $0 \leq j \leq t/2$. The other possible way, for our ambitious champion, is to construct another class of root cases, namely LS[2]$(2^n - 2, 2^n - 1, 2^{n+1} - 2)$, which resolves the halving conjecture for all t [26]. We note that from this class, large sets are known only for $n = 1, 2, 3$ [13, 24, 29, 31].

7 N-Legged trades

In this section we discuss a generalization of trades recently called N-legged trades. We note that (N,t)-partitionable sets which will be presented in the next section are in fact simple N-legged trades and have been utilized in recursive constructions of large sets in the past.

Let v, k and t be integers such that $v \geq k \geq t \geq 0$ and let X be a v-set. A $T[N](t, k, v)$ *trade* (briefly *N-legged trade*), is a pair $T = (X, \{T_1, T_2, \ldots, T_N\})$ such that for $i \neq j$, $(X, \{T_i, T_j\})$ is a $T(t, k, v)$ trade. For the sake of simplicity, we write $T = (X; T_1, T_2, \ldots, T_N)$. Note that any $\mathrm{LS}[N](t, k, v)$ is in fact a $T[N](t, k, v)$ trade; however the converse is not true. We present some examples. The first two ones are taken from [18].

Example 7.1 Let $X = \{1, 2, \ldots, 7\}$. Let

$$T_1 = \{123, 167, 247, 256, 346, 357\},$$
$$T_2 = \{127, 136, 235, 246, 347, 567\},$$
$$T_3 = \{126, 137, 234, 257, 356, 467\}.$$

Then $(X; T_1, T_2, T_3)$ is a $T[3](2, 3, 7)$ trade.

Example 7.2 Let $X = \{1, 2, \ldots, 8\}$. Let

$$T_1 = \{123, 145, 167, 248, 257, 346, 378, 568\},$$
$$T_2 = \{124, 136, 157, 237, 258, 348, 456, 678\},$$
$$T_3 = \{125, 137, 146, 234, 278, 368, 458, 567\},$$
$$T_4 = \{127, 134, 156, 238, 245, 367, 468, 578\}.$$

Then $(X; T_1, T_2, T_3, T_4)$ is a $T[4](2, 3, 8)$ trade.

Example 7.3 Let $X = \{1, 2, \ldots, 9\}$. Let

$$T_1 = \{124, 138, 157, 169, 237, 259, 268, 349, 356, 458, 467, 789\},$$
$$T_2 = \{129, 136, 145, 178, 235, 248, 267, 347, 389, 469, 568, 579\},$$
$$T_3 = \{123, 148, 159, 167, 249, 256, 278, 346, 358, 379, 457, 689\},$$
$$T_4 = \{126, 135, 147, 189, 234, 258, 279, 369, 378, 459, 468, 567\},$$
$$T_5 = \{128, 137, 149, 156, 239, 246, 257, 345, 368, 478, 589, 679\},$$
$$T_6 = \{125, 134, 168, 179, 238, 247, 269, 359, 367, 456, 489, 578\},$$
$$T_7 = \{127, 139, 146, 158, 236, 245, 289, 348, 357, 479, 569, 678\}.$$

Then $(X; T_1, T_2, \ldots, T_7)$ is a $T[7](2, 3, 9)$ trade. Note that (X, T_i) is a 2-(9,3,1) design and hence we also have an $\mathrm{LS}[9](2, 3, 9)$.

As an analogue of Lemma 3.3, we have the following lemma for N-legged trades.

Lemma 7.4 *A $T[N](t, k, v)$ trade T is also a $T[N](t', k, v)$ trade for any $0 \leq t' < t$.*

Let $T = (X; T_1, T_2, \ldots, T_N)$ be a $T[N](t,k,v)$ trade. If we take $t' = 0$ in this lemma, then we find that the number of blocks in T_i for $1 \leq i \leq N$ is fixed and is called the *volume* of T. By taking $t' = 1$, it turns out that the set of points covered by any T_i is the same. This set is called the *foundation* of T. Trades with distinct blocks are said to be *simple*.

Not much is known about N-legged trades. In [14] some computational results for N-legged trades with $t = 2$ and $k = 3$ were obtained. In [12], 3-legged trades have been studied. Here, we briefly review the results of [12]. First no analogue of Theorem 3.4 for N-legged trades is known. It is shown that the minimum foundation size of a simple $T[3](2,3,v)$ trade is 7. In fact there exists exactly one simple $T[3](2,3,7)$ trade which has volume 6 and is given in Example 7.1. There are exactly 7 nonisomorphic simple $T[3](2,3,8)$ trades with foundation size 8. Three of these are of volume 8, one of volume 10 and five of volume 12. Also the maximum possible volume for simple $T[3](2,3,v)$ trades are found for $v \equiv 1, 3, 4 \pmod 6$ and $v \equiv 2 \pmod 9$.

There are many questions concerning N-legged trades. The main question is about the minimum volume and minimum foundation size of a $T[N](t,k,v)$.

It is possible to extend the definition of product of trades to N-legged trades. However, we think that it is more natural to consider the product operation in the context of (N,t)-partitionable sets. We discuss this matter in the next section.

8 (N,t)-Partitionable sets

A powerful approach for the construction of large sets is obtained through the notion of (N,t)-partitionable sets which was first introduced in [5]. The notion of (N,t)-partitionable sets is a generalization the notion of trades and indeed they are equivalent to simple N-legged trades discussed in the previous section.

Let v, k and t be integers such that $v \geq k \geq t \geq 0$ and let X be a v-set. Let $\mathcal{B}_1, \mathcal{B}_2 \subseteq P_k(X)$. We say that $\mathcal{B}_1$ and $\mathcal{B}_2$ are *t-equivalent* if every t-subset of X appears in the same number of blocks of $\mathcal{B}_1$ and $\mathcal{B}_2$. If there exists a partition of $\mathcal{B} \subseteq P_k(X)$ into N mutually t-equivalent subsets, then $\mathcal{B}$ is called an (N,t)-*partitionable set*. It is easily seen that $\mathcal{B}$ can be used to obtain a $T[N](t,k,v)$ trade. The legs of the trade will be the parts of partition of $\mathcal{B}$. Here is an example.

Example 8.1 This example is from [12]. Let

$$\mathcal{B} = \{123, 124, 125, 136, 137, 145, 146, 157, 167, 234, 237, 248,$$
$$257, 258, 278, 346, 348, 368, 378, 456, 458, 567, 568, 678\}.$$

Consider the following partition of $\mathcal{B}$:

$$T_1 = \{123, 145, 167, 248, 257, 346, 378, 568\},$$
$$T_2 = \{124, 136, 157, 237, 258, 348, 456, 678\},$$
$$T_3 = \{125, 137, 146, 234, 278, 368, 458, 568\}.$$

T_1, T_2 and T_3 are mutually 2-equivalent. Therefore, $\mathcal{B}$ is (3,2)-partitionable set. Note that $(\{1,2,\ldots,8\}; T_1, T_2, T_3)$ is a $T[3](2,3,8)$ trade.

Now we present two important lemmas concerning (N,t)-partitionable sets. The first is a trivial one while the other is unexpected. Let X_1 and X_2 be two disjoint sets and let $\mathcal{B}_i \subseteq P_{k_i}(X_i)$ for $i = 1, 2$. Then from Section 4 recall that

$$\mathcal{B}_1 * \mathcal{B}_2 = \{B_1 \cup B_2 |\ B_1 \in \mathcal{B}_1,\ B_2 \in \mathcal{B}_2\}.$$

Lemma 8.2 [5] (i) t-equivalence implies i-equivalence for all $0 \leq i \leq t$.
(ii) The union of disjoint (N,t)-partitionable sets is again an (N,t)-partitionable set.

Lemma 8.3 [5] Let X_1 and X_2 be two disjoint sets and let $\mathcal{B}_i \subseteq P_{k_i}(X_i)$ for $i = 1, 2$. Suppose that $\mathcal{B}_1$ is (N, t_1)-partitionable. Then

(i) $\mathcal{B}_1 * \mathcal{B}_2$ is (N, t_1)-partitionable.
(ii) If $\mathcal{B}_2$ is also (N, t_2)-partitionable, then $\mathcal{B}_1 * \mathcal{B}_2$ is $(N, t_1 + t_2 + 1)$-partitionable.

We now explain the construction used in Lemma 8.3 (ii). Let $T_1, T_2, \ldots, T_N$ be a partition of $\mathcal{B}_1$ into N mutually t_1-equivalent subsets and let $S_1, S_2, \ldots, S_N$ be a partition of $\mathcal{B}_2$ into N mutually t_2-equivalent subsets. We need to find a partition $R_1, R_2, \ldots, R_N$ of $\mathcal{B}_1 * \mathcal{B}_2$ into N mutually $(t_1 + t_2 + 1)$-equivalent subsets. Consider the partition $T_i * S_j$, $1 \leq i, j \leq N$ of $\mathcal{B}_1 * \mathcal{B}_2$. Let L be a Latin square of order n with entries from $\{1, 2, \ldots, N\}$. Define

$$R_f = \bigcup_{L_{ij}=f} T_i * S_j,$$

for $1 \leq f \leq N$. We give an example taken from [18] to clarify the construction.

Example 8.4 Let

$$\mathcal{B}_1 = \{1, 2, 3\},$$
$$\mathcal{B}_2 = \{45, 46, 47, 56, 57, 67\}.$$

$\mathcal{B}_1$ is $(3,0)$-partitionable with the partition $T_1 = \{1\}$, $T_2 = \{2\}$ and $T_3 = \{3\}$. $\mathcal{B}_2$ is $(3,1)$-partitionable with the partition $S_1 = \{45, 67\}$, $S_2 = \{46, 57\}$ and $S_3 = \{47, 56\}$. Consider the following Latin square:

$$\begin{array}{ccc} 1 & 2 & 3 \\ 3 & 1 & 2 \\ 2 & 3 & 1 \end{array}$$

Then we have

$$R_1 = (T_1 * S_1) \cup (T_2 * S_2) \cup (T_3 * S_3) = \{145, 167, 246, 257, 347, 356\},$$
$$R_2 = (T_1 * S_2) \cup (T_2 * S_3) \cup (T_3 * S_1) = \{146, 157, 247, 256, 345, 367\},$$
$$R_3 = (T_1 * S_3) \cup (T_2 * S_1) \cup (T_3 * S_2) = \{147, 156, 245, 267, 346, 357\}.$$

R_1, R_2 and R_3 provide a partition of $\mathcal{B}_1 * \mathcal{B}_2$ into 3 mutually 2-equivalent subsets. Note that $(\{1, 2, \ldots, 7\}; R_1, R_2, R_3)$ is isomorphic to the unique $T[3](2,3,7)$ trade given in Example 7.1.

The approach of (N,t)-partitionable sets for constructing large sets is based on Lemmas 8.2 and 8.3. Suppose that we are looking for an LS[N](t,k,v) on a v-set X. We try to partition $P_k(X)$ in such a way that each part of the partition is an (N,t)-partitionable set. If this done, then by Lemma 8.2, $P_k(X)$ will be an (N,t)-partitionable set which means that we have obtained an LS[N](t,k,v). We have previously used the approach for large sets of size 2 in Section 5. Here, we give an example of large sets of size 3 taken from [26].

Example 8.5 We construct an LS[3](2, 12, 29) from LS[3](2, 3, 11), LS[3](2, 7, 15), LS[3](2, 8, 16) and LS[3](2, 12, 20). Note that there is known no other construction method for LS[3](2, 12, 29). Let $X = \{1, 2, \ldots, 29\}$ and let $u = 20, v = 8$ and $k = 12$ in Lemma 5.6. Then we have

$$P_{12}(X) = \bigcup_{i=0}^{12} \mathcal{B}_i,$$

where

$$\mathcal{B}_i = P_{12-i}(\{1, \ldots, 20-i\}) * P_i(\{22-i, \ldots, 29\}), \quad 0 \leq i \leq 12.$$

$\mathcal{B}_0$ and $\mathcal{B}_{12}$ are (3,2)-partitionable sets since there exists an LS[3](2, 12, 20). By Theorem 2.2, there exist LS[3](1, 2, 10), LS[3](0, 1, 9), LS[3](1, 6, 14), LS[3](0, 10, 18) and LS[3](1, 11, 19). Using Lemma 8.3, it is an easy task to see that all the remaining $\mathcal{B}_i$ are also (3,2)-partitionable sets. For example, consider $\mathcal{B}_2$ and $\mathcal{B}_3$. Here, $P_{10}(\{1, \ldots, 18\})$ is (3,0)-partitionable since there is an LS[3](0, 10, 18), and $P_2(\{20, \ldots, 29\})$ is (3,1)-partitionable since LS[3](1, 2, 10) exists. Now by Lemma 8.3, $\mathcal{B}_2$ is (3,2)-partitionable. By the existence of an LS[3](2, 3, 11), $P_3(\{19, \ldots, 29\})$ is a (3,2)-partitionable set and so is $\mathcal{B}_3$ by Lemma 8.3. The remaining $\mathcal{B}_i$ are dealt with in similar ways. Hence, by Lemma 8.2, $P_{12}(X)$ is (3,2)-partitionable and so an LS[3](2, 12, 29) is constructed.

The recursive constructions for large sets of size 2 given in Section 5 are easily extended to large sets of any size. More constructions can be found in [2, 3, 27, 25, 34]. Here we present two important constructions by Ajoodani-Namini [3] for large sets of prime size p.

Theorem 8.6 [3] *If there exists an LS[p]($t,k,v-1$), then LS[p]($t+1, pk+i, pv+j$) exist for all $0 \leq j < i \leq p-1$.*

Theorem 8.7 [3, 35] *If there exists an LS[p]($t,k,v-1$), then LS[p]($t, pk+i, pv+j$) exist for all $-p \leq j < i \leq p-1$.*

The above theorems could be utilized to produce a large number of infinite families of large sets. As Ajoodani-Namini [2] has noted, these theorems are unique in design theory in the sense that they impose no further conditions on the parameters. By this, we mean that given any large set (whatever the parameters are), using these theorems one can construct infinite families of large sets. The reason for it is that any large set of size N leads to a large set of size p for any prime divisor p of N. Theorem 8.6 is specially interesting since it proves Teirlinck's theorem [36] on the existence of simple t-designs for all t. It also has the extra merit that one can produce t-designs on point sets which are very small compared to those of Teirlinck's.

9 A linear algebraic approach to trades and designs

In this section we present a linear algebraic approach to the study of trades and designs. By giving two applications we show the usefulness of this representation of trades. Let $0 \leq t \leq k \leq v - t$ and and let X be a v-set. Order $P_t(X)$ and $P_k(X)$ lexicographically (or with any other ordering). Let $W_{tk}(v)$ be a $\binom{v}{t} \times \binom{v}{k}$ (0,1)-matrix whose rows and columns are indexed by the elements of $P_t(X)$ and $P_k(X)$, respectively, and for a t-subset T and a k-subset K, $W_{tk}(v)(T,K) = 1$ if and only if $T \subseteq K$. The matrix $W_{tk}(v)$ is an *inclusion matrix* which is often called the *Wilson matrix* since it was first introduced and used by Wilson [37]. We simply write W_{tk} instead of $W_{tk}(v)$ if there is no risk of confusion.

Given a collection of elements of $P_k(X)$, a $\binom{v}{k}$ column vector F can be associated with it: $F = (f_1, f_2, \ldots, f_{\binom{v}{k}})$ where f_i is the frequency of the ith element of $P_k(X)$ in this collection. Conversely, for a given column vector F of size $\binom{v}{k}$ with nonnegative integers, a collection of the elements of $P_k(X)$ can be associated with it by taking f_i copies of the ith element of $P_k(X)$. Also for a given $S = \{S_1, S_2\}$, where S_1 and S_2 are two disjoint collections of elements each from $P_k(X)$, an integral $\binom{v}{k}$ column vector F can be associated with it whose ith entry is f_i if the ith block of $P_k(X)$ appears f_i times in S_1 or $-f_i$ if the ith block of $P_k(X)$ appears f_i times in S_2.

A $\binom{v}{k}$ column vector F with nonnegative integers represents a t-design if $W_{tk}F = \lambda J$, where λ is a positive integer and J is the all-one vector. If we let negative entries in F, then we have a *signed t-design*.

A $\binom{v}{k}$ integral column vector F is a T(t,k,v) trade if and only if $W_{tk}F = 0$. The positive components of F identify the frequencies of the blocks in T^+ (T^-) and the negative components (sign ignored) identify the frequencies of the blocks in T^- (T^+). If F_1 and F_2 represent two T(t,k,v) trades based on X, then $F_1 + F_2$ as well as nF_1 ($n \in \mathbb{Z}$) are also trades. Therefore, the set of all T(t,k,v) trades forms a free $\mathbb{Z}$-module. It is well known that the rank of W_{tk} is $\binom{v}{t}$ and the null space of W_{tk} that generates all T(t,k,v) trades has dimension $\binom{v}{k} - \binom{v}{t}$ [15, 37].

If F_1 and F_2 represent two t-(v,k,λ) designs based on X, then $F_1 - F_2$ is a trade. Therefore, in principle, all t-(v,k,λ) designs can be generated by using trades and any given signed t-(v,k,λ) design. To do this, we first need to find a signed t-(v,k,λ) design F (which is easy to find) and then combine it with all trades G such that $F + G$ is a nonnegative vector. This shows the importance of trades in the study of t-designs and suggests that, to study t-designs, one may investigate trades.

Different bases have been presented in the literature for the $\mathbb{Z}$-module of trades. A survey is given in [23]. In [20] a triangular basis for trades is given. All trades in this basis are minimal. This paper also gives an algorithm based on this basis to find halving designs in triple systems.

An interesting basis for trades is the so called *standard basis* given in [23] which follows from the basis given in [20]. The $\binom{v}{k} - \binom{v}{t}$ trades of the standard basis constitute the columns of a matrix $M^v_{t,k}$ which has the following block structure:

$$M^v_{t,k} = \begin{bmatrix} I \\ \overline{M^v_{t,k}} \end{bmatrix}.$$

The rows corresponding to I are indexed by the so-called *starting blocks* and the remaining rows by the *non-starting blocks* [20]. This basis has many interesting properties yet to be explored. The standard basis for T(2,3,6) trades is given in Table 1. In this table, the first column shows the starting and non starting blocks, the next five columns show the five trades in the basis and the last column is a signed 2-(6,3,2) design (obtained as described below).

Table 1: The standard basis for T(2,3,6) trades

123	1					0
124		1				0
125			1			0
134				1		0
135					1	0
126	-1	-1	-1			2
136	-1			-1	-1	2
145	-1	-1	-1	-1	-1	3
146	1		1		1	-1
156	1	1		1		-1
234	-1	-1		-1		2
235	-1		-1		-1	2
236	1	1	1	1	1	-2
245	1			1	1	-1
345	1	1	1			-1
246					-1	1
256				-1		1
346			-1			1
356		-1				1
456	-1					1

We present two applications of the standard basis. Note that the feasibility conditions for the existence of a t-(v,k,λ) design given in Section 2 are sufficient for the existence of a signed t-(v,k,λ) design [15, 28, 37]. Signed designs are useful in the study of t-designs. To construct designs using the so-called trade-off method, one starts with a signed design and then tries to eliminate negative entries by adding suitable trades. We show how the standard basis is used to produce a signed design. To find a signed t-(v,k,λ) design, it is enough to sum up all the columns of $M_{t,k}^v$, then subtract it from the vector J and finally divide the resulting vector by a suitable coefficient [28]. Note that all entries in this signed design corresponding to starting blocks are zero. This signed design can sometimes be converted to a t-design by adding a suitable trade. As an example a signed 2-(6,3,2) design obtained by this method is shown in the last column of Table 1.

A halving design is equivalent to a trade F whose entries are ± 1. Therefore, to find a halving design one can use the standard basis and take a combination of columns with coefficients 1 or -1 and then check whether the resulting trade is simple and has no zero entry. This approach is effective since the standard basis has the recursive structure

$$M_{t,k}^v = \begin{bmatrix} I & 0 \\ 0 & I \\ M_{t-1,k-1}^{v-1} & 0 \\ N & M_{t,k}^{v-1} \end{bmatrix},$$

which suggests obtaining a halving t-(v,k,λ) design through extending halving $(t-1)$-$(v-1,k-1,\lambda)$ designs. A detailed description of this method can be found in [13] where it has been successfully used to obtain new 6-(14,7,4) designs.

10 Concluding remarks

In this paper we have presented a rather pedagogical review of the application of trades in constructing halving t-designs. We have also considered the notion of (N,t)-partitionable sets as a generalization of trades and have shown how some powerful recursive constructions can be obtained for large sets of t-designs. We hope that we have been able to draw the attention of the reader to the power of the approach of (N,t)-partitionable sets. There are some open problems which suggest further research in the future. The first problem is to find other binomial identities besides the ones given in Section 5 which correspond to recursive constructions for large sets. The halving conjecture for $t=3$ is another problem which seems to be hard. One may think that the similar problem for large sets of 2-designs of size 3 is more accessible. In order to resolve this problem, one should establish the existence of large sets LS[3]$(2, 3^n+j, 3^{n+1}+2)$ for $j=0,1,2$ and for any $n>3$. Concerning N-legged trades, the main question is to determine the minimum volume and minimum foundation size of a $T[N](t,k,v)$.

References

[1] S. Ajoodani-Namini, All block designs with $b = \binom{v}{k}/2$ exist, *Discrete Math.* **179** (1998), 27–35.

[2] S. Ajoodani-Namini, *Large sets of t-designs*, Ph.D. Thesis, California Institute of Technology, Pasadena (1997).

[3] S. Ajoodani-Namini, Extending large sets of t-designs, *J. Combin. Theory Ser. A* **76** (1996), 139–144.

[4] S. Ajoodani-Namini & G. B. Khosrovshahi, On a conjecture of A. Hartman, in *Combinatorics Advances* (eds. C. J. Colbourn & E. S. Mahmoodian), *Math. Appl.*, 329, Kluwer Acad. Publ., Dordrecht (1995), pp. 1–12.

[5] S. Ajoodani-Namini & G. B. Khosrovshahi, More on halving the complete designs, *Discrete Math.* **135** (1994), 29–37.

[6] W. O. Alltop, Extending t-designs, *J. Combin. Theory Ser. A* **18** (1975), 177–186.

[7] Zs. Baranyai, On the factorizations of the complete uniform hypergraph, in *Infinite and finite sets* (eds. A. Hajnal, R. Rado & V. T. Sós), *Colloq. Math. Soc. Janos Bolyai*, 10, North-Holland, Amsterdam (1975), pp. 91–108.

[8] M. Behbahani, G. B. Khosrovshahi & B. Tayfeh-Rezaie, On the spectrum of simple $T(2,3,v)$ trades, *J. Statist. Plann. Inference* **138** (2008), 2236–2242.

[9] E. J. Billington, Combinatorial trades: A survey of recent results, in *Designs, 2002, Further computational and constructive design theory* (ed. W. D. Wallis), *Math. Appl.*, 563, Kluwer Acad. Publ., Boston (2003), pp. 47–67.

[10] P. J. Cameron, *Parallelisms of Complete Designs*, Lond. Math. Soc. Lecture Note Ser., 23, Cambridge University Press, Cambridge (1976).

[11] D. Donovan, E. S. Mahmoodian, C. Ramsay & A. P. Street, Defining sets in combinatorics: a survey, in *Surveys in combinatorics 2003* (ed. C. D. Wensley), *Lond. Math. Soc. Lecture Note Ser.*, 307, Cambridge University Press, Cambridge (2003), pp. 115–174.

[12] B. Esfahbod, *Studies on 3-legged trades* (in Farsi), M.S. Thesis, University of Tehran, Tehran (2007).

[13] Z. Eslami & G. B. Khosrovshahi, Some new 6-$(14,7,4)$ designs, *J. Combin. Theory Ser. A* **93** (2001), 141–152.

[14] A. D. Forbes, M. J. Grannell & T. S. Griggs, Configurations and trades in Steiner triple systems, *Australas. J. Combin.* **29** (2004), 75–84.

[15] J. E. Graver & W. B. Jurkat, The module structure of integral designs, *J. Combin. Theory Ser. A* **15** (1973), 75–90.

[16] A. Hartman, Halving the complete design, *Ann. Discrete Math.* **34** (1987), 207–224.

[17] A. S. Hedayat, The theory of trade-off for t-designs, in *Coding Theory and Design Theory, Part II, Design theory* (ed. D. Ray-Chaudhuri), *IMA Vol. Math. Appl.*, 21, Springer-Verlag, New York (1990), pp. 101–126.

[18] A. S. Hedayat & G. B. Khosrovshahi, Trades, in *Handbook of Combinatorial Designs* (eds. C. J. Colbourn & J. H. Dinitz), Second edition, Chapman & Hall/CRC Press, Boca Raton (2007), pp. 644–648.

[19] H. L. Hwang, On the structure of (v,k,t) trades, *J. Statist. Plann. Inference* **13** (1986), 179–191.

[20] G. B. Khosrovshahi & S. Ajoodani-Namini, A new basis for trades, *SIAM J. Discrete Math.* **3** (1990), 364–372.

[21] G. B. Khosrovshahi & R. Laue, t-Designs with $t \geq 3$, in *Handbook of Combinatorial Designs* (eds. C. J. Colbourn & J. H. Dinitz), Second edition, Chapman & Hall/CRC Press, Boca Raton (2007), pp. 79–101.

[22] G. B. Khosrovshahi, H. R. Maimani & R. Torabi, On trades: an update, *Discrete Appl. Math.* **95** (1999), 361–376.

[23] G. B. Khosrovshahi & Ch. Maysoori, On the bases for trades, *Linear Alg. Appl.* **226–228** (1995), 731–748.

[24] G. B. Khosrovshahi, M. Mohammad-Noori & B. Tayfeh-Rezaie, Classification of 6-(14,7,4) designs with nontrivial automorphism groups, *J. Combin. Des.* **10** (2002), 180–194.

[25] G. B. Khosrovshahi & B. Tayfeh-Rezaie, Lecture notes on large sets of t-designs, http://math.ipm.ac.ir/tayfeh-r/research.htm.

[26] G. B. Khosrovshahi & B. Tayfeh-Rezaie, Some results on the existence of large sets of t-designs, *J. Combin. Des.* **11** (2003), 144–151.

[27] G. B. Khosrovshahi & B. Tayfeh-Rezaie, Root cases of large sets of t-designs, *Discrete Math.* **263** (2003), 143–155.

[28] G. B. Khosrovshahi & B. Tayfeh-Rezaie, A new proof of a classical theorem in design theory, *J. Combin. Theory Ser. A* **93** (2001), 391–396.

[29] D. L. Kreher & S. P. Radziszowski, The existence of simple 6-(14, 7, 4) designs, *J. Combin. Theory Ser. A* **43** (1986), 237–243.

[30] R. Laue, Halvings on small point sets, *J. Combin. Des.* **7** (1999), 233–241.

[31] R. Laue, A. Betten & E. Haberberger, A new smallest simple 6-design with automorphism group A_4, *Congr. Numer.* **150** (2001), 145–153.

[32] S. S. Magliveras & D. W. Leavitt, Simple 6-(33, 8, 36) designs from $P\Gamma L_2(32)$, in *Computational group theory* (ed. M. D. Atkinson), Academic Press, London (1984), pp. 337–352.

[33] R. Mathon & A. Rosa, On 2-(v, k, λ) designs of small order, in *Handbook of Combinatorial Designs* (eds. C. J. Colbourn & J. H. Dinitz), Second edition, Chapman & Hall/CRC Press, Boca Raton (2007), pp. 25–58.

[34] B. Tayfeh-Rezaie, On the existence of large sets of t-designs of prime sizes, *Des. Codes Cryptogr.* **37** (2005), 143–149.

[35] B. Tayfeh-Rezaie, Some infinite families of large sets of t-designs, *J. Combin. Theory Ser. A* **87** (1999), 239–245.

[36] L. Teirlinck, Nontrivial t-designs without repeated blocks exist for all t, *Discrete Math.* **65** (1987), 301–311.

[37] R. M. Wilson, The necessary conditions for t-designs are sufficient for something, *Util. Math.* **4** (1973), 207–217.

G. B. Khosrovshahi

School of Mathematics
Institute for Research in Fundamental Sciences (IPM)
P.O. Box 19395-5746, Tehran, Iran
and
School of Mathematics, Statistics and Computer Science
University of Tehran, Tehran, Iran
rezagbk@ipm.ir

B. Tayfeh-Rezaie

School of Mathematics
Institute for Research in Fundamental Sciences (IPM)
P.O. Box 19395-5746, Tehran, Iran
tayfeh-r@ipm.ir

Extremal graph packing problems: Ore-type versus Dirac-type

H. A. Kierstead, A. V. Kostochka, and Gexin Yu

Abstract

We discuss recent progress and unsolved problems concerning extremal graph packing, emphasizing connections between Dirac-type and Ore-type problems. Extra attention is paid to coloring, and especially equitable coloring, of graphs.

1 Introduction

An important instance of combinatorial packing problems is that of *graph packing*. We say that n-vertex graphs $G_1, G_2, \ldots, G_k$ *pack*, if there exists an edge-disjoint placement of all these graphs onto the same set of n vertices. By definition, two graphs G_1 and G_2 pack, if G_1 is a subgraph of the complement $\overline{G}_2$ of G_2, or, equivalently, G_2 is a subgraph of the complement $\overline{G}_1$ of G_1. Many basic graph theory problems and concepts can be expressed in a unified (and sometimes more natural) form using the language of graph packing. Here are some relevant examples.

Example 1.1 The problem of existence of a spanning (hamiltonian) cycle in an n-vertex graph G (which is a close relative of the famous Travelling Salesman Problem) is equivalent to the question whether the n-cycle C_n packs with the complement $\overline{G}$ of G.

Example 1.2 The independence number $\alpha(G)$ of an n-vertex graph G is at least k if and only if G packs with the graph $K_k + \overline{K}_{n-k}$ consisting of the k-clique and $n - k$ isolated vertices.

Example 1.3 An n-vertex graph G is k-colorable if and only if G packs with an n-vertex graph that is the union of k cliques.

A proper vertex coloring of a graph is *equitable* if the sizes of its color classes differ by at most one.

Example 1.4 An n-vertex graph G is equitably k-colorable if and only if G packs with the complement $H(n, k)$ of the Turán Graph $T(n, k)$, i.e. with the n-vertex graph whose every component is a complete graph with either $\lfloor n/k \rfloor$ or $\lceil n/k \rceil$ vertices.

Turán-type and Ramsey-type problems also can be naturally stated in the language of graph packing.

Since finding optimal solutions of many graph packing problems is NP-hard, corresponding extremal problems giving sufficient conditions for packing graphs are of great interest. Some well-known theorems of this type can be translated into the language of packing as follows.

Theorem 1.5 (Dirac [24]) *If G is an n-vertex graph and $\Delta(G) \leq n/2 - 1$, then G packs with the cycle C_n of length n.*

Theorem 1.6 (Ore [63]) *If G is an n-vertex graph and $d(x) + d(y) \leq n - 2$ for every edge xy in G, then G packs with the cycle C_n of length n.*

Theorem 1.7 (Hajnal and Szemerédi [31]) *Let G be an n-vertex graph with $\Delta(G) \leq r$. Then G packs with the graph $H(n, r+1)$, whose components are complete graphs with either $\lfloor n/(r+1) \rfloor$ or $\lceil n/(r+1) \rceil$ vertices.*

Ore's theorem motivates considering the notion of *Ore-degree*, $\theta(xy)$, of an edge xy in a graph G as the sum, $d(x)+d(y)$, of the degrees of its ends in G. By definition, the Ore-degree of an edge xy is two greater than the degree of the vertex xy in the line graph of G, and coincides with the degree of xy in the total graph of G. We let the *Ore-degree of a graph* G be $\theta(G) = \max_{xy \in E(G)} \theta(xy)$. Thus, Ore's theorem says that every n-vertex graph G with $n \geq 3$ and $\theta(G) \leq n - 2$ packs with the cycle C_n. Observe that $\theta(G)$ is closely related to the maximum degree of the line graph $L(G)$:

$$\theta(G) = \Delta(L(G)) + 2,$$

and that any bound on $\theta(G)$ is in fact a bound on $\Delta(L(G))$. Observe also that for every graph G,

$$\Delta(G) + \delta^*(G) \leq \theta(G) \leq 2\Delta(G), \tag{1.1}$$

where $\delta^*(G)$ is the minimum *positive* degree of a vertex in G. The left inequality in (1.1) is obtained by considering an edge $xy \in E(G)$ incident with a vertex of the maximum degree.

In view of Dirac's and Ore's theorems, we call upper bounds in terms of maximum degree giving sufficient conditions for packing graphs *Dirac-type bounds* and those in terms of the Ore-degree *Ore-type bounds*.

In this survey, we compare recent progress in Dirac-type and Ore-type bounds for graph packing problems. We discuss the general problem in the next section, packing a graph with a power of a cycle or path in Section 3, graph coloring in Section 4, equitable coloring in Section 5, and equitable list coloring in Section 6.

We use the standard notation. In particular, for a graph G, $|G|$ denotes the order and $\|G\|$ denotes the size of G.

2 General packing results

Some milestone results on extremal graph packing problems were obtained in the seventies. In the same issue of the Journal of Combinatorial Theory (B), fundamental papers by Bollobás and Eldridge [10] and Sauer and Spencer [68] appeared. The papers gave sufficient conditions for packing of two graphs with given average or maximum degree. Some of these results were also obtained by Catlin in his Ph.D. Thesis [16] and in [15]. In particular, Sauer and Spencer [68] proved the following Dirac-type bound.

Theorem 2.1 (Sauer and Spencer [68]) *If G_1 and G_2 are n-vertex graphs and $2\Delta(G_1)\Delta(G_2) < n$, then G_1 and G_2 pack.*

The proof is simple (we will show it for a half-Ore version), but the bound is sharp for even n.

Graph packing problems

Example 2.2 Let G_1 be a perfect matching on n vertices, and G_2 contain $K_{1+n/2}$ and have maximum degree $n/2$. Then $2\Delta(G_1)\Delta(G_2) = 2 \cdot 1 \cdot n/2$ but G_1 and G_2 do not pack, since in any packing of G_1 and G_2 only one end of each edge of G_1 can be placed onto a vertex in the copy of $K_{1+n/2}$ in G_2.

Example 2.3 Let $n \equiv 2 \pmod 4$. Let G_1 be a perfect matching on n vertices, and $G_2 = K_{n/2,n/2}$. Again $2\Delta(G_1)\Delta(G_2) = 2 \cdot 1 \cdot n/2$. Again, G_1 and G_2 do not pack, since in any packing of G_1 and G_2, the ends of each edge of G_1 should be placed into the same partite set of $K_{n/2,n/2}$, but $n/2$ is odd.

Kaul and Kostochka [35] proved that Examples 2.2 and 2.3 are the only examples where the bound of Sauer and Spencer is attained.

The half-Ore version of Theorem 2.1 mentioned above is as follows.

Theorem 2.4 *If G_1 and G_2 are n-vertex graphs and*

$$\theta(G_1)\Delta(G_2) < n, \qquad (2.1)$$

then G_1 and G_2 pack.

Proof. For $i \in [2]$, set $G_i = (V_i, E_i)$, $d_i(v) = d_{G_i}(v)$, and $\Delta_i = \Delta(G_i)$. A packing of G_2 with G_1 will be viewed as a mapping f of V_1 onto V_2 such that if $uv \in E_1$ then $f(u)f(v) \notin E_2$. We argue by induction on $\|G_1\|$. The base step $\|G_1\| = 0$ is trivial, so consider the induction step. Let x be a vertex with the minimum positive degree δ^* in G_1 and xy be any edge incident with x. By the induction hypothesis there exists a packing f of $G_1 - xy$ with G_2. If $f(x)f(y) \notin E_2$, then we are done. Otherwise, we will show that there is some *good* $z \in V - y$ such that the mapping g_z obtained from f by switching the images of x and z is a packing of G_1 with G_2. Suppose $z \in V - y$ is *bad*, i.e., not good. Then there exists an edge $uv \in E_1$ with $g_z(u)g_z(v) \in E_2$. Since f packs $G_1 - xy$ with G_2, at least one, say v, of u,v is in $\{x,z\}$. Then either (1) $xu \in E_1$ and $g_z(x)g_z(u) = f(z)f(u) \in E_2$ or (2) $zu \in E_1$ and $g_z(z)g_z(u) = f(x)f(u) \in E_2$. The number of z for which (1) holds is at most $d_1(x)\Delta_2$: There are at most $d_1(x)$ choices for u and each choice witnesses at most Δ_2 different z. Similarly, the number of z for which (2) holds is at most $d_2(f(x))\Delta_1$. Noting that the choice $z = x, u = y$ has been counted twice, and using both (1.1) and (2.1), the total number of such bad z is at most

$$d_1(x)\Delta_2 + d_2(f(x))\Delta_1 - 1 \leq (\delta^* + \Delta_1)\Delta_2 - 1 \leq \theta(G_1)\Delta_2 - 1 < n - 1 = |V_1 - y|.$$

Thus there exists a good $z \in V_1 - y$, and so g_z is a packing of G_1 with G_2. □

A sharpening of this result was proved in [55]:

Theorem 2.5 *If two n-vertex graphs G_1 and G_2 satisfy $\theta(G_1)\Delta(G_2) \leq n$, then G_1 and G_2 pack, with the following exceptions:*

1. *G_1 is a perfect matching and G_2 either is $K_{n/2,n/2}$ with $n/2$ odd or contains $K_{n/2+1}$;*

2. *G_2 is a perfect matching, and G_1 either is $K_{r,n-r}$ with r odd or contains $K_{n/2+1}$.*

Theorem 2.5 shows that Theorem 2.4 has extremal graphs other than Examples 2.2 and 2.3, but not many. Observe also that the conditions in Theorems 2.5 and 2.4 involve $\Delta(G_2)$. The following Ore-type analogue of Theorem 2.1 was conjectured in [56].

Conjecture 2.6 *If G_1 and G_2 are n-vertex graphs and $\theta(G_1)\theta(G_2) < 2n$, then G_1 and G_2 pack.*

The conjecture looks natural and maybe has a simple proof, but so far we have failed to find any proof.

The main conjecture in the area (strengthening Theorem 2.1) is the following BEC-conjecture:

Conjecture 2.7 *If G_1 and G_2 are n-vertex graphs with maximum degrees Δ_1 and Δ_2, respectively, and $(\Delta_1 + 1)(\Delta_2 + 1) \leq n + 1$, then G_1 and G_2 pack.*

This conjecture was posed by Bollobás and Eldridge [9] (see also [8, 10]) and independently by Catlin [16].

The following examples show that the conjecture is sharp for all values of $\Delta(G_1)$ and $\Delta(G_2)$, if true.

Example 2.8 Let positive integers Δ_1 and Δ_2 be fixed. Let G_1 be the disjoint union of Δ_2 copies of K_{Δ_1+1} and one copy of K_{Δ_1-1}. Let G_2 be the disjoint union of Δ_1 copies of K_{Δ_2+1} and one copy of K_{Δ_2-1}. Then $|V(G_1)| = |V(G_2)| = (\Delta_1 + 1)(\Delta_2 + 1) - 2$. Suppose that G_1 and G_2 pack. Then in this packing, each of the Δ_1 copies of K_{Δ_2+1} in G_2 should intersect all the $\Delta_2 + 1$ components of G_1. But the K_{Δ_1-1}-component of G_1 cannot meet all these Δ_1 copies of K_{Δ_2+1}.

Example 2.9 Let a positive integer Δ_1 and a positive odd integer Δ_2 be fixed. Let G_1 be the disjoint union of Δ_2 copies of K_{Δ_1+1} and one copy of K_{Δ_1-1}. Let G_2 be the disjoint union of $\Delta_1 - 1$ copies of K_{Δ_2+1} and one copy of K_{Δ_2,Δ_2}. Again $|V(G_1)| = |V(G_2)| = (\Delta_1 + 1)(\Delta_2 + 1) - 2$. Suppose that G_1 and G_2 pack. In this packing, each of the $\Delta_1 - 1$ copies of K_{Δ_2+1} in G_2 should intersect all the $\Delta_2 + 1$ components of G_1. This leaves two vertices in each clique of size $\Delta_1 + 1$ in G_1 and K_{Δ_2,Δ_2} in G_2. So we come to Example 2.3 of graphs that do not pack.

Only very special cases of the BEC-conjecture have been proved. In particular, the Hajnal-Szemerédi Theorem (Theorem 1.7) on equitable colorings verifies the conjecture in the case when G_2 is the disjoint union of cliques of the same size. Aigner and Brandt [1] and independently (for huge n) Alon and Fisher [2] settled the conjecture in the case $\Delta_1 \leq 2$ (this particular case was conjectured by Sauer and Spencer [68]). Csaba, Shokoufandeh, and Szemerédi [21] proved the BEC-conjecture for $\Delta_1 = 3$ and huge n. Bollobás, Kostochka and Nakprasit [12] showed that although the BEC-conjecture is sharp, if one of the two graphs is sparse, to be precise, d-degenerate for a small d, then much weaker conditions on Δ_1 and Δ_2 imply the existence of a packing. Recall that a graph G is d-*degenerate* if every subgraph G' of G has a vertex of degree (in G') at most d. In this case, the vertices of G can be ordered so that each vertex has fewer than $\text{col}(G) := d + 1$ neighbors that precede it.

Theorem 2.10 (Bollobás, Kostochka and Nakprasit [12]) *Let $d \geq 2$. Let G_1 be a d-degenerate graph of order n and maximum degree Δ_1 and G_2 be a graph of order n and maximum degree at most Δ_2. If*

$$40\Delta_1 \ln \Delta_2 < n \quad \text{and} \quad (2.2)$$

$$40d\Delta_2 < n, \quad (2.3)$$

then there is a packing of G_1 and G_2.

If $\Delta_2 \geq 215$, then $\Delta_2 / \ln \Delta_2 \geq 40$. Therefore, Theorem 2.10 yields that the BEC-conjecture holds if $\Delta_2 \geq 215$ and $\Delta_1 \geq 40d$. Adapting the proof of Theorem 2.10 to control the maximum degree of the union of the two packed graphs, implies the following result on simultaneous packing of many graphs.

Theorem 2.11 (Bollobás, Kostochka and Nakprasit [12]) *Let n, d, Δ and q be positive integers such that $d \geq 2$, $q \leq \frac{n}{1500d^2}$, and $1000d\Delta < \frac{n}{\ln n}$. Let $F_1, \ldots, F_q$ be d-degenerate graphs of order n and maximum degree at most Δ. Then $F_1, \ldots, F_q$ pack.*

For a fixed d, Theorem 2.11 allows packing linearly many (in n) d-degenerate n-vertex graphs of moderate maximum degree. The phenomenon here is that it is easier to pack graphs if the number of vertices is significantly greater than the maximum degrees of the graphs to be packed.

Clearly, restriction (2.3) cannot be weakened by more than 40 times. The following result in [11] shows that (2.2) is also weakest up to a constant factor.

Theorem 2.12 (Bollobás, Kostochka and Nakprasit [11]) *Let k be a positive integer and q be a prime power. Then for every $n \geq q\frac{q^{k+1}-1}{q-1}$, there are graphs $G_1(n,k)$ and $G_2(n,q,k)$ of order n that do not pack and have the following properties.*

(a) $G_1(n,k)$ is a forest with $n - k$ edges and maximum degree at most n/k;

(b) $G_2(n,q,k)$ is a $\frac{q^k-1}{q-1}$-degenerate graph with maximum degree at most $2n/q$.

So, for $l \geq 5$, $q = 3^l$, $k \geq l$, and $n = \frac{3}{2}(3^{k+1} - 1)$, consider the graphs $G_1 = G_1(n,k)$ and $G_2 = G_2(n, 3^l, k)$ of Theorem 2.12. Then (2.3) holds for $l > 4$ (we need $d \geq 2$ in the theorem), so the bounding inequality is (2.2). Furthermore, $\Delta(G_1) \ln \Delta(G_2) \leq \frac{n}{k} \ln n \leq \frac{n}{k}(1 + (k+1) \ln 3) < 2n$, so restriction (2.2) can be possibly weakened only by a constant factor.

Motivated by Theorem 2.10, Theorem 2.1 has very recently been strengthened [44] in terms of *game coloring number*. This parameter is defined by means of the *marking game*, which is played by two players Alice and Bob on a graph G. Bob decides who plays first, and then the players take turns choosing unchosen vertices until all vertices have been chosen. Let L be the order in which the vertices are chosen. The *score* of the game is the least integer s such that every vertex has fewer than s neighbors that precede it in L. The *game coloring number* $\text{gcol}(G)$ is the least integer k such that Alice has a strategy for obtaining a score of at most k regardless of how Bob plays. The new result says:

Theorem 2.13 ([44]) *If two graphs G_1 and G_2 satisfy*

$$(\text{gcol}(G_1) - 1)\Delta(G_2) + (\text{gcol}(G_2) - 1)\Delta(G_1) < n,$$

then they pack.

Theorem 2.13 strengthens Theorem 2.1, since for every graph G,

$$\text{col}(G) \leq \text{gcol}(G) \leq \Delta + 1.$$

Moreover, for some important classes of graphs the upper bound on $\text{gcol}(G)$ can be greatly improved. For instance, Zhu [81] showed that every planar graph G satisfies $\text{gcol}(G) \leq 17$.

In [36], the BEC-conjecture was attacked from a different direction: instead of proving the conjecture for another class of graphs, the following weaker bound for all graphs with high maximum degrees was proved.

Theorem 2.14 ([36]) *Let G_1 and G_2 be n-vertex graphs with maximum degrees Δ_1 and Δ_2, respectively. If $\Delta_1, \Delta_2 \geq 300$ and $(\Delta_1 + 1)(\Delta_2 + 1) \leq 0.6n + 1$, then G_1 and G_2 pack.*

This improves the bound of the Sauer–Spencer Theorem for large Δ_1 and Δ_2 and thus partially answers Problem 4.4 in [32]. Helpful concepts used in the proof are those of critical pairs and of cyclic switchings. Cyclic switchings generalize the ordinary switchings used in the proofs of Theorem 2.4 and the original Sauer–Spencer Theorem.

In all extremal examples (G_1, G_2) for the BEC-conjecture that we know, G_1 contains $\Delta(G_1)$-regular and G_2 contains $\Delta(G_2)$-regular components. This fact and Theorem 2.10 suggest that maybe the following Ore-type analogue of the BEC-conjecture holds.

Conjecture 2.15 ([56]) *If G_1 and G_2 are n-vertex graphs and*

$$(0.5\theta(G_1) + 1)(\Delta(G_2) + 1) \leq n + 1,$$

then G_1 and G_2 pack.

The case of Conjecture 2.15 when G_2 is the union of vertex disjoint triangles (which is the analogue of Corrádi–Hajnal Theorem [18]) was proved independently by Enomoto [26] and Wang [75]. The following extension of the results of Enomoto and Wang is a small step towards Conjecture 2.15.

Theorem 2.16 ([57]) *Each n-vertex graph G with $\theta(G) \leq \frac{4n-8}{3}$ packs with any n-vertex graph H such that each component of H is either C_3, or K_2, or C_5, or $K_4 - e$, or a vertex.*

Yet another way to attack the BEC-conjecture gives the concept of near packing introduced by Eaton [25]. This is an extension of the notion of defective coloring. In a *near packing of degree d*, the copies of the two graphs may overlap so that the maximum degree of the subgraph spanned by the edges common to both copies is at most d. Thus a near packing of degree 0 is an ordinary packing. The following result was proved in [25]:

Theorem 2.17 (Eaton [25]) *Let $0 < a \leq 1$. If G_1 and G_2 are n-vertex graphs, $\Delta(G_1) + \Delta(G_2) \leq n + a - 1$, and $\Delta(G_1) \cdot \Delta(G_2) < an$, then there exists a near packing of G_1 and G_2 of degree d for some integer $d < 2a$. Furthermore, if $(\Delta(G_1) + 1)(\Delta(G_2) + 1) \leq n + 1$, then there exists a near packing of G_1 and G_2 of degree 1.*

Eaton also posed the conjecture below and showed that it is sharp if true:

Conjecture 2.18 (Eaton [25]) *Let G_1 and G_2 be n-vertex graphs and p be a positive real number. If*

$$n \geq \lfloor \frac{\Delta(G_1)}{p} \rfloor (\Delta(G_2) + 1) + \lfloor \frac{\Delta(G_2)}{p} \rfloor (\Delta(G_1) + 1 - p\lfloor \frac{\Delta(G_1)}{p} \rfloor),$$

then there exists a near packing of G_1 and G_2 of degree less than p.

It would be interesting to prove the Ore-version or half-Ore version of Theorem 2.17. For example, is it true that *if $(0.5\theta(G_1) + 1)(\Delta(G_2) + 1) \leq n + 1$, then there exists a near packing of G_1 and G_2 of degree 1?*

Other interesting results on graph packing were obtained by Brandt [14], Csaba [19, 20], Fan and Kierstead [28, 29], Komlós [47], Komlós, Sárközy and Szemerédi [49, 50, 51], Sauer and Wang [69], Wozniak [37, 67], Yap [72, 73, 78] and others. One can look into the 75-page survey [77] of the topic by Wozniak. Some of these results are discussed in the next section. After this paper was accepted, we learned about the following very interesting result by Kühn, Osthus and Treglown [59]: for every fixed graph H and large n, they found asymptotically exact Ore-type conditions for an n-vertex graph G of order divisible by $|H|$ to contain a perfect H-packing of G, i.e., to contain vertex-disjoint copies of H that cover $V(G)$.

3 Packing a graph with a power of a cycle

Let $H = v_1 v_2 \cdots v_n$ be a path or a cycle. An *r-chord* is an edge of the form $v_i v_{i+r}$, where addition is modulo n in the case that H is a cycle. The *r-th power* of H is the graph obtained by adding all i-chords with $i \leq r$. The second power of H is called the *square* of H. Pósa (see [27]) conjectured that every graph on n vertices with minimum degree at least $\frac{2}{3}n$ contains the square of C_n. Seymour [70] strengthened Pósa's Conjecture as follows.

Conjecture 3.1 (Seymour [70]) *Every graph G on n vertices with $\delta(G) \geq \frac{r}{r+1}n$ contains the r-th power of C_n.*

The case $r = 1$ is Dirac's Theorem and the case $r = 2$ is Pósa's Conjecture. Rephrased in terms of packing, Seymour's Conjecture looks like this.

Conjecture 3.2 *The r-th power of C_n packs with every graph G on n vertices with $\Delta(G) \leq \frac{n}{r+1} - 1$.*

Seymour's conjecture implies the packing version of the Hajnal-Szemerédi Theorem, since any $r+1$ consecutive vertices of the r-th power of a cycle induce K_{r+1}. Indeed, Seymour's original motivation for his conjecture was to find an understandable proof of the Hajnal-Szemerédi Theorem. Fan and Kierstead [29] came close to proving

Posa's conjecture, but had to settle for finding the square of a hamiltonian path, albeit with a slightly weaker hypothesis.

Theorem 3.3 (Fan and Kierstead [29]) *Every graph G on n vertices which satisfies $\delta(G) \geq \frac{2n-1}{3}$ contains the square of P_n.*

The following example shows that the theorem is best possible for paths.

Example 3.4 Let G be the complete tripartite graph $K_{t-1,t+1,t+1}$. Then $n := |G| = 3t + 1$ and $\delta(G) = 2t = \frac{2n-2}{3}$. If H is a square path in G, then any three consecutive vertices must be in distinct parts. It follows that G does not contain the square of P_n, since when the last vertex in the small part is used there will still be four unused vertices. Notice also that adding edges inside the small part will not create the square of P_n.

Not only does the weakened hypothesis give the best possible result with respect to paths, but it also implies a strengthened version of the Aigner-Brandt Theorem. Observe that every graph H on n vertices with $\Delta(H) \leq 2$ is contained in the square of P_n. In other words, the square of P_n is *universal* with respect to graphs H on n vertices with $\Delta(H) \leq 2$.

Theorem 3.5 (Fan and Kierstead [29]) *There exists a* universal *graph U on n vertices with $\Delta(U) = 4$ such that (1) U packs with every graph G on n vertices with $\Delta(G) \leq \frac{n-2}{3}$ and (2) U contains every graph H on n vertices with $\Delta(H) \leq 2$.*

Fan and Kierstead [30] also proved:

Theorem 3.6 (Fan and Kierstead [30]) *If G is a graph on n vertices such that $\delta(G) \geq \frac{2}{3}n$, then either (1) G contains the square of C_n or (2) G does not contain the square of C_k for $k > \frac{2}{3}n$, but there exist integers i and j with $i + j = n$ and $i \leq j \leq \frac{2}{3}n$ such that G contains the squares of vertex disjoint copies of C_i and C_j.*

Pósa's Conjecture remains open, but Komlós, Sárközy and Szemerédi [49, 50] have used Szemerédi's Regularity Lemma and their own Blow-Up Lemma [48] to prove it, and the more general conjecture of Seymour, for huge graphs.

Theorem 3.7 (Komlós, Sárközy and Szemerédi [50]) *For every positive integer r there exists an integer N such that for every $n > N$ every graph G on n vertices with $\delta(G) \geq \frac{r}{r+1}n$ contains the r-th power of a hamiltonian cycle.*

It is well known that every balanced bipartite graph G on $n = 2s$ vertices with $\delta(G) \geq \frac{s+1}{2}$ contains C_n. Wang [76] conjectured that if $\delta(G) \geq \frac{s}{2} + 1$ then G is universal with respect to bipartite graphs H on n vertices with $\delta(H) \leq 2$. He showed that if true, this is best possible.

Example 3.8 Let G be the balanced bipartite graph on $n = 2s = 4t + 2$ vertices formed by taking two copies of $K_{t,t+1}$ and joining the two larger parts by a matching with $t + 1$ edges. Then $\delta(G) = t + 1 = \frac{s+1}{2}$, but G does not contain $C_{s+1} + C_{s-1}$.

Define the *ladder* L_n to be the balanced bipartite graph on $n = 2s$ vertices $a_1, b_1, \cdots, a_s, b_s$ such that $a_i b_j \in E(L_n)$ if and only if $|i - j| \leq 1$. It is easy to check that L_n is universal with respect to all bipartite graphs H on n vertices with $\Delta(H) \leq 2$. Czygrinow and Kierstead [22] used the Regularity and Blow-Up lemmas to prove the following strengthening of Wang's conjecture for huge graphs.

Theorem 3.9 (Czygrinow and Kierstead [22]) *There exists an integer N such that every balanced bipartite graph G on $n = 2s > N$ vertices with $\delta(G) \geq \frac{s}{2} + 1$ contains L_n.*

Amar [5] posed the following version of an Ore-type conjecture.

Conjecture 3.10 (Amar [5]) *Every balanced A, B-bigraph G on $n = 2s$ vertices satisfying $d(a) + d(b) \geq s + k$ for all $a \in A$ and $b \in B$ contains a copy of every balanced bipartite graph H on n vertices with k components and $\Delta(H) \leq 2$.*

Bondy and Chvátal [13] proved the case $k = 1$ (prior to the conjecture) and Amar [5] proved the case $k = 2$. Recently Czygrinow, DeBiasio and Kierstead [23] proved the following two theorems. Together they imply Amar's conjecture for sufficiently large graphs. Moreover, Theorem 3.12 shows that the degree bound can be relaxed to $d(a) + d(b) \geq s + 2$ as long as n is large in terms of k.

Theorem 3.11 (Czygrinow, DeBiasio and Kierstead [23]) *For some integer K, every balanced A, B-bigraph G on $n = 2s$ vertices satisfying $d(a) + d(b) \geq s + K$ for all $a \in A$ and $b \in B$ contains L_n.*

Theorem 3.12 (Czygrinow, DeBiasio and Kierstead [23]) *For every integer k there exists an integer N_k such that every balanced A, B-bigraph G on $n = 2s > N_k$ vertices satisfying $d(a) + d(b) \geq s + 2$ for all $a \in A$ and $b \in B$ contains a copy of every balanced bipartite graph H on n vertices with k components and $\Delta(H) \leq 2$. Moreover if $\delta(G) \geq \frac{1}{100k}s$, then G contains L_n.*

4 Coloring

As mentioned in the introduction, various coloring problems are important instances of packing problems. An obvious (but sharp) Dirac-type bound on the (ordinary) chromatic number is

$$\chi(G) \leq \Delta(G) + 1, \tag{4.1}$$

where $\chi(G)$ is the chromatic number of G. Brooks' Theorem below characterizes the graphs for which (4.1) holds with equality.

Theorem 4.1 (Brooks) *If $\chi(G) = \Delta(G) + 1$, then either G contains the complete graph $K_{\Delta(G)+1}$ or $\Delta(G) = 2$ and G contains an odd cycle.*

The counterpart of (4.1) for $\theta(G)$ is

$$\chi(G) \leq \lfloor \theta(G)/2 \rfloor + 1. \tag{4.2}$$

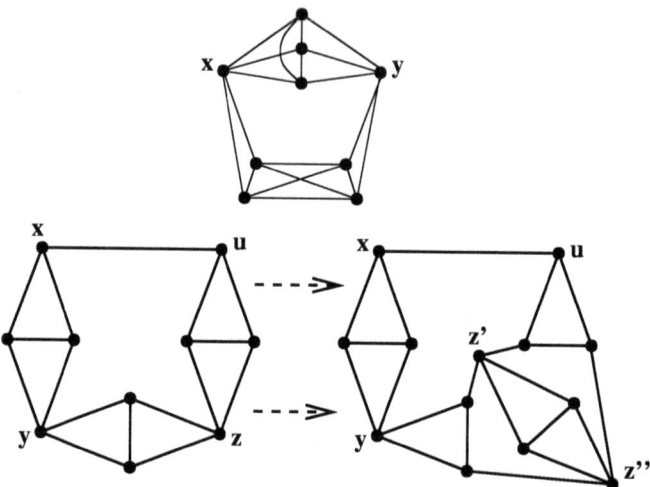

Figure 1: Above: a graph with $\theta = 9$ and $\chi = 5$. Below: two graphs with $\theta = 7$ and $\chi = 4$.

The proof is also obvious and the bound is also attained at complete graphs. However for small odd θ there are more connected graphs for which (4.2) holds with equality.

Example 4.2 Let G be the top graph in Figure 1, i.e., the 9-vertex graph with $V(G) = Q \cup Q'$, $Q \cap Q' = \{x, y\}$, $G[Q] = K_5 - xy$, and $G[Q'] = K_6 - T$, where T is a tree with four leaves, $xy \in E(T)$ and $d(x) = 3 = d(y)$. Since x and y are the only vertices with degree greater than 4 and are not adjacent to each other, $\theta(G) = 9$. Every (proper) 4-coloring of $G[Q]$ assigns x and y the same colors, since $G[N_Q(x) \cap N_Q(y)] = K_3$. On the other hand, every 4-coloring of $G[Q']$ assigns x and y different colors, since $G[N_{Q'}(x) \cup N_{Q'}(y)] = K_4$. Thus $\chi(G) > 4$.

Example 4.3 Let G be a graph with $\theta(G) \leq 7$ and $\chi(G) = 4$ (for example, $G = K_4$). We construct from G a $(3 + |V(G)|)$-vertex graph G' with $\theta(G') \leq 7$ and $\chi(G') = 4$ as follows. Choose a vertex $v \in V(G)$ that has no neighbors of degree 4 (each vertex of degree 4 has this property). Split v into two vertices v_1 and v_2 of degree at most two. Add two new vertices x_v and y_v that are adjacent to v_1, v_2, and to each other. By construction, $\theta(G') = 7$. Suppose that G' has a 3-coloring f. Since both, v_1 and v_2 are adjacent to x_v and y_v, we need $f(v_1) = f(v_2)$. But then f yields a 3-coloring of G, contrary to our assumption. The two bottom graphs in Figure 1 illustrate the idea (with G on the left).

Iterating the idea of Example 4.3 yields infinitely many 2-connected graphs G with $\theta(G) \leq 7$ and $\chi(G) = 4$. In contrast, for graphs with Ore-degree at least 12 (i.e., with chromatic number at least 7), the only extremal connected graphs are complete graphs.

Theorem 4.4 ([41]) *If $7 \leq \chi(G) = \lfloor \theta(G)/2 \rfloor + 1$, then G contains the complete graph $K_{\chi(G)}$.*

We think that the statement of Theorem 4.4 holds also for graphs G with Ore-degree at least 10 but cannot prove it. An even more challenging task would be to describe connected graphs G with $\theta(G) \leq 7$ and $\chi(G) = 4$.

5 Equitable coloring

In several applications of coloring as a partition problem there is an additional requirement that color classes be not too large or be of approximately the same size. Examples are the mutual exclusion scheduling problem [6, 71], scheduling in communication systems [33], construction timetables [46], and round-the-clock scheduling [74]. For other applications in scheduling, partitioning, and load balancing problems, one can look into [7, 58, 71]. A model imposing such a requirement is *equitable coloring*—a proper coloring such that color classes differ in size by at most one. As mentioned in the introduction, equitable coloring is a particular case of the general graph packing problem. Alon and Füredi [2], Pemmaraju [65] and Janson, Łuczak, and Ruciński [34] used equitable colorings to give new bounds on tails of distributions of sums of random variables. Rödl and Ruciński [66] used equitable colorings to give a new proof of the Blow-Up Lemma.

In contrast to ordinary coloring, a graph may have an equitable k-coloring (i.e., an equitable coloring with k colors) but have no equitable $(k+1)$-coloring. Thus, it is natural to look for the minimum number, eq(G), such that for every $k \geq $ eq(G), G has an equitable k-coloring.

Finding eq(G) even for planar graphs G is an NP-hard problem. This motivates a series of extremal problems on equitable colorings. The Dirac-type Theorem 1.7 by Hajnal and Szemerédi [31] in the original language is as follows.

Theorem 5.1 (Hajnal and Szemerédi [31]) *Every graph G with maximum degree at most r has an equitable $(r+1)$-coloring.*

The proof was long and sophisticated. A shorter proof appeared in [38]. Then Kierstead, Kostochka, Mydlarz, and Szemerédi [45] devised an algorithm that in time $O(rn^2)$ finds an equitable $(r+1)$-coloring for any n-vertex graph with maximum degree at most r. It is based on a modification of the proof of Theorem 5.1 in [38]. Here we present a yet shorter and simpler proof of Theorem 5.1.

Proof of Theorem 5.1. Let G be a graph with $\Delta(G) \leq r$. We may assume that $|G|$ is divisible by $r+1$: If $|G| = s(r+1) - p$, where $p \in [r]$, then set $G' = G + K_p$. By construction, $|G'|$ is divisible by $r+1$ and $\Delta(G') \leq r$. Moreover, the restriction of any equitable $(r+1)$-coloring of G' to G is an equitable $(r+1)$-coloring of G. So we may assume $|G| = (r+1)s$.

We argue by induction on $\|G\|$. The base step $\|G\| = 0$ is trivial, so consider the induction step. Let u be a non-isolated vertex. By the induction hypothesis, there exists an equitable $(r+1)$-coloring of $G - E(u)$, where $E(u)$ denotes the set of edges incident with u. We are done unless some color class V contains an edge uv. Since $\Delta(G) \leq r$, some color class T contains no neighbors of u. Moving u to T yields an

$(r+1)$-coloring of G with all classes of size s, except for one *small* class $V^- = V - u$ of size $s-1$ and one *large* class $V^+ = T + u$ of size $s+1$. Such a coloring is called *nearly equitable*.

Given a nearly equitable $(r+1)$-coloring, define an auxiliary digraph $\mathcal{H}$, whose vertices are the color classes, so that UW is a directed edge if and only if some vertex $y \in U$ has no neighbors in W. In this case we say that y *witnesses* edge UW. Let $\mathcal{A}$ be the set of classes from which V^- can be reached in $\mathcal{H}$ and $\mathcal{B}$ be the set of classes not in $\mathcal{A}$. Set $a = |\mathcal{A}|$, $b = |\mathcal{B}|$, $A = \bigcup \mathcal{A}$ and $B = \bigcup \mathcal{B}$. Then $r + 1 = a + b$. Since every vertex $y \in B$ has a neighbor in every class of $\mathcal{A}$,

$$d_A(y) \geq a \text{ for all } y \in B. \qquad (*)$$

Case 0: $V^+ \in \mathcal{A}$. Then there exists a V^+, V^--path $\mathcal{P} = V_1, \ldots, V_k$ in $\mathcal{H}$. Moving each witness y_j of $V_j V_{j+1}$ to V_{j+1} yields an equitable $(r+1)$-coloring of G.

We now argue by a secondary induction on b, whose base step $b = 0$ holds by Case 0. If $V^+ \notin \mathcal{A}$, then $|A| = as - 1$ and $|B| = bs + 1$. Consider the secondary induction step.

A class $W \in \mathcal{A}$ is *terminal* if every $U \in \mathcal{A} - W$ can reach V^- in $\mathcal{H} - W$. Let $\mathcal{A}'$ be the set of terminal classes, $a' = |\mathcal{A}'|$ and $A' = \bigcup \mathcal{A}'$. An edge wz is *solo* if $w \in W \in \mathcal{A}'$, $z \in B$ and $N_W(z) = \{w\}$. Ends of solo edges are *solo* vertices and *solo neighbors* of each other.

If $V^- \in \mathcal{A}'$ then no other class can be in $\mathcal{A}$. Thus in this case $a = 1, b = r$ and by $(*)$, $|E(A, B)| \geq |B| = rs + 1$, a contradiction to the fact that $|E(A, B)| \leq r|A| = r(s-1)$. So, $V^- \notin \mathcal{A}'$.

Suppose that some solo vertex $w \in W \in \mathcal{A}'$ witnesses an edge $WX \in E(\mathcal{H}[A])$. Let $y \in B$ be a solo neighbor of w. Move w to X and y to W. This yields nearly equitable colorings of $G[A + y]$ and $G[B - y]$. Since W is terminal, $X + w$ can reach V^- in $\mathcal{H} - W$. Thus by Case 0, $G[A + y]$ has an equitable a-coloring. By $(*)$, $\Delta(G[B - y]) \leq b - 1$. So by the primary induction hypothesis, $G[B - y]$ has an equitable b-coloring. Combining these equitable colorings yields an equitable $(r+1)$-coloring of G. Thus, we may assume:

Each solo vertex $z \in A'$ witnesses no edges in $\mathcal{H}$, and so $d_A(z) \geq a - 1$. $(**)$

Order $\mathcal{A}$ as $X_0, X_1, \ldots, X_{a-1}$ so that $X_0 = V^-$ and each X_i has a previous out-neighbor.

Case 1: For some $a - b \leq i \leq a - 1$, class X_i is not terminal. This includes the case $a \leq b$. Then some $X_j \in \mathcal{A}'$ cannot reach V^- in $\mathcal{H} - X_i$. So $j > i$ and X_j has no out-neighbors before X_i. In particular, $d_\mathcal{A}^+(X_j) < b$. Then for each $w \in X_j$, $d_A(w) \geq a - b$, and so $d_B(w) < 2b$. Let S be the set of solo vertices in X_j, and $D = X_j \setminus S$. By definition, each $v \in B$ either has a neighbor in S or at least two neighbors in D. Thus $\sum_{w \in X_j} d_B(w) \geq 2|B| - |N_B(S)|$. By $(**)$, $\sum_{w \in S} d_A(w) \geq (a-1)|S|$ and hence $|N_B(S)| \leq b|S|$. Since $|S| + |D| = s$, we have the following contradiction:

$$rs \geq \sum_{w \in X_j} (d_A(w) + d_B(w)) \geq (a-b)|D| + (a-1)|S| + 2|B| - |N_B(S)| \geq$$

$$\geq (a-b)|D| + (a-1)|S| + 2bs + 2 - b|S| = (a+b)|D| + 2 + (a+b-1)|S| > rs.$$

Case 2: All the last b classes $X_{a-b}, \ldots, X_{a-1}$ are terminal. Then $a' \geq b$. For $y \in B$, let $\sigma(y)$ be the number of solo neighbors of y. For each $y \in B$,

$$r \geq d(y) \geq a + d_B(y) + (a' - \sigma(y)) \geq r + 1 + d_B(y) + a' - b - \sigma(y).$$

So $\sigma(y) \geq a' - b + d_B(y) + 1$. Let I be a maximal independent set with $V^+ \subseteq I \subseteq B$. Then $\sum_{y \in I}(d_B(y) + 1) \geq |B| = bs + 1$. Since $a' \geq b$,

$$\sum_{y \in I} \sigma(y) \geq \sum_{y \in I}(a' - b + d_B(y) + 1) \geq s(a' - b) + bs + 1 > a's = |A'|.$$

So some vertex $w \in W \in \mathcal{A}'$ has two solo neighbors y_1 and y_2 in the independent set I.

By the primary induction hypothesis, we can equitably b-color $G[B - y_1]$. Let Y' be the class of y_2 in this coloring. By (**), $d_{B-y_1}(w) \leq r - (a-1) - 1 = b - 1$ and we can move w to some class $U \subseteq B - y_1$. Replacing w with y_1 in W to get W^* and moving w to U yields a new nearly equitable $(r+1)$-coloring of G. Now at least $a+1$ classes, W^*, Y', and all $X \in \mathcal{A}' - W$, can reach V^-. In this case we are done by the secondary induction hypothesis. □

As discussed in Example 2.3, if r is odd, then $K_{r,r}$ has no equitable r-coloring. Chen, Lih and Wu [17] proposed the following common strengthening of Theorem 5.1 and Brooks' theorem.

Conjecture 5.2 (Chen, Lih and Wu [17]) *Let G be a connected graph such that $\Delta(G) = r$. Then G has no equitable r-coloring if and only if either (1) $G = K_{r+1}$, or (2) $r = 2$ and G is an odd cycle, or (3) r is odd and $G = K_{r,r}$.*

Some partial cases of Conjecture 5.2 were proved in [17, 52, 62, 79, 80]. In particular, Chen, Lih and Wu [17] proved that the conjecture holds for $r = 3$:

Theorem 5.3 (Chen, Lih and Wu [17]) *Let G be a connected graph such that $\Delta(G) \leq 3$. Then G has no equitable 3-coloring if and only if $G = K_4$ or $G = K_{3,3}$.*

Brooks' theorem characterizes all graphs with maximum degree r that are r-colorable, since a graph is r-colorable if and only if each of its components is. This is not the case for equitable r-coloring. For example, for each odd $r \geq 3$, the graph consisting of two disjoint copies of $K_{r,r}$ has an equitable r-coloring, but the graph consisting of a copy of $K_{r,r}$ and a copy of K_r does not. This construction can be generalized. Say that a graph H is r-equitable if $|H|$ is divisible by r, H is r-colorable and every r-coloring of H is equitable. If G contains $K_{r,r}$ and $G - K_{r,r}$ is r-equitable, then G does not have an equitable r-coloring. This motivates the study of equitable graphs, i.e., graphs that are r-equitable for some r. It was proved in [40] that there is a good description of the family of all r-equitable graphs; they can all be built from simple examples in a straightforward way.

If an r-colorable graph G has a spanning subgraph whose components are all r-equitable, then G is also r-equitable. We say that an r-equitable graph G is r-reducible if $V(G)$ has a partition $\{V_1, \ldots V_t\}$ into at least two parts such that $G[V_i]$ is r-equitable for each $i \in [t]$; otherwise G is r-irreducible. Clearly K_r is r-irreducible. The reader can see one other 5-irreducible graph F_1 and three other 4-irreducible

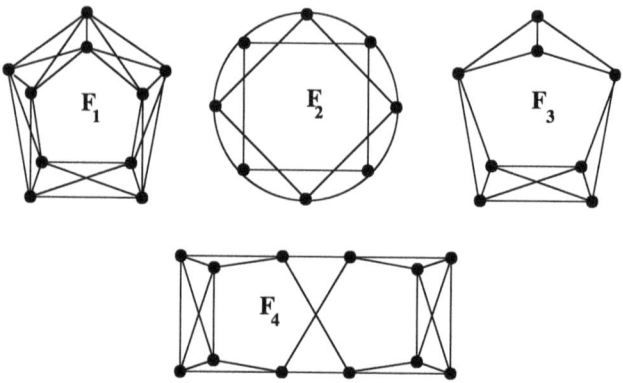

Figure 2: One 5-equitable and three 4-equitable basic graphs.

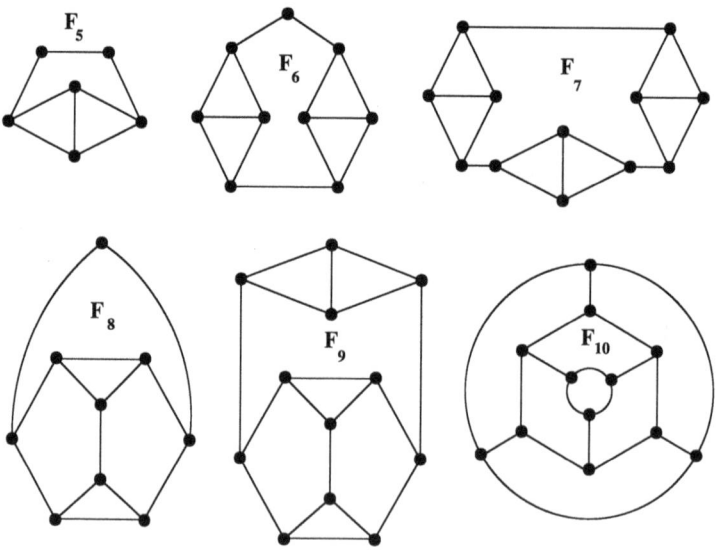

Figure 3: Six 3-equitable basic graphs.

graphs F_2, F_3, F_4 in Figure 2 and six other 3-irreducible graphs $F_5, \ldots, F_{10}$ in Figure 3. Together with K_r, the r-irreducible graphs from this list are the r-basic graphs. An r-decomposition of G is a partition of $V(G)$ into subsets $V_1, \ldots, V_t$ such that each $G[V_i]$ is r-basic. We say that G is r-decomposable if it has an r-decomposition. As was just mentioned, if r is odd and G is r-decomposable, then $G \cup K_{r,r}$ has no equitable r-coloring. It was conjectured in [40] that this is the only obstacle that prevents an r-colorable graph with $\Delta(G) \leq r$ from having an equitable r-coloring.

Conjecture 5.4 ([40]) *Suppose that $r \geq 3$ and G is an r-colorable graph with $\Delta(G) = r$. Then G has no equitable r-coloring if and only if r is odd and there exists $H \subseteq G$ such that $H = K_{r,r}$ and $G - H$ is r-decomposable.*

For $r \geq 6$, this conjecture means that *if an r-colorable graph G with $\Delta(G) \leq r$ has no equitable r-coloring, then r is odd and $V(G)$ can be partitioned into sets $V_0, \ldots, V_t$ such that $G[V_0] = K_{r,r}$ and $G[V_i] = K_r$ for each $i = 1, \ldots, r$.*

A *nearly equitable r-coloring* of a graph G is defined in the above proof of Theorem 5.1. In this case $|G|$ is divisible by r. If r is odd, G contains $K_{r,r}$ and $G - K_{r,r}$ has a nearly equitable r-coloring, then G has an equitable r-coloring, since the small class of one of the components can be combined with the large class of the other. This explains our interest in nearly equitable r-colorings.

Let $\mathcal{G}(r)$ be the class of all graphs G with $\Delta(G) \leq r$ and $\chi(G) \leq r$. Let $\mathcal{G}(r, n)$ be the set of graphs in $\mathcal{G}(r)$ with at most n vertices. The following result implies that Conjectures 5.4 and 5.2 are equivalent.

Theorem 5.5 ([40]) *Let $G \in \mathcal{G}(r)$ with $|G|$ divisible by r. The following are equivalent:*

(A) G is r-decomposable;

(B) G is r-equitable;

(C) G has an equitable r-coloring, but does not have a nearly equitable r-coloring.

Corollary 5.6 ([40]) *For all positive integers r and $n > r$, Conjecture 5.2 holds for all graphs in $\mathcal{G}(r, n)$ if and only if Conjecture 5.4 holds for all graphs in $\mathcal{G}(r, n)$.*

Corollary 5.7 ([40]) *Let $G \in \mathcal{G}(r)$ be r-equitable. Then G has a unique r-decomposition.*

Corollary 5.8 ([40]) *There exists a polynomial time algorithm for deciding if a graph $G \in \mathcal{G}(r)$ is r-equitable.*

Very recently, the case $r = 4$ of Conjecture 5.2 was proved [42].

The following Ore-type analogue of Theorem 5.1 (slightly strengthening a conjecture in [54, 56]) was proved in [39].

Theorem 5.9 ([39]) *For every $r \geq 3$, each graph G with $\theta(G) \leq 2r + 1$ has an equitable $(r+1)$-coloring.*

In particular, the theorem yields that Conjecture 5.2 holds for graphs in which vertices of maximum degree form an independent set. The bounding examples for this theorem (apart from K_{r+1}) are graphs $K_{m,2r-m}$ for every odd $0 < m \leq r$. In the same paper, the following Ore-type analogue of Conjecture 5.2 was proposed.

Conjecture 5.10 ([39]) *Let $r \geq 3$ and G be a connected graph with $\theta(G) \leq 2r$. If G is distinct from K_{r+1} and from $K_{m,2r-m}$ for odd m, then G has an equitable r-coloring.*

Conjecture 5.10 was proved to hold for $r = 3$. In [41] the following common extension of Conjectures 5.4 and 5.10 was posed.

Conjecture 5.11 ([41]) *Suppose that $r \geq 3$. An r-colorable, n-vertex graph G with $\theta(G) \leq 2r$ has no equitable r-coloring, if and only if n is divisible by r, and there exists $W \subseteq G$ such that $W = K_{m,2r-m}$ for some odd m and $G-W$ is r-decomposable.*

It was proved in [41] that Conjecture 5.11 is equivalent to Conjecture 5.10 even for graphs with restricted number of vertices and restricted values of Ore-degree.

Theorem 5.12 ([41]) *Assume that Conjecture 5.10 holds for all graphs with at most n vertices and Ore-degree at most $2r$. Let G be an r-colorable n-vertex graph with $\theta(G) \leq 2r$. Then G has no equitable r-coloring if and only if n is divisible by r and there exists $H \subseteq G$ such that $H = K_{m,2r-m}$ for some odd m and $G - H$ is r-decomposable.*

It follows that Conjecture 5.11 holds for $r = 3$.

6 Equitable list coloring

A list analogue of equitable coloring was introduced by Kostochka, Pelsmajer and West [53]. A *list assignment* L for a graph G assigns to each vertex $v \in V(G)$ a set $L(v)$ of allowable colors. An L-coloring of G is a proper vertex coloring such that for every $v \in V(G)$ the color on v belongs to $L(v)$. Given a k-uniform list assignment L for an n-vertex graph G, the graph G is *equitably L-colorable* if G has an L-coloring where no color is used on more than $\lceil n/k \rceil$ vertices. A graph G is *equitably k-choosable* if G is equitably L-colorable whenever L is a k-uniform list assignment for G.

Because one cannot ensure the appearance of each color, the techniques previously used for ordinary equitable colorings do not work well for equitable list colorings. Nevertheless, Kostochka, Pelsmajer and West [53] suggested that the following analogue of the Hajnal–Szemerédi Theorem holds.

Conjecture 6.1 (Kostochka, Pelsmajer and West [53]) *Every graph G is equitably $(1 + \Delta(G))$-choosable.*

Furthermore, Pelsmajer [64] and independently Lih and Wang [61] confirmed this conjecture for graphs with maximum degree at most 3:

Theorem 6.2 (Pelsmajer [64], Lih and Wang [61]) *If $\Delta(G) \leq 3$, then G is equitably 4-choosable.*

Very recently, the conjecture was proved for graphs with maximum degree 4 [43]. We believe that the following Ore-type analogue of Conjecture 6.1 also holds.

Conjecture 6.3 *Every graph G is equitably $(1 + 0.5\theta(G))$-choosable.*

The result below confirms the conjecture for graphs with small θ and somewhat extends Theorem 6.2.

Theorem 6.4 *If $\theta(G) \leq 6$, then G is equitably 4-choosable.*

Proof Let G be an edge-minimal counterexample to the theorem. In particular, $|L(x)| = 4$ for every $x \in V(G)$. Let v be a vertex of maximum degree in G and $w_1, \ldots, w_{d(v)}$ be the neighbors of v. Since $\theta(G) \leq 6$, $d(v) \leq 5$. If $d(v) \leq 3$, then we are done by Theorem 6.2. So, $d(v) \in \{4, 5\}$.

If $d(v) = 5$, then the component of G containing v is $K_{5,1}$. Let $G_1 = G - v - w_1 - w_2 - w_3$. Since G_1 is a proper subgraph of G, $\theta(G_1) \leq 6$. By the minimality of G, the subgraph G_1 admits some equitable L-coloring f. This means that each color is used on at most $\lceil |V(G_1)|/4 \rceil = \lceil |V(G)|/4 \rceil - 1$ vertices. We extend f to an equitable L-coloring of G as follows: Choose a color $\alpha_0 \in L(v) - f(w_4) - f(w_5)$ as $f(v)$, and then for $i = 1, 2, 3$ choose a color $\alpha_i \in L(w_i) - f(v) - \{f(w_j) : 1 \leq j \leq i-1\}$ as $f(w_i)$. Since each color appears on our four "new" vertices only once, the resulting coloring is equitable.

Suppose now that $d(v) = 4$. Since $\theta(G) \leq 6$, $d(w_i) \leq 2$ for each $i = 1, 2, 3, 4$. For $i = 1, 2, 3, 4$, let u_i be the neighbor of w_i distinct from v, if it exists.

Case 1: u_1 does not exist or $u_1 = w_2$. Consider $G_1 = G - v - w_1 - w_2 - w_3$. By the minimality of G, there exists an equitable L-coloring f of G_1. Extend f to an L-coloring of G as follows: Choose a color $\alpha \in L(v) - f(w_4)$ as $f(v)$, then choose $f(w_3) \in L(w_3) - f(v) - f(u_3)$, $f(w_2) \in L(w_2) - f(v) - f(w_3) - f(u_2)$, and finally $f(w_1) \in L(w_1) - f(v) - f(w_3) - f(w_2)$. Again, since each color appears on our four "new" vertices only once, the resulting coloring is equitable.

So, in what follows, all u_i exist and are distinct from all w_j.

Case 2: $|L(v) - L(w_1)| \geq 2$. Consider $G_1 = G - v - w_1 - w_2 - w_3$ as in Case 1. Let f be an equitable L-coloring of G_1. Choose $f(v) \in L(v) - L(w_1) - f(w_4)$, then $f(w_3) \in L(w_3) - f(v) - f(u_3)$, $f(w_2) \in L(w_2) - f(v) - f(w_3) - f(u_2)$, and finally $f(w_1) \in L(w_1) - f(u_1) - f(w_3) - f(w_2)$. Since $f(v) \notin L(w_1)$, the resulting coloring is proper.

From now on, $|L(v) \cap L(w_i)| \geq 3$ for all $i \in \{1, 2, 3, 4\}$.

Case 3: $|L(v) \cap L(w_1) \cap L(w_2)| \geq 3$. We may assume that

$$|L(v) \cap L(w_1)| \geq |L(v) \cap L(w_2)|. \tag{6.1}$$

Let G_0 be obtained from $G - v - w_3 - w_4$ by identifying w_1 and w_2 into a new vertex w^*. Let $L(w^*) = L(w_2)$. By the minimality of G, the new graph G_0 with the new list L has an equitable L-coloring f.

Subcase 3.1: $f(w^*) \in L(w_1)$. We define $f(w_1) = f(w_2) = f(w^*)$ and let $f(w_3) \in L(w_3) - f(u_3) - f(w^*)$, $f(w_4) \in L(w_4) - f(u_4) - f(w^*) - f(w_3)$, and $f(v) \in L(v) - f(w_3) - f(w_4) - f(w^*)$.

Subcase 3.2: $f(w^*) \notin L(w_1)$. Under the conditions of Case 3, this means that $|L(w_1) \cap L(w_2)| = |L(v) \cap L(w_1) \cap L(w_2)| = 3$ and now (6.1) yields that $f(w*) \notin L(v)$. So, we can let $f(w_2) = f(w*)$, $f(w_3) \in L(w_3) - f(u_3) - f(w^*)$, $f(w_4) \in L(w_4) - f(u_4) - f(w^*) - f(w_3)$, $f(w_1) \in L(w_1) - f(u_1) - f(w_3) - f(w_4)$, and $f(v) \in L(v) - f(w_1) - f(w_3) - f(w_4)$. This finishes Case 3.

If none of Cases 1–3 takes place, then we may assume that $L(v) = \{1, 2, 3, 4\}$ and for $i = 1, 2, 3, 4$, $L(w_i) \cap L(v) = L(v) - i$. Construct the graph G_0 with the list L as in Case 3. Let f be an equitable L-coloring of G_0. If $f(w^*) \neq 1$, then we extend f to the whole of G as in Case 3. So, assume that $f(w^*) = 1$. If $f(u_3) \neq 1$, then let $f(w_3) = f(w_2) = 1$, $f(w_4) \in L(w_4) - f(u_4) - 1$, $f(w_1) \in L(w_1) - f(u_1) - f(w_4)$, and $f(v) \in \{2, 3, 4\} - f(w_4) - f(w_1)$. Finally, suppose $f(u_3) = 1$. In this case, let $f(w_2) = 1$, $f(w_3) \in L(w_3) - L(v)$, $f(w_4) \in L(w_4) - f(u_4) - 1 - f(w_3)$, $f(w_1) \in L(w_1) - f(u_1) - f(w_3) - f(w_4)$, and $f(v) \in \{2, 3, 4\} - f(w_4) - f(w_1)$. □

Acknowledgements

Research of the first author is supported in part by NSA grant H98230-08-1-0069. Research of the second author is supported in part by NSF grant DMS-06-50784 and by grant 06-01-00694 of the Russian Foundation for Basic Research. Research of the third author is partially supported by NSF grant DMS-0852452.

References

[1] M. Aigner & S. Brandt, Embedding arbitrary graphs of maximum degree two, *J. Lond. Math. Soc.* **48** (1993), 39–51.

[2] N. Alon & E. Fischer, 2-factors in dense graphs, *Discrete Math.* **152** (1996), 13–23.

[3] N. Alon & Z. Füredi, Spanning subgraphs of random graphs, *Graphs. Combin.* **8** (1992), 91–94.

[4] N. Alon & J.H. Spencer, *The Probabilistic Method*, Second edition, Wiley, New York (2000).

[5] D. Amar, Partition of a hamiltonian graph into two cycles, *Discrete Math.* **58** (1986), 1–10.

[6] B. Baker & E. Coffman, Mutual exclusion scheduling, *Theor. Comput. Sci.* **162** (1996), 225–243.

[7] J. Blazewicz, K. Ecker, E. Pesch, G. Schmidt & J. Weglarz, *Scheduling computer and manufacturing processes*, Springer, Berlin (2001).

[8] B. Bollobás, *Extremal graph theory*, Academic Press, London–New York (1978).

[9] B. Bollobás & S. E. Eldridge, Maximal matchings in graphs with given maximal and minimal degrees, *Congr. Numer.* **XV** (1976), 165–168.

[10] B. Bollobás & S. E. Eldridge, Packing of graphs and applications to computational complexity, *J. Combin. Theory Ser. B* **25** (1978), 105–124.

[11] B. Bollobás, A. V. Kostochka & K. Nakprasit, On two conjectures on packing of graphs, *Combin. Probab. Comput.* **14** (2005), 723–736.

[12] B. Bollobás, A. V. Kostochka & K. Nakprasit, Packing d-degenerate graphs, *J. Combin. Theory Ser. B* **98** (2008), 85–94.

[13] J. A. Bondy & V. Chvátal, A method in graph theory, *Discrete Math.* **15** (1976), 111–135.

[14] S. Brandt, An extremal result for subgraphs with few edges, *J. Combin. Theory Ser. B* **64** (1995), 288–299.

[15] P. A. Catlin, Subgraphs of graphs I, *Discrete Math.* **10** (1974), 225–233.

[16] P. A. Catlin, *Embedding subgraphs and coloring graphs under extremal degree conditions*, Ph.D. Thesis, Ohio State Univ., Columbus (1976).

[17] B.-L. Chen, K.-W. Lih & P.-L. Wu, Equitable coloring and the maximum degree, *European J. Combin.* **15** (1994), 443–447.

[18] K. Corrádi & A. Hajnal, On the maximum number of independent circuits in a graph, *Acta Math. Acad. Sci. Hung.* **14** (1963), 423–439.

[19] B. Csaba, Approximating the Bollobás–Eldridge–Catlin conjecture for bounded degree graphs, submitted.

[20] B. Csaba, On the Bollobás–Eldridge conjecture for bipartite graphs, *Combin. Probab. Comput.* **16** (2007), 661–691.

[21] B. Csaba, A. Shokoufandeh & E. Szemerédi, Proof of a conjecture of Bollobás and Eldridge for graphs of maximum degree three, *Combinatorica* **23**(1) (2003), 35–72.

[22] A. Czygrinow & H. A. Kierstead, 2-factors in dense bipartite graphs, *Discrete Math.* **257** (2002), 357–369.

[23] A. Czygrinow, L. Debiasio & H. A. Kierstead, On 2-factors in bipartite graphs, submitted.

[24] G. Dirac, Some theorems on abstract graphs, *Proc. Lond. Math. Soc.* **2** (1952), 69–81.

[25] N. Eaton, A near packing of two graphs, *J. Combin. Theory Ser. B* **80** (2000), 98–103.

[26] H. Enomoto, On the existence of disjoint cycles in a graph, *Combinatorica* **18** (1998), 487–492.

[27] P. Erdős, Problem 9, in *Theory of Graphs and Its Applications* (ed. M. Fieldler), Czech. Acad. Sci. Publ., Prague (1964), p.159.

[28] G. Fan & H. A. Kierstead, The square of paths and cycles, *J. Combin. Theory Ser. B* **63** (1995), 55–64.

[29] G. Fan & H. A. Kierstead, Hamiltonian square-paths, *J. Combin. Theory Ser. B* **67** (1996), 167–182.

[30] G. Fan & H. A. Kierstead, Partitioning a graph into two square-cycles, *J. Graph Theory* **23** (1996), 241–256.

[31] A. Hajnal & E. Szemerédi, Proof of a conjecture of Erdős, in *Combinatorial Theory and its Applications* (eds. P. Erdős, A. Rényi & V. T. Sós), Vol. II, North-Holland, New York–London (1970), pp. 601–603.

[32] T. R. Jensen & B. Toft, *Graph coloring problems*, Wiley-Interscience, New York (1995).

[33] S. Irani & V. Leung, Scheduling with conflicts, and applications to traffic signal control, in *Proceedings of the 7th annual ACM-SIAM symposium on discrete algorithms held in Atlanta, GA, 1996*, SIAM, Philadelphia, PA (1996), pp. 85–94.

[34] S. Janson, T. Łuczak & A. Ruciński, *Random Graphs*, Wiley-Interscience, New York (2000).

[35] H. Kaul & A. V. Kostochka, Extremal graphs for a graph packing theorem of Sauer and Spencer, *Combin. Probab. Comput.* **16** (2007), 409–416.

[36] H. Kaul, A. V. Kostochka & G. Yu, On a graph packing conjecture by Bollobás, Eldridge, and Catlin, *Combinatorica* **28** (2008), 469–485.

[37] H. Kheddouci, S. Marshall, J.-F. Sacle & M. Wozniak, On the packing of three graphs, *Discrete Math.* **236** (2001), 197–225.

[38] H. A. Kierstead & A. V. Kostochka, A short proof of the Hajnal–Szemerédi Theorem on equitable coloring, *Combin. Probab. Comput.* **17** (2008), 265–270.

[39] H. A. Kierstead & A. V. Kostochka, An Ore-type theorem on equitable coloring, *J. Combin. Theory Ser. B* **98** (2008), 226–234.

[40] H. A. Kierstead & A. V. Kostochka, Equitable versus nearly equitable coloring and the Chen–Lih–Wu Conjecture, *Combinatorica*, in press.

[41] H. A. Kierstead & A. V. Kostochka, Ore-type versions of Brooks' theorem, *J. Combin. Theory Ser. B* **99** (2009), 298–305.

[42] H. A. Kierstead & A. V. Kostochka, The Chen–Lih–Wu Conjecture holds for $r = 4$, in preparation.

[43] H. A. Kierstead & A. V. Kostochka, Graphs with maximum degree 4 are equitably 5-choosable, in preparation.

[44] H. A. Kierstead & A. V. Kostochka, Efficient graph packing via game coloring, *Combin. Probab. Comput.*, in press.

[45] H. A. Kierstead, A. V. Kostochka, M. Mydlarz & E. Szemerédi, A fast algorithm for equitable coloring, *Combinatorica*, in press.

[46] F. Kitagawa & H. Ikeda, An existential problem of a weight-controlled subset and its application to school timetable construction, *Discrete Math.* **72** (1988), 195–211.

[47] J. Komlós, Tiling Turán theorems, *Combinatorica* **20** (2000), 203–218.

[48] J. Komlós, G. Sárközy & E. Szemerédi, Blow-Up Lemma, *Combinatorica* **17** (1997), 109–123.

[49] J. Komlós, G. Sárközy & E. Szemerédi, On the Pósa–Seymour conjecture, *J. Graph Theory* **29** (1998), 167–176.

[50] J. Komlós, G. Sárközy & E. Szemerédi, Proof of the Seymour's conjecture for large graphs, *Ann. Comb.* **1** (1998), 43–60.

[51] J. Komlós, G. Sárközy & E. Szemerédi, Spanning trees in dense graphs, *Combin. Probab. Comput.* **10** (2001), 397–416.

[52] A. V. Kostochka & K. Nakprasit, On equitable Δ-coloring of graphs with low average degree, *Theoret. Comput. Sci.* **349** (2005), 82–91.

[53] A. V. Kostochka, M. J. Pelsmajer & D. B. West, A list analogue of equitable coloring, *J. Graph Theory* **44** (2003), 166–177.

[54] A. V. Kostochka & G. Yu, Extremal problems on packing of graphs, in *Oberwolfach reports*, 1, (2006), pp. 55–57.

[55] A. V. Kostochka & G. Yu, An Ore-type analogue of the Sauer–Spencer Theorem, *Graphs Combin.* **23** (2007), 419–424.

[56] A. V. Kostochka & G. Yu, Ore-type graph packing problems, *Combin. Probab. Comput.* **16** (2007), 167–169.

[57] A. V. Kostochka & G. Yu, Ore-type conditions implying 2-factors consisting of short cycles, *Discrete Math.*, in press.

[58] J. Krarup & D. de Werra, Chromatic optimisation: Limitations, objectives, uses, references, *European J. Oper. Res.* **11** (1982), 1–19.

[59] D. Kühn, D. Osthus & A. Treglown, An Ore-type theorem for perfect packings in graphs, submitted.

[60] K.-W. Lih, The equitable coloring of graphs, in *Handbook of Combinatorial Optimization* (eds. D.-Z. Du & P. Pardalos), Vol. 3, Kluwer, Dordrecht (1998), pp. 543–566.

[61] K.-W. Lih & W.-F. Wang, Equitable list coloring of graphs, *Taiwanese J. Math.* **8** (2004), 747–759.

[62] K.-W. Lih & P.-L. Wu, On equitable coloring of bipartite graphs, *Discrete Math.* **151** (1996), 155–160.

[63] O. Ore, Note on Hamilton circuits, *Amer. Math. Monthly* **67** (1960), 55.

[64] M. Pelsmajer, Equitable list-coloring for graphs of maximum degree 3, *J. Graph Theory* **47** (2004), 1–8.

[65] S. V. Pemmaraju, Equitable coloring extends Chernoff–Hoeffding bounds, in *Approximation, randomization and combinatorial optimization (Berkeley, CA, 2001)*, Springer, Berlin (2001), 285–296.

[66] V. Rödl & A. Ruciński, Perfect matchings in ϵ-regular graphs and the Blow-Up Lemma, *Combinatorica* **19** (1999), 437–452.

[67] J.-F. Sacle & M. Wozniak, A note on packing of three forests, *Discrete Math.* **164** (1997), 265–274.

[68] N. Sauer & J. Spencer, Edge disjoint placement of graphs, *J. Combin. Theory Ser. B* **25** (1978), 295–302.

[69] N. Sauer & H. Wang, Packing of three copies of a graph, *J. Graph Theory* **21** (1996), 71–80.

[70] P. Seymour, Problem section, in *Combinatorics: Proceedings of the British Combinatorial Conference, 1973* (eds. T. P. McDonough & V. C. Mavron), Cambridge University Press, Cambridge (1974), pp. 201–202.

[71] B. F. Smith, P. E. Bjorstad & W. D. Gropp, *Domain decomposition. Parallel multilevel methods for elliptic partial differential equations*, Cambridge University Press, Cambridge (1996).

[72] S. Teo & H. P. Yap, Packing two graphs of order n having total size at most $2n - 2$, *Graphs Combin.* **6** (1990), 197–205.

[73] S. Teo & H. P. Yap, Two theorems on packings of graphs, *European J. Combin.* **8** (1987), 199–207.

[74] A. Tucker, Perfect graphs and an application to optimizing municipal services, *SIAM Review* **15** (1973), 585–590.

[75] H. Wang, On the maximum number of independent cycles in a graph, *Discrete Math.* **205** (1999), 183–190.

[76] H. Wang, On 2-factors of a bipartite graph, *J. Graph Theory* **31** (1999), 101–106.

[77] M. Wozniak, Packing of graphs, *Dissertationes Math. (Rozprawy Mat.)* **362** (1997), 1–78.

[78] H. P. Yap, Packing of graphs — a survey, *Discrete Math.* **72** (1988), 395–404.

[79] H. P. Yap & Y. Zhang, The equitable Δ-colouring conjecture holds for outerplanar graphs, *Bull. Inst. Math. Acad. Sin.* **25** (1997), 143–149.

[80] H.-P. Yap & Y. Zhang, Equitable colourings of planar graphs, *J. Combin. Math. Combin. Comp.* **27** (1998), 97–105.

[81] X. Zhu, Refined activation strategy for the marking game, *J. Combin. Theory Ser. B* **98** (2008), 1–18.

Department of Mathematics and Statistics,
Arizona State University,
Tempe, AZ 85287, USA
kierstead@asu.edu

Department of Mathematics,
University of Illinois,
Urbana, IL, 61801, USA and
Sobolev Institute of Mathematics,
Novosibirsk, 630090, Russia
kostochk@math.uiuc.edu

Department of Mathematics,
College of William and Mary
Williamsburg, VA 23187, USA
gyu@wm.edu

Embedding large subgraphs into dense graphs

Daniela Kühn and Deryk Osthus

Abstract

What conditions ensure that a graph G contains some given spanning subgraph H? The most famous examples of results of this kind are probably Dirac's theorem on Hamilton cycles and Tutte's theorem on perfect matchings. Perfect matchings are generalized by perfect F-packings, where instead of covering all the vertices of G by disjoint edges, we want to cover G by disjoint copies of a (small) graph F. It is unlikely that there is a characterization of all graphs G which contain a perfect F-packing, so as in the case of Dirac's theorem it makes sense to study conditions on the minimum degree of G which guarantee a perfect F-packing.

The Regularity Lemma of Szemerédi and the Blow-up Lemma of Komlós, Sárközy and Szemerédi have proved to be powerful tools in attacking such problems and quite recently, several long-standing problems and conjectures in the area have been solved using these. In this survey, we give an outline of recent progress (with our main emphasis on F-packings, Hamiltonicity problems and tree embeddings) and describe some of the methods involved.

1 Introduction, overview and basic notation

In this survey, we study the question of when a graph G contains some given large or spanning graph H as a subgraph. Many important problems can be phrased in this way: one example is Dirac's theorem, which states that every graph G on $n \geq 3$ vertices with minimum degree at least $n/2$ contains a Hamilton cycle. Another example is Tutte's theorem on perfect matchings which gives a characterization of all those graphs which contain a perfect matching (so H corresponds to a perfect matching in this case). A result which gives a complete characterization of all those graphs G which contain H (as in the case of Tutte's theorem) is of course much more desirable than a sufficient condition (as in the case of Dirac's theorem). However, for most H that we consider, it is unlikely that such a characterization exists as the corresponding decision problems are usually NP-complete. So it is natural to seek simple sufficient conditions. Here we will focus mostly on degree conditions. This means that G will usually be a dense graph and that we have to restrict H to be rather sparse in order to get interesting results. We will survey the following topics:

- a generalization of the matching problem, which is called the F-packing or F-tiling problem (here the aim is to cover the vertices of G with disjoint copies of a fixed graph F instead of disjoint edges);

- Hamilton cycles (and generalizations) in graphs, directed graphs and hypergraphs;

- large subtrees of graphs;

- arbitrary subgraphs H of bounded degree;

- Ramsey numbers of sparse graphs.

A large part of the progress in the above areas is due to the Regularity Lemma of Szemerédi [125] and the Blow-up Lemma of Komlós, Sárközy and Szemerédi [83]. Roughly speaking, the former states that one can decompose an arbitrary large dense graph into a bounded number of random-like graphs. The latter is a powerful tool for embedding spanning subgraphs H into such random-like graphs. In the final section we give a formal statement of these results and describe in detail an application to a special case of the F-packing problem. We hope that readers who are unfamiliar with these tools will find this a useful guide to how they can be applied.

There are related surveys in the area by Komlós and Simonovits [89] (some minor updates were added later in [88]) and by Komlós [80]. However, much has happened since these were written and the emphasis is different in each case. So we hope that the current survey will be a useful complement and update to these. In particular, as the title indicates, our focus is mainly on embedding large subgraphs and we will ignore other aspects of regularity/quasi-randomness. There is also a recent survey on F-packings (and so-called F-decompositions) by Yuster [130], which is written from a computational perspective.

2 Packing small subgraphs in graphs

2.1 F-packings in graphs of large minimum degree

Given two graphs F and G, an F-packing in G is a collection of vertex-disjoint copies of F in G. (Alternatively, this is often called an F-tiling.) F-packings are natural generalizations of graph matchings (which correspond to the case when F consists of a single edge). An F-packing in G is called *perfect* if it covers all vertices of G. In this case, we also say that G contains an *F-factor* or a *perfect F-matching*. If F has a component which contains at least 3 vertices then the question whether G has a perfect F-packing is difficult from both a structural and algorithmic point of view: Tutte's theorem characterizes those graphs which have a perfect F-packing if F is an edge but for other connected graphs F no such characterization is known. Moreover, Hell and Kirkpatrick [61] showed that the decision problem of whether a graph G has a perfect F-packing is NP-complete if and only if F has a component which contains at least 3 vertices. So, as mentioned earlier, this means that it makes sense to search for degree conditions which ensure the existence of a perfect F-packing. The fundamental result in the area is the Hajnal-Szemerédi theorem:

Theorem 2.1 (Hajnal and Szemerédi [55]) *Every graph whose order n is divisible by r and whose minimum degree is at least $(1 - 1/r)n$ contains a perfect K_r-packing.*

The minimum degree condition is easily seen to be best possible. (The case when $r = 3$ was proved earlier by Corrádi and Hajnal [30].) The result is often phrased in terms of colourings: any graph G whose order is divisible by k and with $\Delta(G) \leq k-1$ has an equitable k-colouring, i.e. a colouring with colour classes of equal size. (So $k := n/r$ here.) Theorem 2.1 raises the question of what minimum degree condition forces a perfect F-packing for arbitrary graphs F. The following result gives a general bound.

Theorem 2.2 (Komlós, Sárközy and Szemerédi [86]) *For every graph F there exists a constant $C = C(F)$ such that every graph G whose order n is divisible by $|F|$ and whose minimum degree is at least $(1 - 1/\chi(F))n + C$ contains a perfect F-packing.*

This confirmed a conjecture of Alon and Yuster [9], who had obtained the above result with an additional error term of εn in the minimum degree condition. As observed in [9], there are graphs F for which the above constant C cannot be omitted completely (e.g. $F = K_{s,s}$ where $s \geq 3$ and s is odd). Thus one might think that this settles the question of which minimum degree guarantees a perfect F-packing. However, we shall see that this is *not* the case. There are graphs F for which the bound on the minimum degree can be improved significantly: we can often replace $\chi(F)$ by a smaller parameter. For a detailed statement of this, we define the *critical chromatic number* $\chi_{cr}(F)$ of a graph F as

$$\chi_{cr}(F) := (\chi(F) - 1)\frac{|F|}{|F| - \sigma(F)},$$

where $\sigma(F)$ denotes the minimum size of the smallest colour class in an optimal colouring of F. (We say that a colouring of F is *optimal* if it uses exactly $\chi(F)$ colours.) So for instance a k-cycle C_k with k odd has $\chi_{cr}(C_k) = 2 + 2/(k-1)$. Note that $\chi_{cr}(F)$ always satisfies $\chi(F) - 1 < \chi_{cr}(F) \leq \chi(F)$ and equals $\chi(F)$ if and only if for every optimal colouring of F all the colour classes have equal size. The critical chromatic number was introduced by Komlós [81]. He (and independently Alon and Fischer [8]) observed that for *any* graph F it can be used to give a lower bound on the minimum degree that guarantees a perfect F-packing.

Proposition 2.3 *For every graph F and every integer n that is divisible by $|F|$ there exists a graph G of order n and minimum degree $\lceil(1 - 1/\chi_{cr}(F))n\rceil - 1$ which does not contain a perfect F-packing.*

Given a graph F, the graph G in the proposition is constructed as follows: write $k := \chi(F)$ and let $\ell \in \mathbb{N}$ be arbitrary. G is a complete k-partite graph with vertex classes $V_1, \ldots, V_k$, where $|V_1| = \sigma(F)\ell - 1$, $n = \ell|F|$ and the sizes of $V_2, \ldots, V_k$ are as equal as possible. Then any perfect F-packing would consist of ℓ copies of F. On the other hand, each such copy would contain at least $\sigma(F)$ vertices in V_1, which is impossible.

Komlós also showed that the critical chromatic number is the parameter which governs the existence of *almost* perfect packings in graphs of large minimum degree. (More generally, he also determined the minimum degree which ensures that a given fraction of vertices is covered.)

Theorem 2.4 (Komlós [81]) *For every graph F and every $\gamma > 0$ there exists an integer $n_0 = n_0(\gamma, F)$ such that every graph G of order $n \geq n_0$ and minimum degree at least $(1 - 1/\chi_{cr}(F))n$ contains an F-packing which covers all but at most γn vertices of G.*

By making V_1 slightly smaller in the previous example, it is easy to see that the minimum degree bound in Theorem 2.4 is also best possible. Confirming a conjecture of Komlós [81], Shokoufandeh and Zhao [121, 122] subsequently proved that the number of uncovered vertices can be reduced to a constant depending only on F.

We [96] proved that for any graph F, either its critical chromatic number or its chromatic number is the relevant parameter which governs the existence of perfect packings in graphs of large minimum degree. The classification depends on a parameter which we call the *highest common factor* of F.

This is defined as follows for non-bipartite graphs F. Given an optimal colouring c of F, let $x_1 \leq x_2 \leq \cdots \leq x_\ell$ denote the sizes of the colour classes of c. Put $\mathcal{D}(c) := \{x_{i+1} - x_i \mid i = 1, \ldots, \ell - 1\}$. Let $\mathcal{D}(F)$ denote the union of all the sets $\mathcal{D}(c)$ taken over all optimal colourings c. We denote by $\mathrm{hcf}(F)$ the highest common factor of all integers in $\mathcal{D}(F)$. If $\mathcal{D}(F) = \{0\}$ we set $\mathrm{hcf}(F) := \infty$. Note that if all the optimal colourings of F have the property that all colour classes have equal size, then $\mathcal{D}(F) = \{0\}$ and so $\mathrm{hcf}(F) \neq 1$ in this case. In particular, if $\chi_{cr}(F) = \chi(F)$, then $\mathrm{hcf}(F) \neq 1$. So, for example, odd cycles of length at least 5 have $\mathrm{hcf} = 1$ whereas complete graphs have $\mathrm{hcf} \neq 1$.

The definition can be extended to bipartite graphs F. For connected bipartite graphs, we always have $\mathrm{hcf}(F) \neq 1$, but for disconnected bipartite graphs the definition also takes into account the relative sizes of the components of F (see [96]).

We proved that, in Theorem 2.2, one can replace the chromatic number by the critical chromatic number if $\mathrm{hcf}(F) = 1$. (A much simpler proof of a weaker result can be found in [93].)

Theorem 2.5 (Kühn and Osthus [96]) *Suppose that F is a graph with $\mathrm{hcf}(F) = 1$. Then there exists a constant $C = C(F)$ such that every graph G whose order n is divisible by $|F|$ and whose minimum degree is at least $(1 - 1/\chi_{cr}(F))n + C$ contains a perfect F-packing.*

Note that Proposition 2.3 shows that the result is best possible up to the value of the constant C. A simple modification of the examples in [8, 81] shows that there are graphs F for which the constant C cannot be omitted entirely. Moreover, it turns out that Theorem 2.2 is already best possible up to the value of the constant C if $\mathrm{hcf}(F) \neq 1$. To see this, for simplicity assume that $k := \chi(F) \geq 3$ and $n = k\ell|F|$ for some $\ell \in \mathbb{N}$ and let G be a complete k-partite graph with vertex classes $V_1, \ldots, V_k$, where $|V_1| := \ell|F| - 1$, $|V_2| := \ell|F| + 1$ and $|V_i| = \ell|F|$ for $i \geq 3$. Consider any F-packing $F_1, \ldots, F_t$ in G. Let G_i be the graph obtained from G by removing $F_1, \ldots, F_i$. So $G = G_0$. If $t := \mathrm{hcf}(F) \neq 1$, then the vertex classes V_i^1 of G_1 still have property that $|V_1^1| - |V_k^1| \not\equiv 0$ modulo t. More generally, this property is preserved for all G_i, so the original F-packing cannot cover all the vertices in $V_1 \cup V_k$.

One can now combine Theorems 2.2 and 2.5 (and the corresponding lower bounds which are discussed in detail in [96]) to obtain a complete answer to the question of which minimum degree forces a perfect F-packing (up to an additive constant). For this, let

$$\chi^*(F) := \begin{cases} \chi_{cr}(F) & \text{if } \mathrm{hcf}(F) = 1; \\ \chi(F) & \text{otherwise.} \end{cases}$$

Also let $\delta(F, n)$ denote the smallest integer k such that every graph G whose order n is divisible by $|F|$ and with $\delta(G) \geq k$ contains a perfect F-packing.

Theorem 2.6 (Kühn and Osthus [96]) *For every graph F there exists a constant $C = C(F)$ such that*

$$\left(1 - \frac{1}{\chi^*(F)}\right) n - 1 \leq \delta(F, n) \leq \left(1 - \frac{1}{\chi^*(F)}\right) n + C.$$

The constant C appearing in Theorems 2.5 and 2.6 is rather large since it is related to the number of partition classes (clusters) obtained by the Regularity Lemma. It would be interesting to know whether one can take e.g. $C = |F|$ (this holds for large n in Theorem 2.2). Another open problem is to characterize all those graphs F for which $\delta(F, n) = \lceil (1 - 1/\chi^*(F))n \rceil$. This is known to be the case for complete graphs by Theorem 2.1 and all graphs with at most 4 vertices (see Kawarabayashi [71] for a proof of the case when F is a K_4 minus an edge and a discussion of the other cases). If n is large, this is also known to hold for cycles (this follows from Theorem 5.3 below) and for the case when F is a complete graph minus an edge [29] (the latter was conjectured in [71]).

2.2 Ore-type degree conditions

Recently, a simpler proof (based on an inductive argument) of the Hajnal–Szemerédi theorem was found by Kierstead and Kostochka [78]. Using similar methods, they subsequently strengthened this to an Ore-type condition [79]:

Theorem 2.7 (Kierstead and Kostochka [79]) *Let G be a graph whose order n is divisible by r. If $d(x) + d(y) \geq 2(1 - 1/r)n - 1$ for all pairs $x \neq y$ of nonadjacent vertices, then G has a perfect K_r-packing.*

Equivalently, if a graph G whose order is divisible by k satisfies $d(x) + d(y) \leq 2k - 1$ for every edge xy, then G has an equitable k-colouring. (So $k := n/r$.) Recently, together with Treglown [99], we proved an Ore-type analogue of Theorem 2.6 (but with a linear error term εn instead of the additive constant C). The result in this case turns out to be genuinely different: again, there are some graphs F for which the degree condition depends on $\chi(F)$ and some for which it depends on $\chi_{cr}(F)$. However, there are also graphs F for which it depends on a parameter which lies strictly between $\chi_{cr}(F)$ and $\chi(F)$. This parameter in turn depends on how many additional colours are necessary to extend colourings of neighbourhoods of certain vertices of F to a colouring of F. It is an open question whether the linear error term in [99] can be reduced to a constant one.

2.3 r-partite versions

Also, it is natural to consider r-partite versions of the Hajnal-Szemerédi theorem. For this, given an r-partite graph G, let $\delta'(G)$ denote the minimum over all vertex classes W of G and all vertices $x \notin W$ of the number of neighbours of x in W. The obvious question is what value of $\delta'(G)$ ensures that G has a perfect K_r-packing. The following (surprisingly difficult) conjecture is implicit in [101]. Fischer [39] originally made a stronger conjecture which did not include the 'exceptional' graph $\Gamma_{r,n}$ defined below.

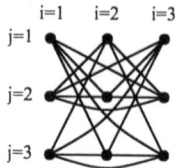

Figure 1: The graph $\Gamma_3 = \Gamma_{3,1}$ in Conjecture 2.8

Conjecture 2.8 *Suppose that $r \geq 2$ and that G is an r-partite graph with vertex classes of size n. If $\delta'(G) \geq (1 - 1/r)n$, then G has a perfect K_r-packing unless both r and n are odd and $G = \Gamma_{r,n}$.*

To define the graph $\Gamma_{r,n}$, we first construct a graph Γ_r: its vertices are labelled g_{ij} with $1 \leq i,j \leq r$. We have an edge between g_{ij} and $g_{i'j'}$ if $i \neq i'$, $j \neq j'$ and $j \leq r-2$ or $j' \leq r-2$. We also have an edge if $i \neq i'$ and we have either $j = j' = r-1$ or $j = j' = r$ (see Fig. 1). $\Gamma_{r,n}$ is then obtained from Γ_r by replacing each vertex with an independent set of size n/r and replacing each edge with a complete bipartite graph.

To see that $\Gamma_{r,n}$ has no perfect K_r-packing when both r and n are odd, let W_ℓ denote the set of vertices of $\Gamma_{r,n}$ which correspond to a vertex of Γ_r with $j = \ell$. Note that every copy of K_r which covers a vertex in $W_1 \cup \cdots \cup W_{r-2}$ has to contain at least 2 vertices in W_{r-1} or at least 2 vertices in W_r. So in order to cover all vertices in $W_1 \cup \cdots \cup W_{r-2}$ we can only use copies of K_r which contain exactly 2 vertices in W_{r-1} or exactly 2 vertices in W_r. But since $|W_{r-1}| = |W_r| = n$ is odd this means that it is impossible to cover all vertices of $\Gamma_{r,n}$ with vertex-disjoint copies of K_r. (Note that the argument uses only that n is odd, but we cannot have that n is odd and r is even.)

A much simpler example which works for all r and n but which gives a weaker bound when r and n are odd is obtained as follows: choose a set A which has less than $(1 - 1/r)n$ vertices in each vertex class and include all edges which have at least one endpoint in A. For large n, the case $r = 3$ of Conjecture 2.8 was solved by Magyar and Martin [101] and the case $r = 4$ by Martin and Szemerédi [102], both using the Regularity Lemma (the case $r = 2$ is elementary). Johansson [67] had earlier proved an approximate version of the case $r = 3$. Csaba and Mydlarz [33] proved a result which implies that Conjecture 2.8 holds approximately when r is large (and n large compared to r). Generalizations to packings of arbitrary graphs were considered in [63, 103, 132]. A variant of the problem (where one considers usual minimum degree $\delta(G)$) was considered by Johansson, Johansson and Markström [68]. They solved the case $r = 3$ and gave bounds for the case $r > 3$. This problem is related to bounding the so-called 'strong chromatic number'.

2.4 Hypergraphs

(Perfect) F-packings have also been investigated for the case when F is a uniform hypergraph. Unsurprisingly, the hypergraph problem turns out to be much more difficult than the graph problem. There are two natural notions of (minimum) degree of the 'dense' hypergraph G. Firstly, one can consider the vertex degree. Secondly, given an r-uniform hypergraph G and an $(r-1)$-tuple W of vertices in G, the degree of W is defined to be the number of hyperedges which contain W. This notion of degree is called *collective degree* or *co-degree*. In contrast to the graph case, even the minimum collective degree which ensures a perfect matching (i.e. when F consists of a single edge) is not easy to determine. Rödl, Ruciński and Szemerédi [118] gave a precise solution to this problem; the answer turns out to be close to $n/2$. This improved upon bounds of [94, 115]. An r-partite version (which is best possible for infinitely many values of n) was proved by Aharoni, Georgakopoulos and Sprüssel [3]. The minimum vertex degree which forces the existence of a perfect matching is unknown. It is natural to make the following conjecture (a related r-partite version is conjectured in [3]).

Conjecture 2.9 *For all integers r and all $\varepsilon > 0$ there is an integer $n_0 = n_0(r, \varepsilon)$ so that the following holds for all $n \geq n_0$ which are divisible by r: if G is an r-uniform hypergraph on n vertices whose minimum vertex degree is at least*

$$(1 - (1 - 1/r)^{r-1} + \varepsilon)\binom{n}{r-1},$$

then G has a perfect matching.

The following construction gives a corresponding lower bound: let V be a set of n vertices and let $A \subseteq V$ be a set of less than n/r vertices and include as hyperedges all r-tuples with at least one vertex in A. The case $r = 3$ of the conjecture was proved recently by Han, Person and Schacht [56].

A hypergraph analogue of Theorem 2.6 currently seems out of reach. So far, the only hypergraph F (apart from the single edge) for which the approximate minimum collective degree which forces a perfect F-packing has been determined is the 3-uniform hypergraph with 4 vertices and 2 edges [95]. Pikhurko [113] gave bounds on the minimum collective degree which forces the complete 3-uniform hypergraph on 4 vertices. In the same paper, he also shows that if $\ell \geq r/2$ and G is an r-uniform hypergraph where every ℓ-tuple of vertices is contained in at least $(1/2 + o(1))\binom{n}{r-\ell}$ hyperedges, then G has a perfect matching, which is best possible up to the $o(1)$-term. This result is rather surprising in view of the fact that Conjecture 2.9 (which corresponds to the case when $\ell = 1$) has a rather different form. Further results on this question are also proved in [56].

3 Trees

One of the earliest applications of the Blow-up Lemma was the solution by Komlós, Sárközy and Szemerédi [82] of a conjecture of Bollobás on the existence of given bounded degree spanning trees. The authors later relaxed the condition of bounded degree to obtain the following result.

Theorem 3.1 (Komlós, Sárközy and Szemerédi [87]) *For any $\gamma > 0$ there exist constants $c > 0$ and n_0 with the following properties. If $n \geq n_0$, T is a tree of order n with $\Delta(T) \leq cn/\log n$, and G is a graph of order n with $\delta(G) \geq (1/2 + \gamma)n$, then T is a subgraph of G.*

The condition $\Delta(T) \leq cn/\log n$ is best possible up to the value of c. (The example which is given in [87] to show this, is a random graph G with edge probability 0.9 and a tree of depth 2 whose root has degree close to $\log n$.)

It is an easy exercise to see that every graph of minimum degree at least k contains any tree with k edges. The following classical conjecture would imply that we can replace the minimum degree condition by one on the average degree.

Conjecture 3.2 (Erdős and Sós [37]) *Every graph of average degree greater than $k-1$ contains any tree with k edges.*

This is trivially true for stars. (On the other hand, stars also show that the bound is best possible in general.) It is also trivial if one assumes an extra factor of 2 in the average degree. It has been proved for some special classes of trees, most notably those of diameter at most 4 [105]. The conjecture is also true for 'locally sparse' graphs – see Sudakov and Vondrak [124] for a discussion of this.

The following result proves (for large n) a related conjecture of Loebl. An approximate version was proved earlier by Ajtai, Komlós and Szemerédi [5].

Theorem 3.3 (Zhao [131]) *There is an integer n_0 so that every graph G on $n \geq n_0$ vertices which has at least $n/2$ vertices of degree at least $n/2$ contains all trees with at most $n/2$ edges.*

This would be generalized by the following conjecture.

Conjecture 3.4 (Komlós and Sós) *Every graph G on n vertices which has at least $n/2$ vertices of degree at least k contains all trees with k edges.*

Again, the conjecture is trivially true (and best possible) for stars. Piguet and Stein [111] proved an approximate version for the case when k is linear in n and n is large. Cooley [26], as well as Hladký and Piguet [62], proved an exact version for this case. All of these proofs are based on the Regularity Lemma. As with Conjecture 3.2, there are several results on special cases which are not based on the Regularity Lemma. For instance, Piguet and Stein proved it for trees of diameter at most 5 [112].

4 Hamilton cycles

4.1 Classical results for graphs and digraphs

As mentioned in the introduction, the decision problem of whether a graph has a Hamilton cycle is NP-complete, so it makes sense to ask for degree conditions which ensure that a graph has a Hamilton cycle. One such result is the classical theorem of Dirac.

Theorem 4.1 (Dirac [36]) *Every graph on $n \geq 3$ vertices with minimum degree at least $n/2$ contains a Hamilton cycle.*

For an analogue in directed graphs it is natural to consider the *minimum semidegree* $\delta^0(G)$ of a digraph G, which is the minimum of its minimum outdegree $\delta^+(G)$

and its minimum indegree $\delta^-(G)$. (Here a directed graph may have two edges between a pair of vertices, but in this case their directions must be opposite.) The corresponding result is a theorem of Ghouila-Houri [45].

Theorem 4.2 (Ghouila-Houri [45]) *Every digraph on n vertices with minimum semidegree at least $n/2$ contains a Hamilton cycle.*

In fact, Ghouila-Houri proved the stronger result that every strongly connected digraph of order n, where every vertex has total degree at least n, has a Hamilton cycle. (When referring to paths and cycles in directed graphs we always mean that these are directed, without mentioning this explicitly.) All of the above degree conditions are best possible. Theorems 4.1 and 4.2 were generalized to a degree condition on pairs of vertices for graphs as well as digraphs:

Theorem 4.3 (Ore [110]) *Suppose that G is a graph with $n \geq 3$ vertices such that every pair $x \neq y$ of nonadjacent vertices satisfies $d(x) + d(y) \geq n$. Then G has a Hamilton cycle.*

Theorem 4.4 (Woodall [129]) *Let G be a strongly connected digraph on $n \geq 2$ vertices. If $d^+(x) + d^-(y) \geq n$ for every pair $x \neq y$ of vertices for which there is no edge from x to y, then G has a Hamilton cycle.*

There are many generalizations of these results. The survey [46] gives an overview for undirected graphs and the monograph [10] gives a discussion of directed versions. Below, we describe some recent progress on degree conditions for Hamilton cycles, much of which is based on the Regularity Lemma.

4.2 Hamilton cycles in oriented graphs

Thomassen [127] raised the natural question of determining the minimum semidegree that forces a Hamilton cycle in an *oriented graph* (i.e. in a directed graph that can be obtained from a simple undirected graph by orienting its edges). Thomassen initially believed that the correct minimum semidegree bound should be $n/3$ (this bound is obtained by considering a 'blow-up' of an oriented triangle). However, Häggkvist [52] later gave the following construction which gives a lower bound of $\lceil (3n-4)/8 \rceil - 1$ (see Fig. 2). For n of the form $n = 4m + 3$ where m is odd, we construct G on n vertices as follows. Partition the vertices into 4 parts A, B, C, D, with $|A| = |C| = m$, $|B| = m+1$ and $|D| = m+2$. Each of A and C spans a regular tournament; B and D are joined by a bipartite tournament (i.e. an orientation of the complete bipartite graph) which is as regular as possible. We also add all edges from A to B, from B to C, from C to D and from D to A. Since every path which joins two vertices in D has to pass through B, it follows that every cycle contains at least as many vertices from B as it contains from D. As $|D| > |B|$ this means that one cannot cover all the vertices of G by disjoint cycles. This construction can be extended to arbitrary n (see [74]). The following result exactly matches this bound and improves earlier ones of several authors, e.g. [52, 54, 76].

Theorem 4.5 (Keevash, Kühn and Osthus [74]) *There exists an integer n_0 so that any oriented graph G on $n \geq n_0$ vertices with minimum semidegree $\delta^0(G) \geq \frac{3n-4}{8}$ contains a Hamilton cycle.*

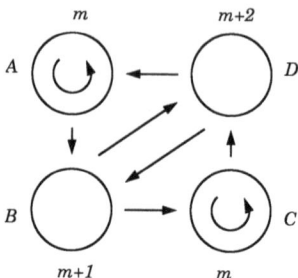

Figure 2: An extremal example for Theorem 4.5

The proof of this result is based on some ideas in [76]. Häggkvist [52] also made the following conjecture which is closely related to Theorem 4.5. Given an oriented graph G, let $\delta(G)$ denote the minimum degree of G (i.e. the minimum number of edges incident to a vertex) and set $\delta^*(G) := \delta(G) + \delta^+(G) + \delta^-(G)$.

Conjecture 4.6 (Häggkvist [52]) *Every oriented graph G on n vertices with $\delta^*(G) > (3n-3)/2$ contains a Hamilton cycle.*

(Note that this conjecture does not quite imply Theorem 4.5 as it results in a marginally greater minimum semidegree condition.) In [76], Conjecture 4.6 was verified approximately, i.e. if $\delta^*(G) \geq (3/2 + o(1))n$, then G has a Hamilton cycle (note this implies an approximate version of Theorem 4.5). The same methods also yield an approximate version of Theorem 4.4 for oriented graphs.

Theorem 4.7 (Kelly, Kühn and Osthus [76]) *For every $\alpha > 0$ there exists an integer $n_0 = n_0(\alpha)$ such that every oriented graph G of order $n \geq n_0$ with $d^+(x) + d^-(y) \geq (3/4 + \alpha)n$ whenever G does not contain an edge from x to y contains a Hamilton cycle.*

The above construction of Häggkvist shows that the bound is best possible up to the term αn. It would be interesting to obtain an exact version of this result.

Note that Theorem 4.5 implies that every sufficiently large regular tournament on n vertices contains at least $n/8$ edge-disjoint Hamilton cycles. (To verify this, note that in a regular tournament, all in- and outdegrees are equal to $(n-1)/2$. We can then greedily remove Hamilton cycles as long as the degrees satisfy the condition in Theorem 4.5.) It is the best bound so far towards the following conjecture of Kelly (see e.g. [10]).

Conjecture 4.8 (Kelly) *Every regular tournament on n vertices can be partitioned into $(n-1)/2$ edge-disjoint Hamilton cycles.*

A result of Frieze and Krivelevich [43] states that every dense ε-regular digraph contains a collection of edge-disjoint Hamilton cycles which covers almost all of its edges. This implies that the same holds for almost every tournament. Together with a lower bound by McKay [104] on the number of regular tournaments, it is easy to see that the above result in [43] also implies that almost every regular tournament contains a collection of edge-disjoint Hamilton cycles which covers almost all of its edges. Thomassen made the following conjecture which replaces the assumption of regularity by high connectivity.

Conjecture 4.9 (Thomassen [128]) *For every $k \geq 2$ there is an integer $f(k)$ so that every strongly $f(k)$-connected tournament has k edge-disjoint Hamilton cycles.*

The following conjecture of Jackson is also closely related to Theorem 4.5 – it would imply a much better degree condition for regular oriented graphs.

Conjecture 4.10 (Jackson [66]) *For $d > 2$, every d-regular oriented graph G on $n \leq 4d + 1$ vertices is Hamiltonian.*

The disjoint union of two regular tournaments on $n/2$ vertices shows that this would be best possible. An undirected analogue of Conjecture 4.10 was proved by Jackson [65]. It is easy to see that every tournament on n vertices with minimum semidegree at least $n/4$ has a Hamilton cycle. In fact, for tournaments T of large order n with minimum semidegree at least $n/4 + \varepsilon n$, Bollobás and Häggkvist [18] proved the stronger result that (for fixed k) T even contains the kth power of a Hamilton cycle. It would be interesting to find corresponding degree conditions which ensure this for arbitrary digraphs and for oriented graphs.

4.3 Degree sequences forcing Hamilton cycles in directed graphs

For undirected graphs, Dirac's theorem is generalized by Chvátal's theorem [22] that characterizes all those degree sequences which ensure the existence of a Hamilton cycle in a graph: suppose that the degrees of the graph are $d_1 \leq \cdots \leq d_n$. If $n \geq 3$ and $d_i \geq i+1$ or $d_{n-i} \geq n-i$ for all $i < n/2$ then G is Hamiltonian. This condition on the degree sequence is best possible in the sense that for any degree sequence violating this condition there is a corresponding graph with no Hamilton cycle. Nash-Williams [109] raised the question of a digraph analogue of Chvátal's theorem quite soon after the latter was proved: for a digraph G it is natural to consider both its outdegree sequence $d_1^+, \ldots, d_n^+$ and its indegree sequence $d_1^-, \ldots, d_n^-$. Throughout, we take the convention that $d_1^+ \leq \cdots \leq d_n^+$ and $d_1^- \leq \cdots \leq d_n^-$ without mentioning this explicitly. Note that the terms d_i^+ and d_i^- do not necessarily correspond to the degree of the same vertex of G.

Conjecture 4.11 (Nash-Williams [109]) *Suppose that G is a strongly connected digraph on $n \geq 3$ vertices such that for all $i < n/2$*

(i) $d_i^+ \geq i+1$ or $d_{n-i}^- \geq n-i$,

(ii) $d_i^- \geq i+1$ or $d_{n-i}^+ \geq n-i$.

Then G contains a Hamilton cycle.

It is even an open problem whether the conditions imply the existence of a cycle through any pair of given vertices (see [12]). It is easy to see that one cannot omit the condition that G is strongly connected. The following example (which is a straightforward generalization of the corresponding undirected example) shows that the degree condition in Conjecture 4.11 would be best possible in the sense that for all $n \geq 3$ and all $k < n/2$ there is a non-Hamiltonian strongly connected digraph G on n vertices which satisfies the degree conditions except that $d_k^+, d_k^- \geq k+1$ are replaced by $d_k^+, d_k^- \geq k$ in the kth pair of conditions. To see this, take an independent set I of size $k < n/2$ and a complete digraph K of order $n-k$. Pick a set X of k vertices of K and add all possible edges (in both directions) between I and X. The digraph G thus obtained is strongly connected, not Hamiltonian and

$$\underbrace{k,\ldots,k}_{k \text{ times}}, \underbrace{n-1-k,\ldots,n-1-k}_{n-2k \text{ times}}, \underbrace{n-1,\ldots,n-1}_{k \text{ times}}$$

is both the out- and indegree sequence of G. In contrast to the undirected case there exist examples with a similar degree sequence to the above but whose structure is quite different (see [98]). In [98], the following approximate version of Conjecture 4.11 for large digraphs was proved.

Theorem 4.12 (Kühn, Osthus and Treglown [98]) *For every $\alpha > 0$ there exists an integer $n_0 = n_0(\alpha)$ such that the following holds. Suppose G is a digraph on $n \geq n_0$ vertices such that for all $i < n/2$*

- $d_i^+ \geq i + \alpha n$ or $d_{n-i-\alpha n}^- \geq n-i$,
- $d_i^- \geq i + \alpha n$ or $d_{n-i-\alpha n}^+ \geq n-i$.

Then G contains a Hamilton cycle.

Theorem 4.12 was derived from a result in [74] on the existence of a Hamilton cycle in an oriented graph satisfying a 'robust' expansion property.

The following weakening of Conjecture 4.11 was posed earlier by Nash-Williams [108]. It would yield a digraph analogue of Pósa's theorem which states that a graph G on $n \geq 3$ vertices has a Hamilton cycle if its degree sequence $d_1 \leq \cdots \leq d_n$ satisfies $d_i \geq i+1$ for all $i < (n-1)/2$ and if additionally $d_{\lceil n/2 \rceil} \geq \lceil n/2 \rceil$ when n is odd [114]. Note that Pósa's theorem is much stronger than Dirac's theorem but is a special case of Chvátal's theorem.

Conjecture 4.13 (Nash-Williams [108]) *Let G be a digraph on $n \geq 3$ vertices such that $d_i^+, d_i^- \geq i+1$ for all $i < (n-1)/2$ and such that additionally $d_{\lceil n/2 \rceil}^+, d_{\lceil n/2 \rceil}^- \geq \lceil n/2 \rceil$ when n is odd. Then G contains a Hamilton cycle.*

The previous example shows the degree condition would be best possible in the same sense as described there. The assumption of strong connectivity is not necessary in Conjecture 4.13, as it follows from the degree conditions. Theorem 4.12 immediately implies an approximate version of Conjecture 4.13.

It turns out that the conditions of Theorem 4.12 even guarantee the digraph G to be *pancyclic*, i.e. G contains a cycle of length t for all $t = 2,\ldots,n$. Thomassen [126] as well as Häggkvist and Thomassen [53] gave degree conditions which imply that every digraph with minimum semidegree $> n/2$ is pancyclic. The latter bound can also be deduced directly from Theorem 4.2. The complete bipartite digraph whose vertex class sizes are as equal as possible shows that the bound is best possible. For oriented graphs the minimum semidegree threshold which guarantees pancyclicity turns out to be $(3n-4)/8$ (see [77]).

4.4 Powers of Hamilton cycles in graphs

The following result is a common extension (for large n) of Dirac's theorem and the Hajnal-Szemerédi theorem. It was originally conjectured (for all n) by Seymour.

Theorem 4.14 (Komlós, Sárközy and Szemerédi [85]) *For every $k \geq 1$ there is an integer n_0 so that every graph G on $n \geq n_0$ vertices and with $\delta(G) \geq \frac{k}{k+1}n$ contains the kth power of a Hamilton cycle.*

Complete $(k+1)$-partite graphs whose vertex classes have almost (but not exactly) equal size show that the minimum degree bound is best possible. Previous to this, a large number of partial results had been proved (see e.g. [100] for a history of the problem). Very recently, Levitt, Sarközy and Szemerédi [100] gave a proof of the case $k = 2$ which avoids the use of the Regularity lemma, resulting in a much better bound on n_0. Their proof is based on a technique introduced by Rödl, Ruciński and Szemerédi [117] for hypergraphs. The idea of this method (as applied in [100]) is first to find an 'absorbing' path P^2: roughly, P^2 is the second power of a path P which, given any vertex x, has the property that x can be inserted into P so that $P \cup x$ still induces the second power of a path. The proof of the existence of P^2 is heavily based on probabilistic arguments. Then one finds the second power Q^2 of a path which is almost spanning in $G - P^2$. One can achieve this by repeated applications of the Erdős-Stone theorem. One then connects up Q^2 and P^2 into the second power of a cycle and finally uses the absorbing property of P^2 to incorporate the vertices left over so far.

4.5 Hamilton cycles in hypergraphs

It is natural to ask whether one can generalize Dirac's theorem to uniform hypergraphs. There are several possible notions of a hypergraph cycle. One generalization of the definition of a cycle in a graph is the following one. An r-uniform hypergraph C is a *cycle of order n* if there a exists a cyclic ordering $v_1, \ldots, v_n$ of its n vertices such that every consecutive pair $v_i v_{i+1}$ lies in a hyperedge of C and such that every hyperedge of C consists of consecutive vertices. Thus the cyclic ordering of the vertices of C induces a cyclic ordering of its hyperedges. A cycle is *tight* if every r consecutive vertices form a hyperedge. A cycle of order n is *loose* if all pairs of consecutive edges (except possibly one pair) have exactly one vertex in common. (So every tight cycle contains a spanning loose cycle but a cycle might not necessarily contain a spanning loose cycle.) There is also the even more general notion of a *Berge-cycle*, which consists of a sequence of vertices where each pair of consecutive vertices is contained in a common hyperedge.

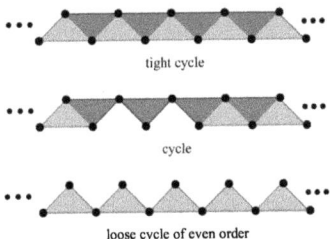

tight cycle

cycle

loose cycle of even order

A *Hamilton cycle* of a uniform hypergraph G is a subhypergraph of G which is a cycle containing all its vertices. Theorem 4.15 gives an analogue of Dirac's theorem for tight hypergraph cycles, while Theorem 4.16 gives an analogue for 3-uniform (loose) cycles.

Theorem 4.15 (Rödl, Ruciński and Szemerédi [117]) *For all $r \in \mathbb{N}$ and $\alpha > 0$ there is an integer $n_0 = n_0(r, \alpha)$ such that every r-uniform hypergraph G with $n \geq n_0$ vertices and minimum degree at least $n/2 + \alpha n$ contains a tight Hamilton cycle.*

Theorem 4.16 (Han and Schacht [57]; Keevash, Kühn, Mycroft and Osthus [73]) *For all $r \in \mathbb{N}$ and $\alpha > 0$ there is an integer $n_0 = n_0(\alpha)$ such that every r-uniform hypergraph G with $n \geq n_0$ vertices and minimum degree at least $n/(2r-2) + \alpha n$ contains a loose Hamilton cycle.*

Both results are best possible up to the error term αn. In fact, if the minimum degree is less than $\lceil n/(2r-2) \rceil$, then we cannot even guarantee any Hamilton cycle in an r-uniform hypergraph. The case $r = 3$ of Theorems 4.15 and 4.16 was proved earlier in [116] and [95] respectively. The result in [57] also covers the notion of an r-uniform ℓ-cycle for $\ell < r/2$ (here we ask for consecutive edges to intersect in precisely ℓ vertices). Hamiltonian Berge-cycles were considered by Bermond et al. [11].

5 Bounded degree spanning subgraphs

Bollobás and Eldridge [17] as well as Catlin [21] made the following very general conjecture on embedding graphs. If true, this conjecture would be a far-reaching generalization of the Hajnal-Szemerédi theorem (Theorem 2.1).

Conjecture 5.1 (Bollobás and Eldridge [17], Catlin [21]) *If G is a graph on n vertices with $\delta(G) \geq \frac{\Delta n - 1}{\Delta + 1}$, then G contains any graph H on n vertices with maximum degree at most Δ.*

The conjecture has been proved for graphs H of maximum degree at most 2 [4, 7] and for large graphs of maximum degree at most 3 [34]. Recently, Csaba [31] proved it for bipartite graphs H of arbitrary maximum degree Δ, provided the order of H is sufficiently large compared to Δ. In many applications of the Blow-up Lemma, the graph H is embedded into G by splitting H up into several suitable parts and applying the Blow-up Lemma to each of these parts (see e.g. the example in Section 7). It is not clear how to achieve this for H as in Conjecture 5.1, as H may be an 'expander'. So the proofs in [31, 34] rely on a variant of the Blow-up Lemma which is suitable for embedding such 'expander graphs'. Also, Kaul, Kostochka and Yu [70] showed (without using the Regularity Lemma) that the conjecture holds if we increase the minimum degree condition to $\frac{\Delta n + 2n/5 - 1}{\Delta + 1}$.

Theorem 2.2 suggests that one might replace Δ in Conjecture 5.1 with $\chi(H) - 1$, resulting in a smaller minimum degree bound for some graphs H. This is far from being true in general (e.g. let H be a 3-regular bipartite expander and let G be the union of two cliques which have equal size and are almost disjoint). However, Bollobás and Komlós conjectured that this does turn out to be true if we restrict our attention to a certain class of 'non-expanding' graphs. This conjecture was recently confirmed in [15]. The bipartite case was proved earlier by Abbasi [1].

Theorem 5.2 (Böttcher, Schacht and Taraz [15]) *For every $\gamma > 0$ and all integers $r \geq 2$ and Δ, there exist $\beta > 0$ and n_0 with the following property. Every graph G of order $n \geq n_0$ and minimum degree at least $(1 - 1/r + \gamma)n$ contains every r-chromatic graph H of order n, maximum degree at most Δ and bandwidth at most βn as a subgraph.*

Here the *bandwidth* of a graph H is the smallest integer b for which there exists an enumeration $v_1, \ldots, v_{|H|}$ of the vertices of H such that every edge $v_i v_j$ of H satisfies $|i - j| \leq b$. Note that kth powers of cycles have bandwidth $2k$, so Theorem 5.2 implies an approximate version of Theorem 4.14. (Actually, this is only the case if n is a multiple of $k+1$, as otherwise the kth power of a Hamilton cycle fails to be $(k+1)$-colourable. But [15] contains a more general result which allows for a small number of vertices of colour $k+2$.) A further class of graphs having small bandwidth and bounded degree are planar graphs with bounded degree [13]. (See [92, 97] for further results on embedding planar graphs in graphs of large minimum degree.) Note that the discussion in Section 2 implies that the minimum degree bound in Theorem 5.2 is approximately best possible for certain graphs H but not for all graphs. Abbasi [2] showed that there are graphs H for which the linear error term γn in Theorem 5.2 is necessary. One might think that one could reduce the error term to a constant for graphs of bounded bandwidth. However, this turns out to be incorrect. (We grateful to Peter Allen for pointing this out to us.)

Alternatively, one can try to replace the bandwidth assumption in Theorem 5.2 with a less restrictive parameter. For instance, Csaba [32] gave a minimum degree condition on G which guarantees a copy of a 'well-separated' graph H in G. Here a graph with n vertices is α-*separable* if there is a set S of vertices of size at most αn so that all components of $H - S$ have size at most αn. It is easy to see that every graph with n vertices and bandwidth at most βn is $\sqrt{\beta}$-separable. (Moreover large trees are α-separable for $\alpha \to 0$ but need not have small bandwidth, so considering separability is less restrictive than bandwidth.)

Here is another common generalization of Dirac's theorem and the triangle case of Theorem 2.1 (i.e. the Corrádi-Hajnal theorem). It proves a conjecture by El-Zahar (actually, El-Zahar made the conjecture for all values of n, this is still open).

Theorem 5.3 (Abbasi [1]) *There exists an integer n_0 so that the following holds. Suppose that G is a graph on $n \geq n_0$ vertices and $n_1, \ldots, n_k \geq 3$ are such that*

$$\sum_{i=1}^{k} n_i = n \quad \text{and} \quad \delta(G) \geq \sum_{i=1}^{k} \lceil n_i/2 \rceil.$$

Then G has k vertex-disjoint cycles whose lengths are $n_1, \ldots, n_k$.

Note that $\sum_{i=1}^{k} \lceil n_i/2 \rceil = \sum_{i=1}^{k}(1 - 1/\chi_{cr}(C_i))n_i$, where C_i denotes a cycle of length n_i. This suggests the following more general question (which was raised by Komlós [81]): Given $t \in \mathbb{N}$, does there exists an $n_0 = n_0(t)$ such that whenever $H_1, \ldots, H_k$ are graphs which each have at most t vertices and which together have $n \geq n_0$

vertices and whenever G is a graph on n vertices with minimum degree at least $\sum_i (1-1/\chi_{cr}(H_i))|H_i|$, then there is a set of vertex-disjoint copies of $H_1, \ldots, H_k$ in G? In this form, the question has a negative answer by (the lower bound in) Theorem 2.6, but it would be interesting to find a common generalization of Theorems 2.6 and 5.3.

It is also natural to ask corresponding questions for oriented and directed graphs. As in the case of Hamilton cycles, the questions appear much harder than in the undirected case and again much less is known. Keevash and Sudakov [75] recently obtained the following result which can be viewed as an oriented version of the $\Delta = 2$ case of Conjecture 5.1.

Theorem 5.4 (Keevash and Sudakov [75]) *There exist constants c, C and an integer n_0 so that whenever G is an oriented graph on $n \geq n_0$ vertices with minimum semidegree at least $(1/2 - c)n$ and whenever $n_1, \ldots, n_t$ are so that $\sum_{i=1}^{t} n_i \leq n - C$, then G contains disjoint cycles of length $n_1, \ldots, n_t$.*

In the case of triangles (i.e. when all the $n_i = 3$), they show that one can choose $C = 3$ (one cannot take $C = 0$). Their paper ([75]) also contains a discussion of related open questions for tournaments and directed graphs. Similar questions were also raised earlier by Song [123]. For instance, given t, what is the smallest integer $f(t)$ so that all but a finite number of $f(t)$-connected tournaments T satisfy the following: Let n be the number of vertices of T and let $\sum_{i=1}^{t} n_i = n$. Then T contains disjoint cycles of length $n_1, \ldots, n_t$.

6 Ramsey Theory

The Regularity Lemma can often be used to show that the Ramsey numbers of sparse graphs H are small. (The *Ramsey number* $R(H)$ of H is the smallest $N \in \mathbb{N}$ such that for every 2-colouring of the complete graph on N vertices one can find a monochromatic copy of H.) In fact, the first result which demonstrated the use of the Regularity Lemma in extremal graph theory was the following result of Chvátal, Rödl, Szemerédi and Trotter [23], which states that graphs of bounded degree have linear Ramsey numbers:

Theorem 6.1 (Chvátal, Rödl, Szemerédi and Trotter [23]) *For all $\Delta \in \mathbb{N}$ there is a constant $C = C(\Delta)$ so that every graph H with maximum degree $\Delta(H) \leq \Delta$ and n vertices satisfies $R(H) \leq Cn$.*

The constant C arising from the original proof (based on the Regularity Lemma) is quite large. The bound was improved in a series of papers. Recently, Fox and Sudakov [40] showed that $R(H) \leq 2^{4\chi(H)\Delta} n$ (the bipartite case was also proved independently by Conlon [24]). For bipartite graphs, a construction from [49] shows that this bound is best possible apart from the value of the absolute constant $4 \cdot 2$ appearing in the exponent.

Theorem 6.1 was recently generalized to hypergraphs [27, 28, 106, 64] using hypergraph versions of the Regularity Lemma. Subsequently, Conlon, Fox and Sudakov [25] gave a shorter proof which gives a better constant and does not rely on the Regularity Lemma.

Embedding large subgraphs

One of the most famous conjectures in Ramsey theory is the Burr-Erdős conjecture on d-degenerate graphs, which generalizes Theorem 6.1. Here a graph G is *d-degenerate* if every subgraph has a vertex of degree at most d. In other words, G has no 'dense' subgraphs.

Conjecture 6.2 (Burr and Erdős [19]) *For every d there is a constant $C = C(d)$ so that every d-degenerate graph H on n vertices satisfies $R(H) \leq Cn$.*

It has been proved in many special cases (see e.g. the introduction of [41] for a recent overview). Also, Kostochka and Sudakov [91] proved that it is 'approximately' true:

Theorem 6.3 (Kostochka and Sudakov [91]) *For every d there is a constant $C = C(d)$ so that every d-degenerate graph H on n vertices satisfies $R(H) \leq 2^{C(\log n)^{2d/(2d+1)}} n$.*

The exponent '$2d/(2d+1)$' of the logarithm was improved to '$1/2$' in [41]. All the results in [24, 40, 41, 91] rely on variants of the same probabilistic argument, which was first applied to special cases of Conjecture 6.2 in [90]. To give an idea of this beautiful argument, we use a simple version to give a proof of the following density result (which is implicit in several of the above papers): it implies that bipartite graphs H whose maximum degree is logarithmic in their order have polynomial Ramsey numbers. (The logarithms in the statement and the proof are binary.)

Theorem 6.4 *Suppose that $H = (A', B', E')$ is a bipartite graph on $n \geq 2$ vertices and $\Delta(H) \leq \log n$. Suppose that $m \geq n^8$. Then every bipartite graph $G = (A, B, E)$ with $|A| = |B| = m$ and at least $m^2/8$ edges contains a copy of H. In particular, $R(H) \leq 2n^8$.*

An immediate corollary is that the Ramsey number of a d-dimensional cube Q_d is polynomial in its number $n = 2^d$ of vertices (this fact was first observed in [120] based on an argument similar to that in [90]). The best current bound of $R(Q_d) \leq d2^{2d+5}$ is given in [40]. Burr and Erdős [19] conjectured that the bound should actually be linear in $n = 2^d$.

Proof Write $\Delta := \log n$. Let $b_1, \ldots, b_s$ be a sequence of $s := 2\Delta$ not necessarily distinct vertices of B, chosen uniformly and independently at random and write $S := \{b_1, \ldots, b_s\}$. Let $N(S)$ denote the set of common neighbours of vertices in S. Clearly, $S \subseteq N(a)$ for every $a \in N(S)$. So Jensen's inequality implies that

$$\mathbb{E}(|N(S)|) = \sum_{a \in A} \mathbb{P}(a \in N(S)) = \sum_{a \in A} \left(\frac{|N(a)|}{m}\right)^s = \frac{\sum_{a \in A}(d(a))^s}{m^s}$$

$$\geq \frac{m \left(\frac{\sum_{a \in A} d(a)}{m}\right)^s}{m^s} \geq \frac{m\left((m^2/8)/m\right)^s}{m^s} = \frac{m}{8^s} \geq \frac{n^8}{n^6} = n^2.$$

We say that a subset $W \subseteq A$ is *bad* if it has size Δ and its common neighbourhood $N(W)$ satisfies $|N(W)| < n$. Now let Z denote the number of bad subsets W of $N(S)$. Note that the probability that a given set $W \subseteq A$ lies in $N(S)$ equals

$(|N(W)|/m)^s$ (since the probability that it lies in the neighbourhood of a fixed vertex $b \in B$ is $|N(W)|/m$). So

$$\mathbb{E}Z = \sum_{W \text{ bad}} \mathbb{P}(W \subseteq N(S)) \leq \binom{m}{\Delta}\left(\frac{n}{m}\right)^s \leq m^\Delta \left(\frac{n}{m}\right)^s = \left(\frac{n^2}{m}\right)^\Delta \leq (1/2)^\Delta < 1.$$

So $\mathbb{E}(|N(S)| - Z) \geq n^2 - 1 \geq n$ and hence there is a choice of S with $|N(S)| - Z \geq n$. By definition, we can delete a vertex from every bad W contained in $N(S)$ to obtain a set $T \subseteq N(S)$ with $|T| \geq n$ so that every subset $W \subseteq T$ with $|W| = \Delta$ satisfies $|N(W)| \geq n$. Clearly we can now embed H: first embed A' arbitrarily into T and then embed the vertices of B' one by one into B, using the property that T has no bad subset.

The bound on $R(H)$ can be derived as follows: consider any 2-colouring of the complete graph on $2n^8$ vertices. Partition its vertices arbitrarily into two sets A and B of size n^8 and then apply the main statement to the subgraph of G induced by the colour class having the most edges between A and B. □

Note that the proof immediately shows that the bound on the maximum degree of H can be relaxed: all we need is the property that every subgraph of H has a vertex $b \in B'$ of low degree. In the proof of (the bipartite case) of Theorem 6.3, this is exploited as follows: roughly speaking one carries out the above argument twice (of course with different parameters than the above). The first time we consider a random subset $S \subseteq B$ and the second time we consider a smaller random subset $S' \subseteq T$.

For some types of sparse graphs H, one can give even more precise estimates for $R(H)$ than the ones which follow from the above results. For instance, Theorem 3.3 has an immediate application to the Ramsey number of trees.

Corollary 6.5 *There is an integer n_0 so that if T_n is a tree on $n \geq n_0$ vertices then $R(T_n) \leq 2n - 2$.*

Indeed, to derive Corollary 6.5 from Theorem 3.3, consider a 2-colouring of a complete graph K_{2n-2} on $2n - 2$ vertices, yielding a red graph G_r and a blue graph G_b. Order the vertices x_i according to their degree (in ascending order) in G_r. If x_{n-1} has degree at least $n - 1$ in G_r, then we can apply Theorem 3.3 to find a red copy of T in G_r. If not, we can apply it to find a blue copy of T in G_b. For even n, the bound is best possible (let T be a star and let G_b and G_r be regular of the same degree) and proves a conjecture of Burr and Erdős [20]. For odd n, they conjectured that the answer is $2n - 3$. Similarly, the Komlós-Sós conjecture (Conjecture 3.4) would imply that $R(T_n, T_m) \leq n + m - 2$, where T_n and T_m are trees on n and m vertices respectively. Of course, Corollary 6.5 is not best possible for every tree. For instance, in the case when the tree is a path, Gerencsér and Gyarfas [44] showed that $R(P_n, P_n) = \lfloor (3n-2)/2 \rfloor$. Further recent results on Ramsey numbers of paths and cycles (many of which rely on the Regularity Lemma) can be found e.g. in [51, 38]. Hypergraph versions (i.e. Ramsey numbers of tight cycles, loose cycles and Berge-cycles) were considered e.g. in [58, 59, 50].

7 A sample application of the Regularity and Blow-up Lemma

In order to illustrate the details of the Regularity method for those not familiar with it, we now prove Theorem 2.2 for the case when $H := C_4$ and when we replace the constant C in the minimum degree condition with a linear error term.

Theorem 7.1 *For every $0 < \eta < 1/2$ there exists an integer n_0 such that every graph G whose order $n \geq n_0$ is divisible by 4 and whose minimum degree is at least $n/2 + \eta n$ contains a perfect C_4-packing.*

(Note that Theorem 7.1 also follows from Theorems 5.2 and 5.3.) We start with the formal definition of ε-regularity. The *density* of a bipartite graph $G = (A, B)$ with vertex classes A and B is

$$d_G(A, B) := \frac{e_G(A, B)}{|A||B|}.$$

We also write $d(A, B)$ if this is unambiguous. Given $\varepsilon > 0$, we say that G is ε-*regular* if for all sets $X \subseteq A$ and $Y \subseteq B$ with $|X| \geq \varepsilon |A|$ and $|Y| \geq \varepsilon |B|$ we have $|d(A, B) - d(X, Y)| < \varepsilon$. Given $d \in [0, 1)$, we say that G is (ε, d)-*superregular* if all sets $X \subseteq A$ and $Y \subseteq B$ with $|X| \geq \varepsilon |A|$ and $|Y| \geq \varepsilon |B|$ satisfy $d(X, Y) > d$ and, furthermore, if $d_G(a) > d|B|$ for all $a \in A$ and $d_G(b) > d|A|$ for all $b \in B$. Moreover, we will denote the neighbourhood of a vertex x in a graph G by $N_G(x)$. Given disjoint sets A and B of vertices of G, we write $(A, B)_G$ for the bipartite subgraph of G whose vertex classes are A and B and whose edges are all the edges of G between A and B.

Szemerédi's Regularity Lemma [125] states that one can partition the vertices of every large graph into a bounded number of 'clusters' so that most of the pairs of clusters induce ε-regular bipartite graphs. Proofs are also included in [16] and [35]. Algorithmic proofs of the Regularity Lemma were given in [6, 42]. There are also several versions for hypergraphs (in fact, all the results in Section 4.5 are based on some hypergraph version of the Regularity Lemma). The first so-called 'strong' versions for r-uniform hypergraphs were proved in [48] and [107, 119].

Lemma 7.2 (Szemerédi [125]) *For all $\varepsilon > 0$ and all integers k_0 there is an $N = N(\varepsilon, k_0)$ such that for every graph G on $n \geq N$ vertices there exists a partition of $V(G)$ into $V_0, V_1, \ldots, V_k$ such that the following holds:*

- $k_0 \leq k \leq N$ and $|V_0| \leq \varepsilon n$,

- $|V_1| = \cdots = |V_k| =: m$,

- *for all but εk^2 pairs $1 \leq i < j \leq k$ the graph $(V_i, V_j)_G$ is ε-regular.*

Unfortunately, the constant N appearing in the lemma is very large, Gowers [47] showed that it has at least a tower-type dependency on ε. We will use the following degree form of Szemerédi's Regularity Lemma which can be easily derived from Lemma 7.2.

Lemma 7.3 (Degree form of the Regularity Lemma) *For all $\varepsilon > 0$ and all integers k_0 there is an $N = N(\varepsilon, k_0)$ such that for every number $d \in [0,1)$ and for every graph G on $n \geq N$ vertices there exist a partition of $V(G)$ into $V_0, V_1, \ldots, V_k$ and a spanning subgraph G' of G such that the following holds:*

- $k_0 \leq k \leq N$ and $|V_0| \leq \varepsilon n$,
- $|V_1| = \cdots = |V_k| =: m$,
- $d_{G'}(x) > d_G(x) - (d+\varepsilon)n$ for all vertices $x \in G$,
- for all $i \geq 1$ the graph $G'[V_i]$ is empty,
- for all $1 \leq i < j \leq k$ the graph $(V_i, V_j)_{G'}$ is ε-regular and has density either 0 or $> d$.

The sets V_i ($i \geq 1$) are called *clusters*, V_0 is called the *exceptional set* and G' is called the *pure graph*.

Sketch of proof of Lemma 7.3 To obtain a partition as in Lemma 7.3, apply Lemma 7.2 with parameters d, ε', k_0' satisfying $1/k_0', \varepsilon' \ll \varepsilon, d, 1/k_0$ to obtain clusters $V_1', \ldots, V_{k'}'$ and an exceptional set V_0'. (Here $a \ll b < 1$ means that there is an increasing function f such that all the calculations in the argument work as long as $a \leq f(b)$.) Let $m' := |V_1'| = \cdots = |V_{k'}'|$. Now delete all edges between pairs of clusters which are not ε'-regular and move any vertices into V_0' which were incident to at least $\varepsilon n/10$ (say) of these deleted edges. Secondly, delete all (remaining) edges between pairs of clusters whose density is at most $d+\varepsilon'$. Consider such a pair (V_i', V_j') of clusters. For every vertex $x \in V_i'$ which has more than $(d+2\varepsilon')m'$ neighbours in V_j' mark all but $(d+2\varepsilon')m'$ edges between x and V_j'. Do the same for the vertices in V_j' and more generally for all pairs of clusters of density at most $d+\varepsilon'$. It is easy to check that in total this yields at most $\varepsilon' n^2$ marked edges. Move all vertices into V_0' which are incident to at least $\varepsilon n/10$ of the marked edges. Thirdly, delete any edges within the clusters. Finally, we need to make sure that the clusters have equal size again (as we may have lost this property during the deletion process). This can be done by splitting up the clusters into smaller subclusters (which contain almost all the vertices and have equal size) and moving a small number of further vertices into V_0'. A straightforward calculation shows that the new exceptional set V_0 has size at most εn as required. □

The *reduced graph* R is the graph whose vertices are $1, \ldots, k$ and in which i is joined to j whenever the bipartite subgraph $(V_i, V_j)_{G'}$ of G' induced by V_i and V_j is ε-regular and has density $> d$. Thus ij is an edge of R if and only if G' has an edge between V_i and V_j. Roughly speaking, the following result states that R almost 'inherits' the minimum degree of G.

Proposition 7.4 *If $0 < 2\varepsilon \leq d \leq c/2$ and $\delta(G) \geq cn$ then $\delta(R) \geq (c-2d)|R|$.*

Proof Consider any vertex i of R and pick $x \in V_i$. Then every neighbour of x in G' lies in $V_0 \cup \bigcup_{j \in N_R(i)} V_j$. Thus $(c-(d+\varepsilon))n \leq d_{G'}(x) \leq d_R(i)m + \varepsilon n$ and so $d_R(i) \geq (c-2d)n/m \geq (c-2d)|R|$ as required. □

The proof of Proposition 7.4 is a point where it is important that R was defined using the graph G' obtained from Lemma 7.3 and not using the partition given by Lemma 7.2.

In our proof of Theorem 7.1 the reduced graph R will contain a Hamilton path P. Recall that every edge ij of $P \subseteq R$ corresponds to the ε-regular bipartite subgraph $(V_i, V_j)_{G'}$ of G' having density $> d$. The next result shows that by removing a small number of vertices from each cluster (which will be added to the exceptional set V_0) we can guarantee that the edges of P even correspond to superregular pairs.

Proposition 7.5 *Suppose that $4\varepsilon < d \leq 1$ and that P is a Hamilton path in R. Then every cluster V_i contains a subcluster $V_i' \subseteq V_i$ of size $m - 2\varepsilon m$ such that $(V_i', V_j')_{G'}$ is $(2\varepsilon, d - 3\varepsilon)$-superregular for every edge $ij \in P$.*

Proof We may assume that $P = 1 \ldots k$. Given any $i < k$, the definition of regularity implies that there are at most εm vertices $x \in V_i$ such that $|N_{G'}(x) \cap V_{i+1}| \leq (d - \varepsilon)m$. Similarly, for each $i > 1$ there are at most εm vertices $x \in V_i$ such that $|N_{G'}(x) \cap V_{i-1}| \leq (d - \varepsilon)m$. Let V_i' be a subset of size $m - 2\varepsilon m$ of V_i which contains none of the above vertices (for all $i = 1, \ldots, k$). Then $V_1', \ldots, V_k'$ are as required. □

Of course, in Proposition 7.5 it is not important that P is a Hamilton path. One can prove an analogue whenever P is a subgraph of R of bounded maximum degree. We will also use the following special case of the Blow-up Lemma of Komlós, Sárközy and Szemerédi [83]. It implies that dense superregular pairs behave like complete bipartite graphs with respect to containing bounded degree graphs as subgraphs, i.e. if the superregular pair has vertex classes V_i and V_j then any bounded degree bipartite graph on these vertex classes is a subgraph of this superregular pair. An algorithmic version of the Blow-up Lemma was proved by the same authors in [84]. A hypergraph version was recently proved by Keevash [72].

Lemma 7.6 (Blow-up Lemma, bipartite case) *Given $d > 0$ and $\Delta \in \mathbb{N}$, there is a positive constant $\varepsilon_0 = \varepsilon_0(d, \Delta)$ such that the following holds for every $\varepsilon < \varepsilon_0$. Given $m \in \mathbb{N}$, let G^* be an (ε, d)-superregular bipartite graph with vertex classes of size m. Then G^* contains a copy of every subgraph H of $K_{m,m}$ with $\Delta(H) \leq \Delta$.*

Proof of Theorem 7.1 We choose further positive constants ε and d as well as $n_0 \in \mathbb{N}$ such that
$$1/n_0 \ll \varepsilon \ll d \ll \eta < 1/2.$$
(In order to simplify the exposition we will not determine these constants explicitly.) We start by applying the degree form of the Regularity Lemma (Lemma 7.3) with parameters ε, d and $k_0 := 1/\varepsilon$ to G to obtain clusters $V_1, \ldots, V_k$, an exceptional set V_0, a pure graph G' and a reduced graph R. Thus $k := |R|$ and
$$\delta(R) \geq (1/2 + \eta - 2d)k \geq (1 + \eta)k/2 \tag{7.1}$$
by Proposition 7.4. So R contains a Hamilton path P (this follows e.g. from Dirac's theorem). By relabelling if necessary we may assume that $P = 1 \ldots k$. Apply Proposition 7.5 to obtain subclusters $V_i' \subseteq V_i$ of size $m - 2\varepsilon m =: m'$ such that for

every edge $i(i + 1) \in P$ the bipartite subgraph $(V_i', V_{i+1}')_{G'}$ of G' induced by V_i' and V_{i+1}' is $(2\varepsilon, d/2)$-superregular. Note that the definition of ε-regularity implies that $(V_i', V_j')_{G'}$ is still 2ε-regular of density at least $d - \varepsilon \geq d/2$ whenever ij is an edge of R. We add all those vertices of G that are not contained in some V_i' to the exceptional set V_0. Moreover, if k is odd then we also add all the vertices in V_k' to V_0. We still denote the reduced graph by R, its number of vertices by k and the exceptional set by V_0. Thus

$$|V_0| \leq \varepsilon n + 2\varepsilon n + m \leq 4\varepsilon n.$$

Let M denote the perfect matching in P. So M consists of the edges $12, 34, \ldots, (k-1)k$. The Blow-up Lemma would imply that for every odd i the bipartite graph $(V_i', V_{i+1}')_{G'}$ contains a perfect C_4-packing, provided that 2 divides m'. So we have already proved that G contains a C_4-packing covering *almost* all of its vertices (this can also be easily proved without the Regularity Lemma). In order to obtain a perfect C_4-packing, we have to incorporate the exceptional vertices.

To make it simpler to deal with divisibility issues later on, for every odd i we will now choose a set X_i of 7 vertices of G which we can put in any of V_i' and V_{i+1}' without destroying the superregularity of $(V_i', V_{i+1}')_{G'}$. More precisely, (7.1) implies that the vertices i and $i+1$ of R have a common neighbour, j say. Recall that both $(V_i', V_j')_{G'}$ and $(V_{i+1}', V_j')_{G'}$ are 2ε-regular and have density at least $d/2$. So almost all vertices in V_j' have at least $(d/2 - 2\varepsilon)m'$ neighbours in both V_i' and V_{i+1}'. Let $X_i \subseteq V_j'$ be a set of 7 such vertices. Clearly, we may choose the sets X_i disjoint for distinct odd i. Remove all the vertices in $X_1 \cup X_3 \cup \cdots \cup X_{k-1} =: X$ from the clusters they belong to. By removing at most $|X|k \leq 7k^2$ further vertices and adding them to the exceptional set we may assume that the subclusters $V_i'' \subseteq V_i'$ thus obtained satisfy $|V_1''| = \cdots = |V_k''| =: m''$. (The vertices in X are not added to V_0.) Note that we now have

$$|V_0| \leq 4\varepsilon n + 7k^2 \leq 5\varepsilon n.$$

Consider any vertex $x \in V_0$. Call an odd i *good for* x if x has at least $\eta^2 m''$ neighbours in both V_i'' and V_{i+1}'' (in the graph G'). Then the number g_x of good indices satisfies

$$(1+\eta)n/2 \leq d_{G'}(x) - |V_0| - |X| \leq 2g_x m'' + (k/2 - g_x)(1+\eta^2)m'' \leq 2g_x m'' + (1+\eta^2)n/2,$$

which shows that $g_x \geq \eta k/8 = \eta |M|/4$. Since $|V_0|/(\sqrt{\varepsilon}m'') \leq \eta|M|/4$, this implies that we can assign each $x \in V_0$ to an odd index i which is good for x in such a way that to each odd i we assign at most $\sqrt{\varepsilon}m''$ exceptional vertices. Now consider any matching edge $i(i+1) \in M$. Add each exceptional vertex assigned to i to V_i' or V_{i+1}' so that the sizes of the sets $V_i^* \supseteq V_i''$ and $V_{i+1}^* \supseteq V_{i+1}''$ obtained in this way differ by at most 1. It is easy to check that the bipartite subgraph $(V_i^*, V_{i+1}^*)_{G'}$ of G' is still $(2\sqrt{\varepsilon}, d/8)$-superregular.

Since the vertices in X_i can be added to any of V_i^* and V_{i+1}^* without destroying the superregularity of $(V_i^*, V_{i+1}^*)_{G'}$, we could now apply the Blow-up Lemma to find a C_4-packing of $G'[V_i^* \cup V_{i+1}^* \cup X_i]$ which covers all but at most 3 vertices (and so altogether these packings would form a C_4-packing of G covering all but at most $3k$ vertices of G). To ensure the existence of a perfect C_4-packing, we need to make $|V_i^* \cup V_{i+1}^* \cup X_i|$ divisible by 4 for every odd i. We will do this for every

$i = 1, 3, \ldots, k-1$ in turn by shifting the remainders mod 4 along the path P. More precisely, suppose that $|V_1^* \cup V_2^* \cup X_1| \equiv a \mod 4$ where $0 \le a < 4$. Choose a disjoint copies of C_4, each having 1 vertex in V_2^*, 2 vertices in V_3^* and 1 vertex in V_4^*. Remove the vertices in these copies from the clusters they belong to and still denote the subclusters thus obtained by V_i^*. (Each such copy of C_4 can be found greedily using that both $(V_2^*, V_3^*)_{G'}$ and $(V_3^*, V_4^*)_{G'}$ are still $2\sqrt{\varepsilon}$-regular and have density at least $d/8$. Indeed, to find the first copy, pick any vertex $x \in V_2^*$ having at least $(d/8 - 2\sqrt{\varepsilon})|V_3^*|$ neighbours in V_3^*. The regularity of $(V_2^*, V_3^*)_{G'}$ implies that almost all vertices in V_2^* can play the role of x. The regularity of $(V_3^*, V_4^*)_{G'}$ now implies that its bipartite subgraph induced by the neighbourhood of x in V_3^* and by V_4^* has density at least $d/8 - 2\sqrt{\varepsilon}$. So there are many vertices $y \in V_4^*$ which have at least 2 neighbours in $N_{G'}(x) \cap V_3^*$. Then x and y together with 2 such neighbours form a copy of C_4.) Now $|V_1^* \cup V_2^* \cup X_1|$ is divisible by 4. Similarly, by removing at most 3 further copies of C_4, each having 1 vertex in V_4^*, 2 vertices in V_5^* and 1 vertex in V_6^* we can achieve that $|V_3^* \cup V_4^* \cup X_3|$ is divisible by 4. Since $n = |G|$ is divisible by 4 we can continue in this way to achieve that $|V_i^* \cup V_{i+1}^* \cup X_i|$ is divisible by 4 for every odd i.

Recall that before we took out all these copies of C_4, for every odd i the sizes of V_i^* and V_{i+1}^* differed by at most 1. Thus now these sizes differ (crudely) by at most 7. But every vertex $x \in X_i$ can be added to both V_i^* and V_{i+1}^* without destroying the superregularity. Add the vertices from X_i to V_i^* and V_{i+1}^* in such a way that the sets $V_i^\circ \supseteq V_i^*$ and $V_{i+1}^\circ \supseteq V_{i+1}^*$ thus obtained have equal size. (This size must be even since $|V_i^* \cup V_{i+1}^* \cup X_i|$ is divisible by 4.) It is easy to check that $(V_i^\circ, V_{i+1}^\circ)_{G'}$ is still $(3\sqrt{\varepsilon}, d/9)$-superregular. Thus we can apply the Blow-up Lemma (Lemma 7.6) to obtain a perfect C_4-packing in $(V_i^\circ, V_{i+1}^\circ)_{G'}$. The union of all these packings (over all odd i) together with the C_4's we have chosen before form a perfect C_4-packing of G. □

Acknowledgements

We would like to thank Demetres Christofides, Nikolaos Fountoulakis and Andrew Treglown for their comments on an earlier version of this manuscript.

References

[1] S. Abbasi, The solution of the El-Zahar problem, Ph.D. Thesis, Rutgers University (1998).

[2] S. Abbasi, How tight is the Bollobás–Komlós Conjecture?, *Graphs Combin.* **16** (2000), 129–137.

[3] R. Aharoni, A. Georgakopoulos & P. Sprüssel, Perfect matchings in r-partite r-graphs, *European J. Combin.* **30** (2009), 39–42.

[4] M. Aigner & S. Brandt, Embedding arbitrary graphs of maximum degree two, *J. Lond. Math. Soc.* **48** (1993), 39–51.

[5] M. Ajtai, J. Komlós & E. Szemerédi, On a conjecture of Loebl, in *Graph theory, Combinatorics, and Algorithms, Vol. 1, 2 (Kalamazoo, MI, 1992)*, Wiley-Interscience, New York, (1995) pp. 1135–1146.

[6] N. Alon, R. A. Duke, H. Lefmann, V. Rödl & R. Yuster, The algorithmic aspects of the Regularity Lemma, *J. Algorithms* **16** (1994), 80–109.

[7] N. Alon & E. Fischer, 2-factors in dense graphs, *Discrete Math.* **152** (1996), 13–23.

[8] N. Alon & E. Fischer, Refining the graph density condition for the existence of almost K-factors, *Ars Combin.* **52** (1999), 296–308.

[9] N. Alon & R. Yuster, H-factors in dense graphs, *J. Combin. Theory Ser. B* **66** (1996), 269–282.

[10] J. Bang-Jensen & G. Gutin, *Digraphs: Theory, Algorithms and Applications*, Springer (2000).

[11] J. C. Bermond, A. Germa, M. C. Heydemann & D. Sotteau, Hypergraphes hamiltoniens, *Prob. Comb. Théorie Graph Orsay* **260** (1976), 39–43.

[12] J. C. Bermond & C. Thomassen, Cycles in digraphs — a survey, *J. Graph Theory* **5** (1981), 1–43.

[13] J. Böttcher, K. P. Pruessmann, A. Taraz & A. Würfl, Bandwidth, treewidth, separators, expansion, and universality, in *Proceedings of the TGGT conference, Electron. Notes Discrete Math.*, 31, (2008) pp. 91–96.

[14] J. Böttcher, M. Schacht & A. Taraz, Spanning 3-colourable subgraphs of small bandwidth in dense graphs, *J. Combin. Theory Ser. B* **98** (2008), 752–777.

[15] J. Böttcher, M. Schacht & A. Taraz, Proof of the bandwidth conjecture of Bollobás and Komlós, *Math. Ann.* **343** (2009), 175–205.

[16] B. Bollobás, *Modern Graph Theory, Graduate Texts in Mathematics*, 184, Springer-Verlag (1998).

[17] B. Bollobás & S. E. Eldridge, Packings of graphs and applications to computational complexity, *J. Combin. Theory Ser. B* **25** (1978), 105–124.

[18] B. Bollobás & R. Häggkvist, Powers of Hamilton cycles in tournaments, *J. Combin. Theory Ser. B* **50** (1990), 309–318.

[19] A. Burr & P. Erdős, On the magnitude of generalized Ramsey numbers for graphs, in *Infinite and Finite Sets I, Colloq. Math. Soc. Janos Bolyai*, 10, North-Holland, Amsterdam (1975), pp. 214–240.

[20] A. Burr & P. Erdős, Extremal Ramsey theory for graphs, *Utilitas Math.* **9** (1976), 247–258.

[21] P. A. Catlin, *Embedding subgraphs and coloring graphs under extremal degree conditions*, Ph.D. Thesis, Ohio State Univ., Columbus (1976).

[22] V. Chvátal, On Hamilton's ideals, *J. Combin. Theory Ser. B* **12** (1972), 163–168.

[23] V. Chvátal, V. Rödl, E. Szemerédi & W. T. Trotter, Jr., The Ramsey number of a graph with a bounded maximum degree, *J. Combin. Theory Ser. B* **34** (1983), 239–243.

[24] D. Conlon, Hypergraph packing and sparse bipartite Ramsey numbers, preprint.

[25] D. Conlon, J. Fox & B. Sudakov, Ramsey numbers of sparse hypergraphs, *Random Structures Algorithms*, in press.

[26] O. Cooley, Proof of the Loebl–Komlós–Sós conjecture for large dense graphs, *Discrete Math.*, in press.

[27] O. Cooley, N. Fountoulakis, D. Kühn & D. Osthus, 3-uniform hypergraphs of bounded degree have linear Ramsey numbers, *J. Combin. Theory Ser. B* **98** (2008), 484–505.

[28] O. Cooley, N. Fountoulakis, D. Kühn & D. Osthus, Embeddings and Ramsey numbers of sparse k-uniform hypergraphs, *Combinatorica*, in press.

[29] O. Cooley, D. Kühn & D. Osthus, Perfect packings with complete graphs minus an edge, *European J. Combin.* **28** (2007), 2143–2155.

[30] K. Corrádi & A. Hajnal, On the maximal number of independent circuits in a graph, *Acta Math. Acad. Sci. Hungar.* **14** (1963), 423–439.

[31] B. Csaba, On the Bollobás–Eldridge conjecture for bipartite graphs, *Combin. Probab. Comput.* **16** (2007), 661–691.

[32] B. Csaba, On embedding well-separable graphs, *Discrete Math.* **308** (2008), 4322–4331.

[33] B. Csaba & M. Mydlarz, Approximate multipartite version of the Hajnal–Szemerédi theorem, preprint.

[34] B. Csaba, A. Shokoufandeh & E. Szemerédi, Proof of a conjecture of Bollobás and Eldridge for graphs of maximum degree three, *Combinatorica* **23** (2003), 35–72.

[35] R. Diestel, *Graph Theory*, Third edition, *Graduate Texts in Mathematics*, 173, Springer-Verlag (2005).

[36] G. A. Dirac, Some theorems on abstract graphs, *Proc. Lond. Math. Soc.* **2** (1952), 69–81.

[37] P. Erdős, Extremal problems in graph theory,, in *Theory of Graphs and its Applications (Proc. Sympos. Smolenice, 1963)* Publ. House Czechoslovak Acad. Sci., Prague (1964), pp. 29–36.

[38] A. Figaj & T. Łuczak, The Ramsey number for a triple of long even cycles *J. Combin. Theory Ser. B* **97** (2007), 584–596.

[39] E. Fischer, Variants of the Hajnal–Szemerédi Theorem, *J. Graph Theory* **31** (1999), 275–282.

[40] J. Fox & B. Sudakov, Density theorems for bipartite graphs and related Ramsey-type results, *Combinatorica*, in press.

[41] J. Fox & B. Sudakov, Two remarks on the Burr–Erdős conjecture, *European J. Combin.*, in press.

[42] A. Frieze & R. Kannan, A simple algorithm for constructing Szemerédi's regularity partition, *Electron. J. Combin.* **6** (1999), R17.

[43] A. Frieze & M. Krivelevich, On packing Hamilton cycles in ε-regular graphs, *J. Combin. Theory Ser. B* **94** (2005), 159–172.

[44] L. Gerencsér & A. Gyarfas, On Ramsey-type problems, *Ann. Univ. Sci. Budapest Eötvös Sect. Math.* **10** (1967), 167–170.

[45] A. Ghouila-Houri, Une condition suffisante d'existence d'un circuit hamiltonien, *C. R. Acad. Sci. Paris* **25** (1960), 495–497.

[46] R. Gould, Advances on the hamiltonian problem: A survey, *Graphs Combin.* **19** (2003), 7–52.

[47] W. T. Gowers, Lower bounds of tower type for Szemerédi's uniformity lemma, *Geom. Funct. Anal.* **7** (1997), 322–337.

[48] W. T. Gowers, Hypergraph regularity and the multidimensional Szemerédi theorem, *Ann. of Math.* **166** (2007), 897–946.

[49] R. Graham, V. Rödl & A. Ruciński, On bipartite graphs with linear Ramsey numbers, *Combinatorica* **21** (2001), 199–209.

[50] A. Gyárfás, J. Lehel, G. N. Sarközy & R. Schelp, Monochromatic Hamiltonian Berge-cycles in colored complete uniform hypergraphs, *J. Combin. Theory Ser. B* **98** (2008), 342–358.

[51] A. Gyárfás, M. Ruszinkó, G. N. Sarközy & E. Szemerédi, Three color Ramsey numbers for paths, *Combinatorica* **27** (2007), 35–69.

[52] R. Häggkvist, Hamilton cycles in oriented graphs, *Combin. Probab. Comput.* **2** (1993), 25–32.

[53] R. Häggkvist & C. Thomassen, On pancyclic digraphs, *J. Combin. Theory Ser. B* **20** (1976), 20–40.

[54] R. Häggkvist & A. Thomason, Oriented Hamilton cycles in oriented graphs, in *Combinatorics, Geometry and Probability*, Cambridge University Press, Cambridge (1997), pp. 339–353.

[55] A. Hajnal & E. Szemerédi, Proof of a conjecture of Erdős, in *Combinatorial Theory and its Applications (Vol. 2)* (eds. P. Erdős, A. Rényi and V. T. Sós), *Colloq. Math. Soc. Janos Bolyai*, 4, North-Holland, Amsterdam (1970), pp. 601–623.

[56] H. Hàn, Y. Person & M. Schacht, On perfect matchings in uniform hypergraphs with large minimum vertex degree, preprint.

[57] H. Hàn & M. Schacht, Dirac-type results for loose Hamilton cycles in uniform hypergraphs, preprint.

[58] P. E. Haxell, T. Łuczak, Y. Peng, V. Rödl, A. Ruciński, M. Simonovits & J. Skokan, The Ramsey number for hypergraph cycles I, *J. Combin. Theory Ser. A* **113** (2006), 67–83.

[59] P. E. Haxell, T. Łuczak, Y. Peng, V. Rödl, A. Ruciński & J. Skokan, The Ramsey number for hypergraph cycles II, *Combin. Probab. Comput.*, in press.

[60] P. Hell & D. G. Kirkpatrick, Scheduling, matching and colouring, in *Algebraic methods in graph theory, Vol. I, II (Szeged, 1978) Colloq. Math. Soc. János Bolyai*, 25, North-Holland, Amsterdam (1981), pp. 273–279.

[61] P. Hell & D. G. Kirkpatrick, On the complexity of general graph factor problems, *SIAM J. Comput.* **12** (1983), 601–609.

[62] J. Hladký & D. Piguet, Loebl–Komlós–Sós Conjecture: dense case, preprint.

[63] J. Hladký & M. Schacht, Note on bipartite graph tilings, preprint.

[64] Y. Ishigami, Linear Ramsey numbers for bounded-degree hypergraphs, preprint.

[65] B. Jackson, Hamilton cycles in regular 2-connected graphs, *J. Combin. Theory Ser. B* **29** (1980), 27–46.

[66] B. Jackson, Long paths and cycles in oriented graphs, *J. Graph Theory* **5** (1981), 245–252.

[67] R. Johansson, Triangle-factors in a balanced blown-up triangle, *Discrete Math.* **211** (2000), 249–254.

[68] A. Johansson, R. Johansson & K. Markström, Factors of r-partite graphs, preprint.

[69] V. Kann, Maximum bounded H-matching is MAX SNP-complete, *Inform. Process. Lett.* **49** (1994), 309–318.

[70] H. Kaul, A. Kostochka & G. Yu, On a graph packing conjecture of Bollobás, Eldridge, and Catlin, *Combinatorica* **28** (2008), 469–485.

[71] K. Kawarabayashi, K_4^--factors in a graph, *J. Graph Theory* **39** (2002), 111–128.

[72] P. Keevash, A hypergraph blow-up lemma, preprint.

[73] P. Keevash, D. Kühn, R. Mycroft & D. Osthus, Loose Hamilton cycles in hypergraphs, preprint.

[74] P. Keevash, D. Kühn & D. Osthus, An exact minimum degree condition for Hamilton cycles in oriented graphs, *J. Lond. Math. Soc.* **79** (2009), 144–166.

[75] P. Keevash & B. Sudakov, Triangle packings and 1-factors in oriented graphs, preprint.

[76] L. Kelly, D. Kühn & D. Osthus, A Dirac-type result on Hamilton cycles in oriented graphs, *Combin. Probab. Comput.* **17** (2008), 689–709.

[77] L. Kelly, D. Kühn & D. Osthus, Cycles of given length in oriented graphs, preprint.

[78] H. Kierstead & A. Kostochka, A short proof of the Hajnal–Szemerédi Theorem on equitable coloring, *Combin. Probab. Comput.* **17** (2008), 265–270.

[79] H. Kierstead & A. Kostochka, An Ore-type theorem on equitable coloring, *J. Combin. Theory Ser. B* **98** (2008), 226–234.

[80] J. Komlós, The Blow-Up Lemma, *Combin. Probab. Comput.* **8** (1999), 161–176.

[81] J. Komlós, Tiling Turán theorems, *Combinatorica* **20** (2000), 203–218.

[82] J. Komlós, G. N. Sárközy & E. Szemerédi, Proof of a packing conjecture of Bollobás, *Combin. Probab. Comput.* **4** (1995), 241–255.

[83] J. Komlós, G. N. Sárközy & E. Szemerédi, Blow-Up Lemma, *Combinatorica* **17** (1997), 109–123.

[84] J. Komlós, G. N. Sárközy & E. Szemerédi, An algorithmic version of the Blow-Up Lemma, *Random Structures Algorithms* **12** (1998), 297–312.

[85] J. Komlós, G. N. Sárközy & E. Szemerédi, Proof of the Seymour conjecture for large graphs, *Ann. Combin.* **2** (1998), 43–60.

[86] J. Komlós, G. N. Sárközy & E. Szemerédi, Proof of the Alon–Yuster conjecture, *Discrete Math.* **235** (2001), 255–269.

[87] J. Komlós, G. N. Sárközy & E. Szemerédi, Spanning trees in dense graphs, *Combin. Probab. Comput.* **10** (2001), 397–416.

[88] J. Komlós, A. Shokoufandeh, M. Simonovits & E. Szemerédi, The Regularity Lemma and its applications in graph theory, in *Theoretical aspects of computer science (Tehran, 2000), Springer Lecture Notes in Comput. Sci.*, 2292, (2002), pp. 84–112.

[89] J. Komlós & M. Simonovits, Szemerédi's Regularity Lemma and its applications in graph theory, in *Combinatorics, Paul Erdős is Eighty, Vol. 2 (Keszthely, 1993) Bolyai Soc. Mathematical Studies*, 2, János Bolyai Math. Soc., Budapest (1996), pp. 295–352.

[90] A. V. Kostochka & V. Rödl, On graphs with small Ramsey numbers, *J. Graph Theory* **37** (2001), 198–204.

[91] A. V. Kostochka & B. Sudakov, On graphs with small Ramsey numbers, *Combin. Probab. Comput.* **12** (2003), 627–641.

[92] D. Kühn & D. Osthus, Spanning triangulations in graphs, *J. Graph Theory* **49** (2005), 205–233.

[93] D. Kühn & D. Osthus, Critical chromatic number and complexity of perfect packings in graphs, in *Proceedings of the 17th ACM-SIAM Symposium on Discrete Algorithms (SODA 2006)*, ACM, New York (2006), pp. 851–859.

[94] D. Kühn & D. Osthus, Matchings in hypergraphs of large minimum degree, *J. Graph Theory* **51** (2006), 269–280.

[95] D. Kühn & D. Osthus, Loose Hamilton cycles in 3-uniform hypergraphs of large minimum degree, *J. Combin. Theory Ser. B* **96** (2006), 767–821.

[96] D. Kühn & D. Osthus, The minimum degree threshold for perfect graph packings, *Combinatorica* **29** (2009), 65–107.

[97] D. Kühn, D. Osthus & A. Taraz, Large planar subgraphs in dense graphs, *J. Combin. Theory Ser. B* **95** (2005), 263–282.

[98] D. Kühn, D. Osthus & A. Treglown, Hamiltonian degree sequences in digraphs, preprint.

[99] D. Kühn, D. Osthus & A. Treglown, An Ore-type theorem for perfect packings in graphs, preprint.

[100] I. Levitt, G. N. Sárközy & E. Szemerédi, How to avoid using the Regularity Lemma; Pósa's conjecture revisited, *Discrete Math.*, in press.

[101] Cs. Magyar & R. Martin, Tripartite version of the Corrádi–Hajnal Theorem, *Discrete Math.* **254** (2002), 289–308.

[102] R. Martin & E. Szemerédi, Quadripartite version of the Hajnal–Szemerédi Theorem, *Discrete Math.* **308** (2008), 4337–4360.

[103] R. Martin & Y. Zhao, Tiling tripartite graphs with 3-colorable graphs, preprint.

[104] B. McKay, The asymptotic numbers of regular tournaments, Eulerian digraphs and Eulerian oriented graphs, *Combinatorica* **10** (1990), 367–377.

[105] A. McLennan, The Erdős–Sós conjecture for trees of diameter four, *J. Graph Theory* **49** (2005), 291–301.

[106] B. Nagle, S. Olsen, V. Rödl & M. Schacht, On the Ramsey number of sparse 3-graphs, *Graphs Combin.* **24** (2008), 205–228.

[107] B. Nagle, V. Rödl & M. Schacht, The counting lemma for k-uniform hypergraphs, *Random Structures Algorithms* **28** (2006), 113–179.

[108] C. St. J. A. Nash-Williams, Hamilton circuits in graphs and digraphs, in *The many facets of graph theory*, *Springer Verlag Lecture Notes*, 110, Springer Verlag, (1968), pp. 237–243.

[109] C. St. J. A. Nash-Williams, Hamiltonian circuits, *Studies in Math.* **12** (1975), 301–360.

[110] O. Ore, Note on Hamilton circuits, *Amer. Math. Monthly* **67** (1960), 55.

[111] D. Piguet & M. Stein, An approximate version of the Loebl–Komlós–Sós conjecture, preprint.

[112] D. Piguet & M. Stein, The Loebl–Komlós–Sós conjecture for trees of diameter 5 and certain caterpillars, *Electron. J. Combin.* **15** (2008), R106.

[113] O. Pikhurko, Perfect matchings and K_4^3-tilings in hypergraphs of large codegree, *Graphs Combin.* **24** (2008), 391–404.

[114] L. Pósa, A theorem concerning Hamiltonian lines, *Magyar Tud. Akad. Mat. Fiz. Oszt. Kozl.* **7** (1962), 225–226.

[115] V. Rödl, A. Ruciński & E. Szemerédi, Perfect matchings in uniform hypergraphs with large minimum degree, *European J. Combin.* **27** (2006), 1333–1349.

[116] V. Rödl, A. Ruciński & E. Szemerédi, A Dirac-type theorem for 3-uniform hypergraphs, *Combin. Probab. Comput.* **15** (2006), 229–251.

[117] V. Rödl, A. Ruciński & E. Szemerédi, An approximate Dirac theorem for k-uniform hypergraphs, *Combinatorica* **28** (2008), 229–260.

[118] V. Rödl, A. Ruciński & E. Szemerédi, Perfect matchings in large uniform hypergraphs with large minimum collective degree, *J. Combin. Theory Ser. A*, in press.

[119] V. Rödl & J. Skokan, Regularity lemma for k-uniform hypergraphs, *Random Structures Algorithms* **25** (2004), 1–42.

[120] L. Shi, Cube Ramsey numbers are polynomial, *Random Structures Algorithms* **19** (2001), 99–101.

[121] A. Shokoufandeh & Y. Zhao, Proof of a conjecture of Komlós, *Random Structures Algorithms* **23** (2003), 180–205.

[122] A. Shokoufandeh & Y. Zhao, On a tiling conjecture for 3-chromatic graphs, *Discrete Math.* **277** (2004), 171–191.

[123] Z. M. Song, Complementary cycles of all lengths in tournaments, *J. Combin. Theory Ser. B* **57** (1993), 18–25.

[124] B. Sudakov & J. Vondrak, Nearly optimal embedding of trees, preprint.

[125] E. Szemerédi, Regular partitions of graphs, in *Problémes combinatoires et Théorie des Graphes (Colloq. Internat. CNRS, Univ. Orsay, Orsay, 1976) Colloq. Internat. CNRS*, 260, CNRS, Paris (1978), pp. 399–401.

[126] C. Thomassen, An Ore-type condition implying a digraph to be pancyclic, *Discrete Math.* **19** (1977), 85–92.

[127] C. Thomassen, Long cycles in digraphs with constraints on the degrees, in *Surveys in Combinatorics* (ed. B. Bollobás), *Lond. Math. Soc. Lecture Note Ser.*, 38, Cambridge University Press, Cambridge–New York (1979), pp. 211–228.

[128] C. Thomassen, Edge-disjoint hamiltonian paths and cycles in tournaments, *Proc. Lond. Math. Soc.* **45** (1982), 151–168.

[129] D. Woodall, Sufficient conditions for cycles in digraphs, *Proc. Lond. Math. Soc.* **24** (1972), 739–755.

[130] R. Yuster, Combinatorial and computational aspects of graph packing and graph decomposition, *Computer Science Review* **1** (2007), 12–26.

[131] Y. Zhao, Proof of the $(n/2 - n/2 - n/2)$ conjecture for large n, preprint.

[132] Y. Zhao, Bipartite tiling problems, *SIAM J. Discrete Math.*, in press.

Daniela Kühn and Deryk Osthus
School of Mathematics Birmingham University
Edgbaston
Birmingham B15 2TT
UK
{kuehn,osthus}@maths.bham.ac.uk

Counting planar graphs and related families of graphs

Omer Giménez   Marc Noy

Abstract

In this article we survey recent results on the asymptotic enumeration of planar graphs and, more generally, graphs embeddable in a fixed surface and graphs defined in terms of excluded minors. We also discuss in detail properties of random planar graphs, such as the number of edges, the degree distribution or the size of the largest k-connected component. Most of the results we present use generating functions and analytic tools.

Introduction

We consider planar graphs as combinatorial objects, regardless of how many different embeddings they may have in the plane. Let g_n be the number of labelled planar graphs with n vertices, and let R_n be a graph taken uniformly at random among all g_n labelled planar graphs with n vertices. The main questions we are interested in are the following.

(1) How large is g_n asymptotically?

(2) Which are the typical properties of R_n for n large?

(3) What can be said about related families of graphs?

As we are going to see, we have a complete answer to question (1), and significant results concerning (2) and (3). A precise estimate for g_n was obtained by the present authors [38], together with the derivation of limit laws for the basic parameters in random planar graphs. The techniques introduced in [38] have been applied to the analysis of other classes of graphs [12, 35], to the analysis of more advanced parameters [24, 25, 39], and to the uniform sampling of planar graphs [30]. A more probabilistic approach for analysing similar questions is used in [47, 48, 44, 46].

Graph enumeration is a well-established subject in combinatorics. The monograph by Harary and Palmer [40] is mostly devoted to determining the number of unlabelled graphs of various kinds with given number of vertices and edges, and the main tool is Pólya's enumeration theorem [57]. Closer in spirit to this paper is the monograph by Moon [50], a thorough account of the enumeration of labelled trees and parameters in random trees. After the 1970s there is more emphasis on asymptotics, and on the interplay between graph enumeration and the theory of random graphs. A noteworthy example is the enumeration of regular graphs and, more generally, graphs with given degree sequence [15, 69].

The main source and motivation for the results reviewed here is the theory of map enumeration. A map is a connected planar graph together with a particular embedding in the plane. There is a rich and beautiful theory of map enumeration, started by Tutte in the 1960s. The field has grown enormously since then and many classes of maps have been studied, including maps in arbitrary surfaces. Moreover, deep connections with algebra, geometry and physics have been uncovered [42].

Some of the results on maps are key ingredients in the enumeration of planar graphs. This was first realized by Bender, Gao and Wormald [4], and opened the way to many enumerative results that followed. However, we wish to point out that counting planar graphs is more difficult than counting planar maps. First, the generating functions for planar graphs are no longer algebraic; and secondly, we need to root planar graphs both at vertices and at edges, and this implies quite non-trivial relations among the various generating functions involved.

Here is an outline of the paper. Sections 1 to 4 contain background material: decomposition of graphs into 2-connected and 3-connected components; generating functions associated to planar graphs; planar maps and their connection with planar graphs; and a quick introduction to the analytic tools based on singularity analysis. In Section 5 we present the asymptotic enumeration of planar graphs, one of the main results reviewed in this paper.

Sections 6 to 8 are devoted to random graphs. In Section 6 we define the model of random planar graphs and discuss some of the basic parameters, like the number of edges and the number of components. In Section 7 we study the distribution of vertex degrees, and in Section 8 we discuss extremal parameters, with particular emphasis on the size of the largest 2-connected component.

In Section 9 we introduce a general framework for enumerating classes of graphs with given 3-connected components, and for analysing random graphs from these classes. We show a dichotomy between classes similar to planar graphs, in which the largest block is of linear size, and classes similar to series-parallel graphs, in which the largest block is small. Section 10 gives an overview of graphs embeddable in higher surfaces and classes of graphs defined in terms of excluded minors.

Section 11 is devoted to the study of graphs with given average degree and shows examples of graph classes in which the structure depends critically on the average degree. In Section 12 we discuss algorithms for sampling planar graphs uniformly at random, and in Section 13 we review briefly what is known about unlabelled graphs.

Finally, a few words on notation and terminology. All graphs are finite, simple and labelled, and n is the number of vertices. Generating functions are of the exponential type, unless we explicitly state the contrary. Variables x and y mark, respectively, vertices and edges; variable z marks edges in 3-connected graphs and maps. By $[x^n]A(x)$ we mean the coefficient of x^n in the power series $A(x)$. The partial derivatives of $A(x,y)$ with respect to x and y are written $A'(x,y)$ and $A_y(x,y)$. The notation $a_n \sim b_n$ is equivalent to $\lim_{n\to\infty} a_n/b_n = 1$. By a.a.s. we mean asymptotically almost surely, which in our case means a property of random graphs whose probability tends to 1 as n goes to infinity. The probability of an event A is written $\mathbf{P}(A)$, and the expected value of a random variable X is written $\mathbf{E}[X]$.

1 Graph decompositions

We start by recalling the elementary fact that a graph decomposes into disjoint connected components. It is also well known that a connected graph decomposes into blocks, or 2-connected components, along cut vertices. As illustrated in Figure 1, the blocks and the cut vertices are arranged in a tree-like structure. Notice that the blocks can share vertices but not edges, and that a single edge is the smallest block.

Possibly less known is the fact, due to Tutte [64], that a 2-connected graph decomposes into 3-connected components. We present here a brief outline of the theory; for a more formal and comprehensive exposition we refer to [19].

Counting planar graphs

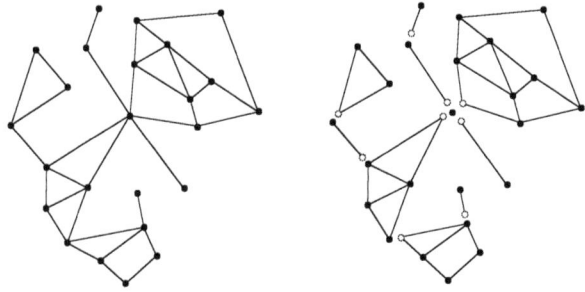

Figure 1: Blocks.

Let G be a 2-connected graph and let $\{x,y\}$ be a 2-cut of G. Let $V_1,\ldots,V_k$ be the vertex sets of the $k \geq 2$ connected components after removing $\{x,y\}$, and let G_i be the subgraph of G induced by $V_i \cup \{x,y\}$. We say that the 2-cut $\{x,y\}$ is *good* if any other 2-cut $\{u,v\}$ of G is contained in some G_i; otherwise we call it *bad*. Good 2-cuts in 2-connected graphs play the role of cut vertices in connected graphs, since they decompose the graph into smaller pieces forming a tree-like structure, as in Figure 2. We proceed to show that the only 2-connected graphs without good 2-cuts are either 3-connected or cycles, represented respectively by black and white vertices in the tree of the figure; grey vertices in the tree correspond to 2-cuts that induce three or more connected components. For this we need a simple lemma.

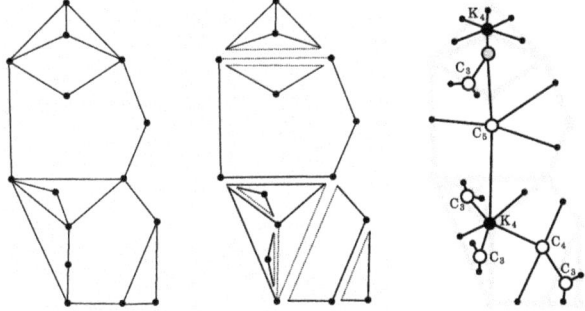

Figure 2: A 2-connected graph, its decomposition along good 2-cuts, and the associated tree structure.

Lemma 1.1 *Let G be a 2-connected graph. A 2-cut $\{x,y\}$ is bad if and only if the following conditions hold:*

(a) vertices x and y are not adjacent;

(b) the removal of $\{x,y\}$ produces exactly two connected graphs G_1 and G_2;

(c) both G_1 and G_2 have cut vertices.

Proof Clearly, if (a), (b) and (c) hold then the 2-cut $\{x,y\}$ is bad, since the cut vertices of G_1 and G_2 form a 2-cut of G that disconnects x from y, as in Figure 3.

To prove the converse, assume that the 2-cut $\{x,y\}$ is bad, and let $\{u,v\}$ be another 2-cut of G where $u \in V_1$ and $v \in V_2$. Note that $\{u,v\}$ disconnects x from y, otherwise either u or v are cut vertices of G, a contradiction since G is 2-connected. Conditions (a), (b) and (c) follow from this observation. $\square$

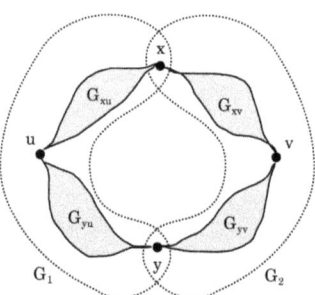

Figure 3: A bad 2-cut $\{x,y\}$.

Let now G be a 2-connected graph without good 2-cuts. Then either G has no 2-cuts at all and it is 3-connected (or a triangle), or else all its 2-cuts are bad. Let $\{x,y\}$ be one of them. The proof of Lemma 1.1 implies that G is as in Figure 3, that is, vertices x, y, u and v divide G into four subgraphs G_{xu}, G_{xv}, G_{yu} and G_{yv}. Consider for instance G_{xu}: it is either a single edge joining u and x, or it contains some other vertices. In this case $\{u,x\}$ is a 2-cut of G, which is bad by assumption. It follows that G_{xu} has a cut vertex. An induction argument shows that G must be a cycle with at least 4 vertices, as was to be proved.

Let us consider again the tree structure obtained by decomposing a graph along good 2-cuts. We say that we have an h-composition when the different pieces are joined along a 3-connected graph H (black nodes in Figure 2); a *series* composition when the subgraphs are joined along a cycle (white nodes); and a *parallel* composition when we have several subgraphs that share a common 2-cut (grey nodes). Parallel compositions can be seen also as joining subgraphs along cocycles, that is, two vertices joined by parallel edges.

The consequence of the previous decompositions is that 3-connected graphs can be considered as the building blocks of all graphs. In particular, this also applies to planar graphs, since it is easy to check that a graph is planar if and only if all its connected, 2-connected and 3-connected components are planar.

2 Generating functions

Generating functions are the key tool we use for enumeration. Let $\mathcal{G}$ be a class of labelled graphs with the property, shared by the class of planar graphs, that a graph is in $\mathcal{G}$ if and only if all its connected, 2-connected and 3-connected components are in $\mathcal{G}$. Let g_n be the number of graphs in $\mathcal{G}$ with n vertices. The associated generating function (GF for short) is

$$G(x) = \sum_{n \geq 0} g_n \frac{x^n}{n!}.$$

Since our graphs are labelled, it is an *exponential* GF. Similarly, let c_n be the number of connected graphs in $\mathcal{G}$ with n vertices, and let $C(x)$ be the associated GF. Then we have

$$G(x) = e^{C(x)}, \qquad (2.1)$$

since the exponential operator applied to GFs corresponds to taking (unordered) connected components. Here and throughout we use the so-called symbolic method for translating combinatorial constructions into equations satisfied by the associated generating functions; see [28] for a thorough exposition.

Let now b_n be the number of 2-connected graphs in $\mathcal{G}$ with n vertices, and let $B(x)$ be the associated GF. The equation linking $C(x)$ and $B(x)$ is a bit more complex than (2.1), but still elementary:

$$xC'(x) = xe^{B'(xC'(x))}, \qquad (2.2)$$

where $C'(x)$ and $B'(x)$ are derivatives. We do not cancel x since we are going to work with $xC'(x) = \sum nc_n x^n/n!$. This is the GF associated to connected graphs *rooted* at a vertex, since there are n ways to select the root vertex. Equation (2.2) reflects the recursive decomposition of rooted connected graphs. The term $\exp(B')$ corresponds to the set of 2-connected components containing the root, and the substitution $B'(xC'(x))$ encodes the recursion.

In order to discuss the situation for 3-connected graphs, we need to enrich our generating functions by taking into account also the number of edges. Let $g_{n,k}$ be the number of graphs in $\mathcal{G}$ with n vertices and k edges, and let

$$G(x,y) = \sum_{n,k} g_{n,k} y^k \frac{x^n}{n!}$$

be the associated bivariate GF. We define similarly $C(x,y)$ and $B(x,y)$ for connected and 2-connected graphs, respectively. Notice that if we set $y = 1$ (that is, we ignore the edges), then we recover the univariate GFs introduced above.

Equations (2.1) and (2.2) extend naturally to

$$G(x,y) = \exp(C(x,y)), \qquad xC'(x,y) = x \exp\left(B'(xC'(x,y),y)\right), \qquad (2.3)$$

where the derivatives are partial derivatives with respect to x. This is because connected and 2-connected components do not share edges.

Next we introduce a class of graphs closely related to 2-connected graphs. A *network* is a graph with two distinguished vertices, called poles, such that the graph obtained by adding an edge between the two poles (if they are not adjacent) is 2-connected. If there is an edge joining the poles, it is called the *root edge*. Moreover, the two poles are not labelled but they are distinguishable. Networks are the key technical device for encoding the decomposition of 2-connected graphs into 3-connected components. Let $D(x,y)$ be the GF associated to networks, where again x and y mark vertices and edges. Then we have

$$2(1+y)B_y(x,y) = x^2(1 + D(x,y)), \qquad (2.4)$$

where B_y denotes derivative with respect to y. The left-hand side corresponds to 2-connected graphs rooted at an oriented edge, and this edge may be included or not; hence the factor $2(1+y)$. The right-hand side corresponds to networks (possibly empty), where the factor x^2 gives labels to the two poles.

We distinguish between three kinds of networks. A network is *series* if it is obtained from a cycle C with a distinguished edge e, whose endpoints become the poles, and every edge different from e is replaced by a network. Equivalently, when removing the root edge if present, the resulting graph is not 2-connected. A network is *parallel* if it is obtained by gluing two or more networks, none of them containing the root edge, along the common poles. Equivalently, when the two poles are a 2-cut of the network. Finally, an h-network is obtained from a 3-connected graph H rooted at an oriented edge, by replacing every edge of H (other than the root) by an arbitrary network. In each of the three cases the edge joining the poles can be added or not. In addition, we must consider the trivial network consisting of a single edge. Trakhtenbrot [60] showed that a network is either series, parallel or an h-network, and Walsh [67] translated this fact into generating functions as we show next.

In order to establish the equation defining $D(x,y)$ we need the GF

$$T(x,z) = \sum_{n,k} t_{n,k} z^k \frac{x^n}{n!},$$

where $t_{n,k}$ is the number of 3-connected graphs in $\mathcal{G}$ with n vertices and k edges. Notice that in this case the we use variable z for marking edges instead of y.

Let $S(x,y)$ be the GF of series networks without root edge. A series network is a sequence of $k \geq 2$ networks where we identify the second pole of the i-th network with the first pole of the $(i+1)$-th network (see Figure 4). We can make this decomposition unique by observing that a series network is the composition of a network which is not series (they are enumerated by $D(x,y) - S(x,y)$) with an arbitrary network. This translates into the equation

$$S(x,y) = x(D(x,y) - S(x,y))D(x,y). \tag{2.5}$$

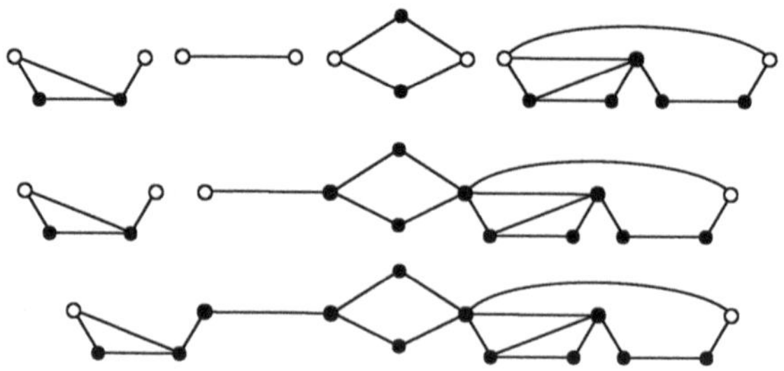

Figure 4: Series composition of networks.

Next we remark that a network is either a series network, or an h-network, or the parallel composition of series and h-networks along the common poles. Series networks are enumerated by the generating function $S(x,y)$ and h-networks by

$2x^{-2}T_z(x, D(x,y))$. The latter corresponds to 3-connected graphs rooted at a directed edge, in which the root is deleted and its endpoints bear no label, and every other edge is replaced by a network. Thus we have

$$D(x,y) = (1+y)\exp\bigl[S(x,y) + 2x^{-2}T_z(x, D(x,y))\bigr] - 1, \qquad (2.6)$$

where the term $(1+y)$ corresponds to adding or not adding an edge between the poles, and the term $\exp(S + 2x^{-2}T_z)$ corresponds to composition of networks formed with series and h-networks. We subtract 1 to remove the graph with two vertices and no edges.

Finally, eliminating $S(x,y)$ from (2.5) and (2.6), we obtain the defining equation for $D = D(x,y)$ as

$$\frac{2}{x^2}T_z(x,D) - \log\left(\frac{1+D}{1+y}\right) + \frac{xD^2}{1+xD} = 0. \qquad (2.7)$$

Summarizing, $T(x,z)$ determines $D(x,y)$ uniquely, which in turn determines $B(x,y)$, which in turn determines $C(x,y)$ and $G(x,y)$. In other words, if we know the GF of 3-connected graphs in the class $\mathcal{G}$ then, to some extent, we know the GF for all graphs in the class. Admittedly, this is via a chain of non-trivial functional-differential equations (2.7), (2.4), (2.3) starting with $T(x,z)$ and ending with $G(x,y)$, but the crucial point is that the former determines the latter.

Let us consider the class of planar graphs, which is the most interesting example for us. The smallest 3-connected planar graphs are K_4, the wheel W_4 with four spokes, and K_5^-, the graph K_5 minus one edge. They can be labelled, respectively, in $1, 15$ and 10 ways. Hence

$$T(x,z) = z^6\frac{x^4}{4!} + \left(15z^8 + 10z^9\right)\frac{x^5}{5!} + \cdots$$

From (2.7) we obtain

$$D(x,y) = y + (y^2+y^3)x + (y^3 + 7y^4 + 6y^5 + y^6)\frac{x^2}{2!} + \cdots$$

For instance, there are 7 planar networks with 4 edges and two labelled vertices (in addition to the two poles), which are depicted in Figure 5.

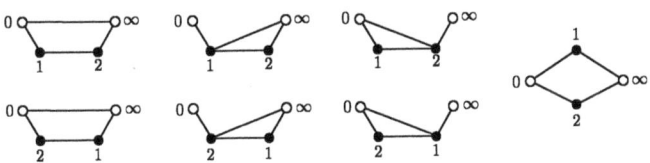

Figure 5: The planar networks with 4 edges and two labelled vertices; the poles are marked 0 and ∞.

Using (2.4) we integrate term by term and obtain

$$B(x,y) = y\frac{x^2}{2!} + y^3\frac{x^3}{3!} + (3y^4+6y^5+y^6)\frac{x^4}{4!} + (12y^5+70y^6+100y^7+45y^8+10y^9)\frac{x^5}{5!} + \cdots$$

For instance, the term $45y^8x^5/5!$ accounts for the 15 labellings of W_4 plus the 30 labellings of the graph obtained by gluing K_4 and a triangle along an edge. Finally, using (2.3) we obtain

$$\begin{aligned} C(x,y) &= x + y\frac{x^2}{2!} + (3y^2+y^3)\frac{x^3}{3!} + (16y^3+15y^4+6y^5+y^6)\frac{x^4}{4!} + \cdots \\ G(x,y) &= x + (1+y)\frac{x^2}{2!} + (1+y)^3\frac{x^3}{3!} + (1+y)^6\frac{x^4}{4!} + ((1+y)^{10}-y^{10})\frac{x^5}{5!} \cdots \end{aligned}$$

For instance, the term $16y^3x^4/4!$ in $C(x,y)$ accounts for the 16 labelled trees on 4 vertices. The last equation corresponds to the fact that all graphs up to 5 vertices are planar with only the exception of K_5, which has a unique labelling.

The moral of the preceding computations is that from the knowledge of $T(x,z)$ we can effectively obtain the first coefficients of $G(x,y)$, that is, the number of planar graphs with a given number of vertices and edges. This applies to any class of graphs satisfying the condition stated at the beginning of this section.

The good news is that we have an explicit expression for $T(x,z)$ in the case of planar graphs. This is discussed in the next section, together with Whitney's theorem and the theory of map enumeration.

3 Maps and graphs

A planar graph can have several non-equivalent embeddings in the plane. For instance, the two drawings in Figure 6 correspond to different embeddings: they cannot be transformed into each other by a homeomorphism of the plane. A connected (unlabelled) planar graph together with a specific embedding is called a *planar map*. From now on we omit the qualifier and speak simply of *maps*. This definition can be made completely rigorous by considering the circular ordering of the edges around each vertex [49]. We remark that maps are usually allowed to have loops and multiple edges, but not in our case.

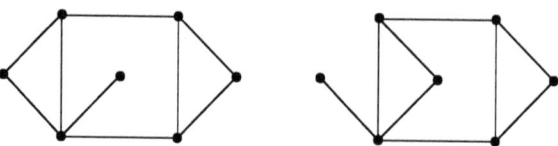

Figure 6: Different embeddings of the same planar graph.

In this paper we consider only rooted maps: an edge $r = \overrightarrow{xy}$ of the map is selected and oriented, and becomes the root edge; x is the root vertex and the face to the right of r (usually chosen to be the outer face) is the root face. Two rooted maps are isomorphic if there is bijection that fixes the root and preserves incidences between vertices, edges and faces. A very useful property of rooting for the purpose of counting is that, as is easily checked, rooted maps do not have non-trivial isomorphisms.

We need at this point the classical theorem of Whitney [68]: a 3-connected planar graph has a unique embedding in the sphere up to homeomorphism, or a unique embedding in the plane up to the choice of the outer face. As a consequence,

counting 3-connected planar graphs is essentially equivalent to counting 3-connected maps. More precisely, if $M_{n,k}$ is the number of 3-connected rooted maps with n vertices and k edges, and $t_{n,k}$ the number of 3-connected graphs as before, then, as observed in [4],
$$n!\, M_{n,k} = 4k\, t_{n,k}.$$
Indeed, both sides count in two ways labelled rooted 3-connected maps. This implies the relation
$$M(x,z) = \sum_{n,k} M_{n,k} z^k x^n = 4z T_z(x,z) \qquad (3.1)$$
between the ordinary GF of maps and the exponential GF of graphs. Hence knowing $M(x,z)$ we have access to $T(x,z)$ which, we recall, is the starting point for the enumeration of planar graphs.

The series $M(x,z)$ was obtained by Mullin and Schellenberg [51]. They showed that
$$M(x,z) = x^2 z^2 \left(\frac{1}{1+xz} + \frac{1}{1+z} - 1 - \frac{(1+U)^2(1+V)^2}{(1+U+V)^3} \right), \qquad (3.2)$$
where $U(x,z)$ and $V(x,z)$ are algebraic functions given by
$$U = xz(1+V)^2, \quad V = z(1+U)^2. \qquad (3.3)$$
Observe that M, being a rational function of algebraic functions, is also algebraic. We remark that the solution in [51] is given according to the number of vertices and faces, but by Euler's formula this also determines the number of edges.

Let us sketch the main ideas from [51]. There is a simple bijection between 2-connected maps (without loops or multiple edges) and quadrangulations. To a map M we associate a quadrangulation Q as follows. For every face f of M, add a new vertex f' and join it to all the vertices incident with f. Now erase all the edges of M to obtain the map Q (see Figure 7). It is not difficult to check that under this bijection 3-connected maps correspond precisely to quadrangulations without separating quadrangles. For counting the latter the approach is similar to that used by Tutte [62] for counting triangulations without separating triangles.

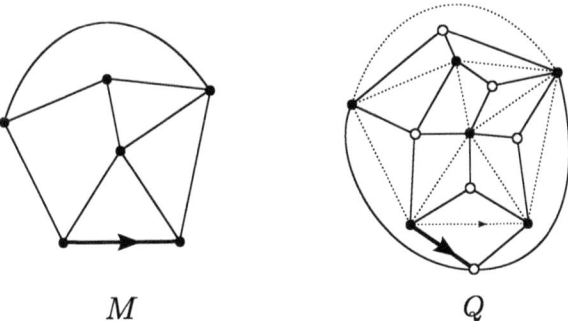

Figure 7: Quadrangulation associated to a 3-connected map.

We close this section with two remarks. The theory of map enumeration was started by Tutte in his famous 'census' papers, including [62, 63]. Many classes of rooted maps have been enumerated since then, but not all of them; for instance, a notable open problem is counting 4-connected maps. They lead invariably to algebraic (ordinary) generating functions. The main reason is that the embedding gives a natural *ordering* on the edges and faces around a vertex, and also on the 2-connected components around a cut vertex. This corresponds to the *sequence* construction, which translates into the operator $1/(1 - A(x))$ for GFs. However the *set* construction, used extensively in Section 2, gives rise to the exponential operator $\exp A(x)$. As a rule, graph enumeration leads to non-algebraic GFs, unless the degree of connectivity makes it equivalent to map enumeration. However, as we see later, from an analytic point the algebraic nature of $M(x,z)$ determines in an essential way the shape of the asymptotic estimates for planar graphs.

The second remark is that maps have been studied also in surfaces of higher genus, leading again to algebraic GFs; see [2, 31] for the enumeration of maps in general surfaces. Later in Section 10 we discuss graphs on general surfaces.

4 Analytic combinatorics

In order to obtain asymptotic estimates and probability limit laws, we need to consider generating functions as analytic functions of complex variables. We enter here the field of *analytic combinatorics*, as developed in the forthcoming book by Flajolet and Sedgewick [28]. In this section we review the principles of the theory and discuss the results needed later.

Let $A(x) = \sum a_n x^n$ be a power series with non-negative coefficients, a natural assumption in enumeration, and suppose $A(x)$ has radius of convergence $\rho > 0$. Thus $A(x)$ defines an analytic function of x in the open disk $|x| < \rho$ of the complex plane. Our goal is to estimate the order of magnitude of the a_n. The first approximation comes from the value of ρ. Indeed, it is well known that

$$\rho^{-1} = \limsup_{n \to \infty} (a_n)^{1/n}.$$

In our applications the lim sup is always a limit and we have

$$a_n \sim w(n)\rho^{-n}, \tag{4.1}$$

where $w(n)$ grows subexponentially, in the sense that $\lim w(n)^{1/n} = 1$. The typical shape we encounter is $w(n) = c \cdot n^\alpha$, where c is a constant and α is a rational number, most often $\alpha = \nu/2$ for some integer ν. The value ρ^{-1} is called the *growth constant* of the a_n.

Pringsheim's theorem guarantees that ρ is a *singularity* of $A(x)$, that is, $A(x)$ cannot be analytically continued in a neighbourhood of ρ. Since $A(x)$ is analytic for $|x| < \rho$, we see that ρ is a singularity of smallest modulus, also called a *dominant singularity*. There might be other singularities of modulus ρ, but this situation does not arise in our context and we can assume from now on that $x = \rho$ is the unique dominant singularity.

The estimate (4.1) tells us that the exponential growth of the a_n is determined by the *location* of the dominant singularity; this is the first principle. The second principle is that the subexponential term $w(n)$ is determined by the *nature* of the dominant singularity, that is, by the behaviour of $A(x)$ near ρ. As an example let us consider the generating function of the Catalan numbers

$$c(x) = \sum \frac{1}{n+1}\binom{2n}{n} x^n = \frac{1 - \sqrt{1-4x}}{2x}. \qquad (4.2)$$

The dominant singularity is at $x = 1/4$, because the function $\sqrt{z}$ of a complex variable z cannot be defined analytically in a neighbourhood of 0. Using Stirling's estimate it follows easily that

$$\frac{1}{n+1}\binom{2n}{n} \sim \frac{1}{\sqrt{\pi}} n^{-3/2} 4^n. \qquad (4.3)$$

Thus the subexponential term is of the form $n^{-3/2}$.

In general we do not have a closed form expression for the coefficients a_n of $A(x)$. However, the theory of *singularity analysis* [27, 28] allows one to obtain precise estimates of the coefficients under rather general conditions on the behaviour of $A(x)$ near the dominant singularity ρ. The following result is a simplified version, adapted to our needs, of the basic 'transfer theorem' from [27]. For $R > \rho > 0$ and $0 < \phi < \pi/2$, define the complex domain

$$\Delta(\phi, \rho, R) = \{x \in \mathbb{C} : x \neq \rho, |x| < R, |\arg(x-\rho)| > \phi\},$$

illustrated in Figure 8.

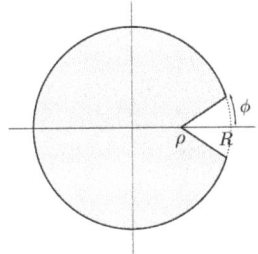

Figure 8: The domain $\Delta(\phi, \rho, R)$.

Theorem 4.1 (Transfer) *Let α be a real number not in $\{0, -1, -2, \dots\}$ and let $R > \rho > 0$. Assume that $f(x)$ is analytic in a domain $\Delta = \Delta(\phi, \rho, R)$.
If, as $x \to \rho$ in Δ,*

$$f(x) \sim (1 - x/\rho)^{-\alpha}$$

then

$$[x^n] f(x) \sim \frac{n^{\alpha-1}}{\Gamma(\alpha)} \rho^{-n}.$$

In combinatorial applications, the analytic conditions of the previous theorem are often easy to establish. The proof of Theorem 4.1 is based on Cauchy's integral formula

$$[x^n]f(x) = \frac{1}{2\pi i}\int_C \frac{f(x)}{x^{n+1}}dx,$$

together with a suitable selection of the contour C enclosing the origin, and a careful analysis of the main contribution in the integral; we refer to [28, Chapter 6] for details.

The main point is that from the knowledge of $f(x)$ near the singularity ρ we can deduce a precise asymptotic estimate for $[x^n]f(x)$. When $f(x)$ is an algebraic function, the Newton-Puiseux theorem provides a representation of $f(x)$ as a series in fractional powers of $1 - x/\rho$. Typically one has a square-root singular expansion, that is, an expansion of $f(x)$ near $x = \rho$ in powers of $\sqrt{1 - x/\rho}$. The trivial example in (4.2) is an exact square root, that is, $c(x)$ satisfies the hypothesis of the theorem with $\alpha = -1/2$, and we get a term $n^{-3/2}$, in accordance with (4.3).

For the algebraic generating functions associated to planar maps one systematically encounters $\alpha = -3/2$ and a corresponding subexponential term $n^{-5/2}$. This kind of universal law for maps is very remarkable and we do not have yet a clear explanation for it [17]. The case of 3-connected maps is discussed in the next section.

The singular expansions we encounter in this paper are always of the form

$$f(x) = f_0 + f_2 X^2 + f_4 X^4 + \cdots + f_{2k} X^{2k} + f_{2k+1} X^{2k+1} + O(X^{2k+2}),$$

where $X = \sqrt{1 - x/\rho}$. That is, $2k + 1$ is the smallest odd integer i such that $f_i \neq 0$. The even powers of X are analytic functions and do not contribute to the asymptotics of $[x^n]f(x)$. The number $e = (2k+1)/2$ is called the *singular exponent*, and by the transfer theorem we obtain the estimate

$$[x^n]f(x) \sim c \cdot n^{-e-1}\rho^{-n},$$

where $c = f_{2k+1}/\Gamma(-e)$.

5 The number of planar graphs

This section is devoted to the asymptotic enumeration of planar graphs. Bender, Gao and Wormald [4] showed that the number b_n of 2-connected planar graphs with n vertices is asymptotically

$$b_n \sim b \cdot n^{-7/2}\delta^n n!, \tag{5.1}$$

where b and δ are well-defined analytic constants and $\delta \approx 26.18412$. The proof is based, as we discuss below, on the decomposition of 2-connected graphs into 3-connected components, and on the enumeration of 3-connected planar maps.

Extending the techniques introduced in [4], the present authors were able to prove the following result [38].

Theorem 5.1 *Let g_n and c_n be, respectively, the number of planar graphs and connected planar graphs with n vertices. Then*

$$g_n \sim g \cdot n^{-7/2}\gamma^n n!, \qquad c_n \sim c \cdot n^{-7/2}\gamma^n n!, \tag{5.2}$$

where g, c and γ are well-defined analytic constants and $\gamma \approx 27.22688$.

Counting planar graphs

The proof of Theorem 5.1 is quite technical and we refer to [38] for details. Here we sketch the main ingredients in the proof. For the remainder of the this section, all graphs are planar.

We start with 3-connected graphs. If $M_n = \sum_k M_{n,k}$ is the number of 3-connected maps with n vertices as in Section 3, then $M(x) = \sum M_n x^n$ has a unique dominant singularity at $r = (7\sqrt{7} - 17)/32$. This value is obtained as the root of the discriminant of the quartic equation (3.3) defining $U(x,1)$.

The singular expansion of $M(x)$ near r is of the form

$$M(x) = \mu_0 + \mu_2 X^2 + \mu_3 X^3 + O(X^4),$$

where $X = \sqrt{1 - x/r}$. Applying Theorem 4.1 we get an estimate $M_n \sim c_1 \cdot n^{-5/2} r^{-n}$, where c_1 is a suitable constant. Equation (3.1) then implies an estimate for the number $t_n = \sum_k t_{n,k}$ of 3-connected planar graphs of the form

$$t_n \sim c_2 \cdot n^{-7/2} r^{-n}.$$

Note that the growth constant for 3-connected graphs is $r^{-1} = (272 + 112\sqrt{7})/27 \approx 21.04904$.

The approach taken in [4] for counting 2-connected planar graphs is rather indirect. From (2.7) it follows a singular expansion

$$D(x,y) = D_0(y) + D_2(y)X^2 + D_3(y)X^3 + O(X^4), \qquad (5.3)$$

where now $X = \sqrt{1 - x/R(y)}$, and $R(y)$ is the dominant singularity of $D(x,y)$ as a function of x. We stress the fact that the singularity $R(y)$ comes from the corresponding singularity of $T(x,z)$, that is, (2.7) does not introduce new singularities. As a consequence the local expansion of $D(x,y)$ near $R(y)$ is of an algebraic nature.

If we set $y = 1$ (ignore the number of edges) we get an expansion

$$D(x,1) = D_0 + D_2 X^2 + D_3 X^3 + O(X^4),$$

in powers of $X = \sqrt{1 - x/R}$, where $R = R(1) \approx 0.03819$. By Theorem 4.1 we get an estimate

$$[x^n]D(x,1) \sim c_3 \cdot n^{-5/2} \delta^n n!,$$

where $\delta = R^{-1}$ is the growth constant in (5.1). Now setting $y = 1$ in Equation (2.4) we obtain the relation

$$4 \sum_k k b_{n,k} = [x^{n-2}]D(x,1).$$

The sum on the left is the unnormalized expected value μ_n of the number of edges in 2-connected graphs, that is, $\sum_k k b_{n,k} = \mu_n b_n$. Applying the results in [6], a limit normal law follows, with $\mu_n \sim \kappa n$ for a computable constant $\kappa \approx 2.2629$. Then we have

$$b_n = \frac{1}{\mu_n} \sum_k k b_{n,k}$$

and the estimate (5.1) follows.

In order to proceed further we need an expression for $B(x,y)$ in terms of $D(x,y)$. This is accomplished in [38] by solving the differential equation (2.4); that is, by finding a primitive of $(1 + D(x,y))/(1+y)$, considered as a function of y. This is a key technical result whose proof relies on algebraic manipulation. We quote the result in full just to give an idea of the shape of the resulting expression.

Lemma 5.2 Let $W(x,z) = z(1 + U(x,z))$, where U is as in (3.3). Then

$$B(x,y) = \beta\left(x, y, D(x,y), W(x, D(x,y))\right), \tag{5.4}$$

where

$$\beta(x,y,z,w) = \frac{x^2}{2}\beta_1(x,y,z) - \frac{x}{4}\beta_2(x,z,w),$$

and

$$\beta_1(x,y,z) = \frac{z(6x - 2 + xz)}{4x} + (1+z)\log\left(\frac{1+y}{1+z}\right) - \frac{\log(1+z)}{2} + \frac{\log(1+xz)}{2x^2};$$

$$\beta_2(x,z,w) = \frac{2(1+x)(1+w)(z+w^2) + 3(w-z)}{2(1+w)^2} - \frac{1}{2x}\log(1 + xz + xw + xw^2)$$
$$+ \frac{1-4x}{2x}\log(1+w) + \frac{1 - 4x + 2x^2}{4x}\log\left(\frac{1 - x + xz + -xw + xw^2}{(1-x)(z + w^2 + 1 + w)}\right).$$

Admittedly, this is rather unwieldy but it has the advantage of being an *explicit* expression for $B(x,y)$, hence also for $B(x)$ setting $y=1$, in terms of known functions $D(x,y)$ and $U(x,z)$. It is remarkable that Chapuy et al. [19] have been able to obtain a completely combinatorial interpretation of all the terms in the preceding expression for $B(x,y)$.

By plugging (5.3) into (5.4) and setting $y = 1$, we obtain a singular expansion

$$B(x) = B_0 + B_2 X^2 + B_4 X^4 + B_5 X^5 + O(X^6), \tag{5.5}$$

again with $X = \sqrt{1 - x/R}$. Once more, (5.4) does not introduce new singularities and R is also the main singularity of $B(x)$. Observe that the singular exponent in (5.3) is 3/2 and not 5/2 as above. The reason is that networks are *rooted* objects, and rooting corresponds to differentiation and decreases the singular exponent by one.

From the previous expansion we can recover (5.1) by using the transfer theorem. But the important point is that now we can analyse $C(x)$, the GF for connected graphs, using Equation (2.2), which we recall is $xC'(x) = xe^{B'(xC'(x))}$. If we set $F(x) = xC'(x)$ for convenience, it can be rewritten as

$$F(x)e^{-B'(F(x))} = x.$$

This means that

$$\psi(u) = ue^{-B'(u)} \tag{5.6}$$

is the functional inverse of $F(x)$. The dominant singularity of ψ is the same as that of $B(x)$, which is equal to R. In order to determine the dominant singularity ρ of $F(x)$, we have to decide which of the following possibilities hold.

1. There exists $\tau \in (0, R)$ (necessarily unique) such that $\psi'(\tau) = 0$. By the implicit function theorem ψ ceases to be invertible at τ, and $\rho = \psi(\tau)$.

2. We have $\psi'(u) \neq 0$ for all $u \in (0, R)$, and there is no obstacle to the analyticity of the inverse function. Then $\rho = \psi(R)$ is the evaluation of $\psi(u)$ at its dominant singularity.

The condition $\psi'(\tau) = 0$ is equivalent to $B''(\tau) = 1/\tau$. Since $B''(u)$ is increasing (the series $B(u)$ has positive coefficients) and $1/u$ is decreasing, we are in case (2) if and only if $B''(R) < 1/R$. Using the explicit expression (5.4) we can check that this is the case, hence $\rho = \psi(R)$.

We have determined $\rho = Re^{-B'(R)} \approx 0.03672841$, and the growth constant for planar graphs $\gamma = \rho^{-1} \approx 27.22688$ in Theorem 5.1. We record for further reference the inverse relation

$$R = F(\rho) = \rho C'(\rho). \tag{5.7}$$

The fact that ψ' does not vanish and $\rho = \psi(R)$ implies that we have to invert ψ precisely at the singularity R in order to obtain the singularity ρ of $F(x)$. It follows that the inverse function $F(x)$ has a singular expansion near ρ in powers of $\sqrt{1 - x/\rho}$ similar to (5.5), although with singular exponent $3/2$ because $F(x) = xC'(x)$ enumerates rooted graphs. By integration we get an expansion for $C(x)$ and using (2.1) we get in turn an expansion for $G(x)$:

$$C(x) = C_0 + C_2 X^2 + C_4 X^4 + C_5 X^5 + O(X^6), \tag{5.8}$$
$$G(x) = G_0 + G_2 X^2 + G_4 X^4 + G_5 X^5 + O(X^6),$$

where now $X = \sqrt{1 - x/\rho}$. Then the transfer theorem gives directly (5.2), where $c = C_5/\Gamma(-5/2)$ and $g = ce^{C_0}$.

Summarizing, the main ingredients for the solution are the following:

1. An explicit expression for the GF of 3-connected graphs $T(x, z)$. In the planar case this comes from the equivalence between 3-connected planar maps and graphs (Whitney's theorem), together with the enumeration of 3-connected maps [51].

2. Equations (2.4) and (2.7) linking 3-connected and 2-connected graphs through networks, first used in [4] for counting 2-connected planar graphs.

3. An explicit solution of the differential equation (2.4). As shown in [38] this can be a difficult technical step.

4. Singularity analysis of the different generating functions involved and transfer theorems, in the spirit of [28].

This basic scheme can be worked out in principle for any class of graphs whenever we know the 3-connected members well enough to get an expression for $T(x, z)$. In Section 9 we develop such a general approach.

6 Random planar graphs

In this section we show how to prove results about random planar graphs using generating functions. But first, what is a random planar graph? We work with the following model, introduced by Denise, Vasconcellos and Welsh [21]. Let $\mathcal{G}_n$ be the set of labelled planar graphs with n vertices, and let $g_n = |\mathcal{G}_n|$ as in the previous section. We let R_n be a graph taken from $\mathcal{G}_n$ at random with the uniform distribution, that is, each graph is taken with probability $1/g_n$. We are interested in typical properties of R_n as $n \to \infty$.

Possibly the most basic parameter is the number of edges $e(R_n)$. Since a planar graph has at most $3n - 6$ edges, the range of this random variable is $[0, 3n]$. It was shown in [21] that the expected value of $e(R_n)$ is at least $3n/2$. This result was sharpened later [16, 36, 55] to

$$\mathbf{P}(1.85\, n < e(R_n) < 2.44\, n) \to 1, \qquad \text{as } n \to \infty.$$

As we show below, $e(R_n)$ is in fact strongly concentrated around $2.21n$.

The first substantial progress on random planar graphs was by McDiarmid, Steger and Welsh [47]. One of their main results is that, for any fixed labelled planar graph H, R_n contains a.a.s. a linear number of disjoint copies of H. This kind of 'density' result holds for many classes of maps; see [5, 34] and the references therein. It was also proved for 2-connected planar graphs in [4].

More precisely, fixing a root vertex r in H, each of these copies is attached to the rest of R_n through a single edge incident with the vertex in the copy corresponding to r (see Figure 9), and moreover the labelling in each copy is order-isomorphic to the labelling of H. Such a copy is called in [47] an *appearance* of H. If we let $f_H(R_n)$ be the number of appearances of H, then $f_H(R_n)$ is linear in n. The precise result from [47] is the following.

Theorem 6.1 *Let H be a planar graph with h vertices. Then there exists α depending only on h such that for n large enough*

$$\mathbf{P}(f_H(R_n) \leq \alpha n) < e^{-\alpha n}.$$

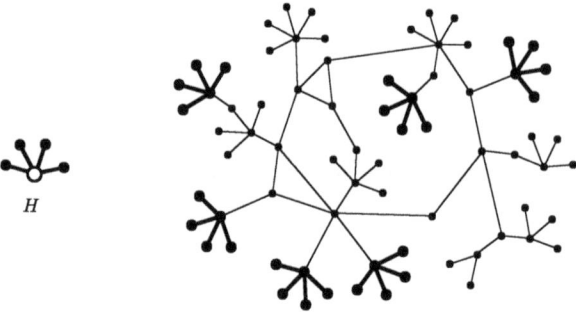

Figure 9: Many copies of H.

In particular, for $k > 0$ let H be a star with k vertices and the root at the center of the star. Then we obtain that a.a.s. R_n has a linear number of vertices of degree k. A noteworthy consequence is that with high probability R_n has exponentially many automorphisms. Indeed, take H as the star with two branches and notice that each copy of H in the above sense produces an automorphism of order two. Since these automorphisms can be composed independently and there are linearly many copies, the result follows.

A useful feature of Theorem 6.1 is that it holds not only for random planar graphs, but more generally for random graphs taken from a class $\mathcal{G}$ satisfying the following conditions: (i) if g_n is the number of (labelled) graphs in $\mathcal{G}$ with n vertices, then $g_n \le c^n n!$ for some constant c; (ii) a graph G is in $\mathcal{G}$ if and only if each component of G is in $\mathcal{G}$; and (iii) for each graph in $\mathcal{G}$, if u and v are in different components then the graph obtained by adding an edge between u and v is also in $\mathcal{G}$. For instance, these conditions are satisfied for the class of graphs avoiding a fixed collection of 2-connected graphs as minors; condition (i) is the only non-trivial one and it holds by a result from [53]. We go back to graph minors later in Section 10.

We come now to the use of analytic methods, which give less general but very precise results. Let us recall that

$$G(x,y) = \sum g_{n,k}\, y^k \frac{x^n}{n!}$$

is the generating function of planar graphs, where x and y mark vertices and edges, respectively. The first observation is that the probability generating function of the random variable $e(R_n)$, the number of edges, is equal to

$$p_n(y) = \sum_k \frac{g_{n,k}}{g_n} y^k = \frac{[x^n]G(x,y)}{[x^n]G(x,1)},$$

since $g_{n,k}/g_n$ is the probability that a random planar graph with n vertices has exactly k edges. In Section 5 we have seen that $g_n \sim g \cdot n^{-7/2} \gamma^n n!$. If we fix $y > 0$ and consider $G(x,y)$ as a function of x then it can be shown, using again singularity analysis, that

$$[x^n]G(x,y) \sim g(y)\, n^{-7/2} \gamma(y)^n n!,$$

where $g(y)$ and $\gamma(y)$ are analytic functions of y (observe that $g(1) = g$ and $\gamma(1) = \gamma$). We see that the dominant singularity $\gamma(y)^{-1}$ moves with y but the exponent $-7/2$ remains unchanged.

If we take into account error terms in the previous expansion, it follows that

$$p_n(y) = \frac{g(y)}{g}\left(\frac{\gamma(y)}{\gamma}\right)^n + O(1/n). \tag{6.1}$$

Thus the probability generating function is asymptotically 'almost' a fixed power of n. An exact power would correspond to a sum of n independent identically distributed random variables. It is well known that in this case the classical Central Limit Theorem implies a normal distribution. The so called Quasi-Powers Theorem [28] shows that from (6.1) one can prove that the limiting distribution is again normal. The precise result proved in [38] is the following.

Theorem 6.2 *The number of edges $e(R_n)$ in a random planar graph with n vertices is asymptotically normal, and the mean μ_n and variance σ_n^2 satisfy*

$$\mu_n \sim \kappa n, \qquad \sigma_n^2 \sim \lambda n, \tag{6.2}$$

where $\kappa \approx 2.21326$ and $\lambda \approx 0.43034$ are well-defined analytic constants.

The same is true, with the same constants, for connected random planar graphs.

Another result from [38] is a central limit theorem for the number of appearances of a planar subgraph, as defined above.

Theorem 6.3 *Let H be a fixed rooted connected planar graph with h vertices. Then the number of appearances $f_H(R_n)$ in a random planar graph is asymptotically normal and the mean μ_n, and variance σ_n^2 satisfy*

$$\mu_n \sim \frac{\rho^h}{h!} n, \qquad \sigma_n^2 \sim \rho n, \qquad (6.3)$$

where $\rho = \gamma^{-1}$ and γ is as in Theorem 5.1.

Notice in particular that the variance does not depend on H at all, and the expected value depends only on h and not on the structure of H. The full statement in [38] provides also large deviations estimates, thus making more precise the statement in Theorem 6.1.

We now turn to a different parameter. It is shown in [47] that the probability that a random planar graph is connected is bounded away from 0 and from 1, and that the number of connected components is stochastically dominated by $1 + X$, where X is a Poisson distribution with mean 1. This is made more precise in [38], where the following is proved.

Theorem 6.4 *The number of connected components in a random planar graph is distributed asymptotically as a $1+X$, where X is a Poisson law of parameter ν, and $\nu \approx 0.037439$ is a well-defined constant. In particular, the probability that a random planar graph is connected tends to $e^{-\nu} \approx 0.96325$.*

The proof uses the fact that $C(x)^k/k!$ is the GF for planar graphs with exactly k components. By taking the k-th power in (5.8) and using $G(x) = \exp C(x)$, we get

$$[x^n]C(x)^k \sim k\, C_0^{k-1}[x^n]C(x), \qquad [x^n]G(x) \sim e^{C_0}[x^n]C(x).$$

Now the probability that a random planar graph has exactly k components is asymptotically

$$\frac{[x^n]C(x)^k/k!}{[x^n]G(x)} \sim \frac{kC_0^{k-1}}{k!} e^{-C_0} = \frac{\nu^{k-1}}{(k-1)!} e^{-\nu},$$

where $\nu = C_0$. This is the Poisson distribution shifted by one, which is consistent, since the number of components is always positive.

As we discuss later in Section 9, the main results about the normal limit law for the number of edges and the Poisson limit law for the number of components can be proved for other classes of graphs, like series-parallel graphs and more generally graphs not containing some particular 3-connected graph graph as a minor, provided we have access to the counting generating functions.

It is natural to ask which additional parameters of planar graphs can be studied using our combinatorial and analytic framework. Clearly such a parameter has to appear, either explicitly or implicitly, in the successive graph decompositions into connected, 2-connected or 3-connected components. For instance, it is not too difficult to study parameters like the number of 2-connected components, the number of cut vertices, or the number of copies of a particular connected or 2-connected component [38]. On the other hand, parameters like the diameter or the number of k-colourings do not seem to fit into our scheme. In the next two sections we analyse several parameters that, although more complex than those of the present section, still can be analysed using combinatorial decompositions and analytic methods.

7 Degree distribution

We are interested in the number of vertices of a given degree in a random labelled planar graph. The first observation is that, following Theorem 6.1, with high probability there is a linear number of vertices of degree k for each $k > 0$. As we see now it is possible to obtain more precise results.

This problem is analysed in [25], and for the simpler class of series-parallel graphs it is proved that for each fixed $k > 0$, the number of vertices of degree k is asymptotically normal with linear mean and variance. The proof proceeds by establishing a system of equations satisfied by the generating functions A_i, enumerating rooted graphs with a root of degree $i = 1, 2, \ldots, k, \infty$, where ∞ means larger than k. The main tool is an extension of the analytic methods developed in [23] for proving central limit theorems.

For planar graphs it is an open problem to prove a normal limit law, but linearity of expectation has been established [24]. For each $k > 0$, the expected number of vertices of degree k in random planar graphs is asymptotically $d_k n$, for computable constants $d_k > 0$. This is equivalent to saying that the probability that a fixed vertex, say vertex with label 1, has degree k tends to a limit d_k as n goes to infinity (just consider for each vertex a variable indicating whether it has degree k). It is shown in [24] that this limit exists and an explicit expression for the probability generating function is obtained.

Theorem 7.1 *For each $k > 0$, the number of vertices of degree k in random planar graphs is asymptotically $d_k n$, where d_k is a computable constant. An explicit expression for $p(w) = \sum_{k \geq 1} d_k w^k$ can be computed and $\sum d_k = 1$.*

Moreover, as k goes to infinity,

$$d_k \sim c \cdot k^{-1/2} q^k \qquad (7.1)$$

for computable constants $c \approx 3.01751$ and $q \approx 0.67345$.

Analogous results are proved for 2- and 3-connected planar graphs, with the same value of q for 2-connected graphs, and $q = \sqrt{7} - 2$ for 3-connected graphs. We remark that the subexponential term $k^{-1/2}$ in the previous estimate is the usual pattern for several classes of planar maps [43]

Table 1 displays the approximate values for small degrees.

	d_1	d_2	d_3	d_4	d_5	d_6
Planar	0.03673	0.16258	0.23544	0.18677	0.12950	0.08618
Planar 2-connected	0	0.17284	0.24812	0.19253	0.13253	0.08798
Planar 3-connected	0	0	0.32749	0.24321	0.15942	0.10104

Table 1: Probabilities of small degrees.

In the remainder of this section, we outline the proof of Theorem 7.1, which is quite technical [24]. A first observation is that the degree distribution is the same for planar graphs and for *connected* planar graphs; an intuitive explanation is that, as we discuss in the next section, the expected number of vertices not in the largest component in a random planar graph is constant. Also, $d_0 = 0$ since the number of isolated vertices follows a discrete Poisson law.

Let $C^\bullet(x,w)$ denote the generating function of rooted connected planar graphs where w marks the degree of the root and the root bears no label. This parameter is additive, in the sense that the degree of the root v is the sum of the degrees of the 2-connected components containing v. If $B^\bullet(x,w)$ denotes the corresponding generating function for rooted 2-connected planar graphs, then we have

$$C^\bullet(x,w) = \exp[B^\bullet(xC'(x),w)].$$

Notice that, unlike Equation (2.2), this is not a recursive equation for $C^\bullet(x,w)$. The reason is that we are recording only the degree of the root.

Since $C'(x)$ is already well known to us, we concentrate on $B^\bullet(x,w)$. Once more, we take as starting point the decomposition of 2-connected graphs into 3-connected components. Let $D(x,y,w)$, $S(x,y,w)$ and $T^\bullet(x,z,w)$ denote, respectively, the GFs of networks, series networks with non-adjacent poles, and 3-connected planar graphs rooted at a directed edge, where w marks the degree of the first pole for networks, and the degree of the tail of the root edge for graphs. The equations linking these GFs are

$$wB_w^\bullet(x,y,w) = xyw \exp\left(S(x,y,w)\right), \qquad (7.2)$$

$$S(x,y,w) = xD(x,y)\left(D(x,y,w) - S(x,y,w)\right), \qquad (7.3)$$

$$D(x,y,w) = (1+yw)\exp\left(S(x,y,w) + \frac{T^\bullet\left(x,D(x,y),\frac{D(x,y,w)}{D(x,y)}\right)}{x^2 D(x,y,w)}\right) - 1. \qquad (7.4)$$

Equations (7.2), (7.3) and (7.4) are the analogues of (2.4), (2.5) and (2.6), respectively, when taking into account the degree of the root. They depend on $D(x,y)$, which we already know, and on $T^\bullet(x,z,w)$, which we do not. Also, observe that the term B_y in (2.4) becomes $B_w^\bullet$ in (7.2), and not $B_y^\bullet$ (otherwise, there would be two roots, the vertex root whose degree is marked by w, and the edge marked by the partial derivative). Thus, an integration step in the spirit of Lemma 5.2 is required to obtain a (quite long) expression for $B^\bullet(x,y,w)$.

It remains to obtain an expression for $T^\bullet(x,z,w)$. As in Section 3, we use the bijection between 3-connected maps and quadrangulations with no separating quadrangles, but now we must take into account the degree of the root vertex. Using classical results on map enumeration [18, 51] we are able to obtain the corresponding generating function $M(x,z,w)$ of 3-connected maps. The final expression is more complex than its counterpart (3.2), and is not reproduced here. An equation similar to (3.1) gives access to $T^\bullet(x,z,w)$.

Notice that $T^\bullet(x,z,1) = zT_z(x,z)$, $B^\bullet(x,y,1) = B'(x,y)$ and $C^\bullet(x,y,1) = C'(x,y)$. Hence a singular exponent $3/2$ is expected after differentiation, and this is indeed the case. The singular expansion of $T^\bullet(x,z,w)$ for $|w| \leq 1/(\sqrt{7}-2)$ is

$$T^\bullet(x,z,w) = T_0(z,w) + T_2(z,w)X^2 + T_3(z,w)X^3 + O(X^4),$$

where $X = \sqrt{1 - x/r(z)}$. In particular, and this is essential, the dominant singularity $r(z)$ does not depend on w. Using techniques similar to those outlined in

Section 5, it can be shown that the same kind of singular expansion extends to $D^\bullet, B^\bullet$ and $C^\bullet$, for $|w| \le q^{-1} \approx 1.48$. We obtain

$$D^\bullet(x,y,w) = D_0(y,w) + D_2(y,w)X^2 + D_3(y,w)X^3 + O(X^4)$$
$$B^\bullet(x,y,w) = B_0(y,w) + B_2(y,w)X^2 + B_3(y,w)X^3 + O(X^4)$$

where $X = \sqrt{1 - x/R(y)}$, and

$$C^\bullet(x,y,w) = C_0(y,w) + C_2(y,w)X^2 + C_3(y,w)X^3 + O(X^4), \qquad (7.5)$$

where now $X = \sqrt{1 - x/\rho(y)}$. Again the value of w affects neither the nature nor the location of the dominant singularities, only the singular coefficients. We remark that the explicit expressions for the coefficients D_i and B_i are quite involved and it takes several pages of type to write them down [24].

At this point we can set $y = 1$ (edges are no longer necessary) and we work with $C^\bullet(x,w)$. If $C_k^\bullet(x)$ is the GF of rooted connected graphs in which the root has degree k (and bears no label), then we have

$$C^\bullet(x,w) = \sum_k C_k^\bullet(x) w^n.$$

The limit probability d_k that the root vertex has degree k is equal to

$$d_k = \lim_{n\to\infty} \frac{[x^n] C_k(x)}{[x^n] C'(x)}, \qquad (7.6)$$

and this can be estimated from (7.5) by singularity analysis. In fact we get an explicit form for the probability generating function as

$$p(w) = \sum_{k \ge 1} d_k w^k = -e^{B_0(1,w) - B_0(1,1)} B_2(1,w)$$
$$+ e^{B_0(1,w) - B_0(1,1)} \frac{1 + B_2(1,1)}{B_3(1,1)} B_3(1,w),$$

where $B_j(y,w)$, $j = 0, 2, 3$, are as before. We do not want to conceal that the expression for $p(w)$ obtained in [24], as it depends on the B_j, is very involved. However, with the help of Maple we can compute the coefficients d_k and the estimate (7.1).

The same approach works for 2-connected and 3-connected planar graphs. We deduce that there is a linear number of vertices $d_k n$ for any constant degree k in almost all connected, 2-connected or 3-connected planar graphs, where the d_k can be computed explicitly in each case (see Table 1 above).

Finally, by performing singularity analysis with respect to w on the dominant coefficient $C_3(1,w)$ in (7.5), we obtain the asymptotics (7.1). The estimates for the corresponding probabilities of 2-connected and 3-connected planar graphs are of the same type.

8 Extremal parameters

We start by discussing the size $L(n)$ of the largest component in random planar graphs with n vertices. It is shown in [44] that the largest component is 'huge', in the sense that it contains everything except a few vertices. More precisely, if we let $M(n) = n - L(n)$ be the number of vertices *missed* by the largest component, then

$$\mathbf{E}[M(n)] \leq 6 + o(1). \tag{8.1}$$

The proof is based on a simple lemma of great generality. Following [44], call a class $\mathcal{G}$ of graphs *bridge-addable* if whenever G is in $\mathcal{G}$ and e is an edge between different components of G, then $G + e$ is also in $\mathcal{G}$.

Lemma 8.1 *Let $\mathcal{G}$ a bridge-addable class of graphs and let μ_n be the expected number of edges of graphs in $\mathcal{G}$ with n vertices. Then*

$$\mathbf{E}[M(n)] \leq 2\frac{\mu_n}{n}.$$

Clearly planar graphs form a bridge-addable class and $\mu_n \leq 3n - 6$. Hence (8.1) follows directly.

Next we consider the size of the largest block, that is, the largest 2-connected component. This problem has been well studied for planar *maps*. Following the first results on triangulations [7], Gao and Wormald [32] proved that the largest block in a random map with n edges has almost surely $n/3$ edges, with deviations of order $n^{2/3}$. More precisely, if X_n is the size of the largest block, then

$$\mathbf{P}(|X_n - n/3| < \lambda(n)n^{2/3}) \to 1, \quad \text{as } n \to \infty,$$

where $\lambda(n)$ is any function going to infinity with n. More generally, they show similar results for a variety of maps and components, such as finding the size of the largest 3-connected component in a random 2-connected map (3-connected components in maps are defined essentially as in Section 1 and do not correspond necessarily to submaps).

The picture was further clarified by Banderier et al. [1]. They found that the largest 2-connected component in random maps obeys a continuous limit law, which is called by the authors the 'Airy distribution of the map type', and is closely related to a stable law of index $3/2$. The density of this distribution is given by

$$g(x) = 2e^{-2x^3/3}(x\mathrm{Ai}(x^2) - \mathrm{Ai}'(x^2)), \tag{8.2}$$

where $\mathrm{Ai}(x)$ is the Airy function, a particular solution of the differential equation $y'' - xy = 0$. A plot of $g(x)$ is shown in Figure 10. We remark that the left tail (as $x \to -\infty$) decays polynomially fast while the right tail (as $x \to +\infty$) decays exponentially fast.

Here is a simplified version of the main result from [1].

Theorem 8.2 *For a variety of random maps and components, including 2-connected components in arbitrary maps, the size X_n of the largest component satisfies*

$$\mathbf{P}(X_n = \alpha n + xn^{2/3}) \sim n^{-2/3}cg(cx),$$

where α is a centering constant and c is a scaling parameter.

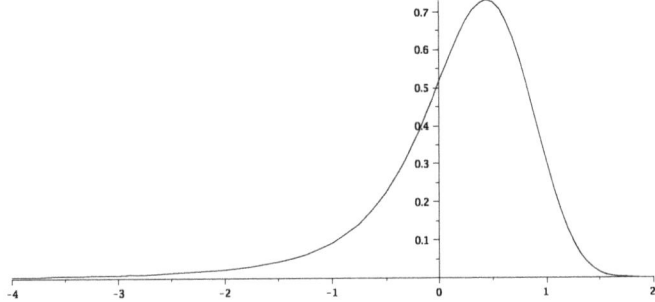

Figure 10: The Airy distribution.

In particular, we have $\mathbf{E}[X_n] \sim cn$. The parameter c quantifies in some sense the dispersion of the distribution (not the variance, since the second moment does not exist). The proof is based on estimating coefficients of high powers of generating functions using contour integration and the saddle point method. However, far from being a standard application of the method, it requires a deep and very careful analysis due to the presence of coalescing saddle points.

A noteworthy feature of [1] is that it shows that an Airy law is to be expected under a general combinatorial-analytic scheme. In the language of maps, let $\mathcal{K} \subset \mathcal{M}$ be two classes of maps and assume a map of $\mathcal{M}$ is made recursively from a map $\kappa \in \mathcal{K}$ (the core) by substituting every edge of κ (or every corner, or every triangle, depending on the kind of map) by elements of $\mathcal{M}$. The associated generating functions $M(z) = \sum M_n z^n$ and $K(z) = \sum K_n z^n$ then satisfy an equation

$$M(z) = K(H(z)),$$

where usually $H(z)$ is a simple variation of $M(z)$. For instance, for the basic case where the core of a map is the 2-connected component containing the root edge, we have $H(z) = z(1 + M(z))^2$, corresponding to the fact that if the core has k edges, then there are $2k$ corners at which to place additional maps (possibly empty). To introduce the parameter 'size of the core' is now a simple matter. If variable u marks this parameter then clearly

$$M(z,u) = K(uH(z)). \tag{8.3}$$

This is called a composition scheme, and for the families of maps analysed in [1] it is *critical*, in the sense that $H(\rho_H) = \rho_K$, where ρ_H and ρ_K are the corresponding dominant singularities. Moreover, both have singular exponents 3/2; that is, $H(z) = h_0 + h_2 Z^2 + h_3 Z^3 + O(Z^4)$, $Z = \sqrt{1 - z/\rho_H}$, and similarly for $K(z)$. In this situation the analysis of the core size can be expanded to cover the size X_n of the largest component (2-connected component in the former example) and, as shown in [1], X_n satisfies a limit Airy law as in Theorem 8.2.

Now we turn our attention to planar graphs instead of maps. Recall that the decomposition of a rooted connected graph into blocks is encoded in Equation (2.2). There is a technical issue when defining the core in this case, since the root may belong to several blocks and there is no way to single out one of them. For simplicity of exposition, we consider here the core as being the set of blocks containing the root.

Let then u be a new variable that marks the sum of the sizes of blocks containing the root vertex. Then (2.2) becomes

$$xC'(x,u) = xe^{B'(uxC'(x))}.$$

This follows the scheme (8.3) with $M(x,u) = xC'(x,u)$, $H(x) = xC'(x)$, and $K(x) = xe^{B'(x)}$. Equation (5.7) shows that the scheme is critical, and (5.5), (5.8) show that the singular exponents are $3/2$, since by differentiating the exponent decreases by 1. The theory developed in [1] applies and we obtain the following result [39], where R is the dominant singularity of $B(x)$ and the B_i are the singular coefficients in (5.5).

Theorem 8.3 Let X_n be size of the largest block in random planar graphs with n vertices. Then

$$\mathbf{P}(X_n = \alpha n + xn^{2/3}) \sim n^{-2/3} cg(cx),$$

where

$$\alpha = \frac{R - 2B_4}{R} \approx 0.95982, \qquad c = \left(\frac{-2R}{15B_5}\right)^{2/3} \approx 128.35169,$$

and $g(x)$ is as in (8.2).

Moreover, the size of the second largest block is $O(n^{2/3})$.

There is a technical argument needed to go from the size of the core to the size of the largest block. The distribution of the core size is bimodal: there is a positive probability p of having a small core, and probability $1 - p$ of having a large core of linear size. One shows (see the proof of Theorem 7 in [1]) that the probability of having a large core which is not the largest block is negligible, and the proof can be adapted to planar graphs.

A similar result holds for 3-connected components in random planar graphs, although the definition of the core in this case is more technical. In essence it says that, in the decomposition of 2-connected graphs described in Section 1, there is a 3-connected graph of linear size a.a.s.

Our next extremal parameter is the maximum degree, denoted Δ_n for graphs with n vertices. It has been well studied for trees. A classical result of Moon [50] says that for labelled trees

$$\mathbf{E}[\Delta_n] \sim \frac{\log(n)}{\log\log(n)}. \tag{8.4}$$

For planar graphs the following has been proved [46].

Theorem 8.4 Let Δ_n be the maximum degree in random planar graphs with n vertices. Then with high probability

$$c\log(n) \leq \Delta_n \leq C\log(n),$$

for suitable constants $0 < c < C$.

The proof is combinatorial and extends to more general classes of graphs.

It is conjectured in [24] that

$$\mathbf{E}[\Delta_n] \sim \frac{\log(n)}{\log(1/q)},$$

where q is as in (7.1). An analogous result, and even a limit law for Δ_n, is proved in [33] for planar maps counted according to the number of edges. A rough indication for the validity of the conjecture is that, according to (7.1), the number of vertices of degree k is of order $k^{-1/2}q^k n$, and this becomes constant when k is $\log(n)/\log(1/q)$. In contrast, the number of vertices of degree k in trees is of order $n/k!$, leading to the estimate (8.4).

We conclude this section with a brief remark about the diameter. Very interesting results have been obtained for some classes of maps. For instance, the diameter in planar quadrangulations is of order $\Theta(n^{1/4})$ [20], a result intimately connected with the search for continuous limits of large random planar maps. However for planar *graphs* the problem is open.

9 A general framework for enumeration

After a long detour through random graphs in the last three sections, we return to enumeration. The material from this section is taken mostly from [39].

Call a class of labelled graphs $\mathcal{G}$ *closed* if the following condition holds: a graph is in $\mathcal{G}$ if and only if its connected, 2-connected and 3-connected components are in $\mathcal{G}$. A closed family is completely determined by its 3-connected members. The basic example is the class of planar graphs, but there are others, especially minor-closed classes whose excluded minors are 3-connected; see the next section.

The point of view we adopt here is to take an *arbitrary* family $\mathcal{T}$ of 3-connected graphs, and *define* a closed class $\mathcal{G}$ whose 3-connected members are those in $\mathcal{T}$. Actually, the family $\mathcal{T}$ cannot be too large if we wish to apply the analytic tools introduced formerly; this means that the number of graphs in $\mathcal{T}$ of size n is at most $c^n n!$ for some constant c.

If a class $\mathcal{G}$ is closed, we have seen in Section 2 that there are equations linking the GFs associated to the numbers of graphs of each kind: $T(x,z)$ for 3-connected graphs; $D(x,y)$ for networks; $B(x,y)$ for 2-connected graphs; $C(x,y)$ for connected graphs; $G(x,y)$ for all graphs. Suppose we know $T(x,z)$, either explicitly or through some functional equation. Suppose further that we know its dominant singularity $r(z)$ for fixed z, and that $T(x,z)$ is either analytic or has a singular expansion at $r(z)$ with singular exponent e, usually 3/2 or 5/2. Then we ask:

> How does the singular structure of $T(x,z)$ affect the singular structure of $B(x,y)$ and $C(x,y)$, and thus the asymptotic estimates of the counting sequences and the properties of random graphs in the class $\mathcal{G}$?

Let us discuss two basic examples. As usual, g_n, c_n, b_n are the numbers of arbitrary, connected and 2-connected graphs in $\mathcal{G}$. If $\mathcal{T}$ is the family of 3-connected planar graphs, then $\mathcal{G}$ is the class of planar graphs and

$$g_n \sim g n^{-7/2} \gamma^n n!, \qquad c_n \sim c n^{-7/2} \gamma^n n!, \qquad b_n \sim b n^{-7/2} \delta^n n!,$$

with $\gamma \approx 27.23$ and $\delta \approx 26.18$. If $\mathcal{T}$ is empty, then $\mathcal{G}$ is the class of series-parallel graphs (since the only networks are series and parallel) and

$$g_n \sim g_1 n^{-5/2} \gamma_1^n n!, \qquad c_n \sim c_1 n^{-5/2} \gamma_1^n n!, \qquad b_n \sim b_1 n^{-5/2} \delta_1^n n!,$$

with $\gamma_1 \approx 9.07$ and $\delta_1 \approx 7.81$ [12].

The reason for this qualitative difference in the asymptotics lies in the two possible cases discussed after Equation (5.6). For planar graphs we are in case (2) and the singularity ρ of $C(x)$ comes from the singularity R of $B(x)$. But for series-parallel graphs we are in case (1) and ρ comes from a branch point. This explains the difference in the asymptotics of c_n and g_n. A similar situation occurs for the dominant singularity R of $B(x)$. It may come from the singularities of $T(x,z)$, as in the planar case, or from a branch point when solving the Equation (2.7) for networks.

This is the main dichotomy: the subexponential terms are either $n^{-7/2}$ or $n^{-5/2}$. There is also a *critical* case, when two possible sources of singularities coincide. In this situation the subexponential term becomes $n^{-8/3}$.

Theorem 9.1 *Let $\mathcal{T}$ be a family of 3-connected graphs with generating function $T(x,z)$, and let $\mathcal{G}$ be the associated closed family.*

(1) If either $T(x,z)$ is analytic or has singular exponent $e < 2$ then $B(x), C(x)$ and $G(x)$ have singular exponent $3/2$ and we have the estimates

$$b_n \sim b n^{-5/2} R^{-n} n!, \qquad c_n \sim c n^{-5/2} \rho^{-n} n!, \qquad g_n \sim g n^{-5/2} \rho^{-n} n!$$

(2) If $T(x,z)$ has singular exponent $5/2$, then one of the following holds:

(2.1) $B(x), C(x)$ and $G(x)$ have singular exponent $5/2$;

(2.2) $B(x)$ has singular exponent $5/2$, and $C(x)$ and $G(x)$ have singular exponent $3/2$;

(2.3) $B(x), C(x)$ and $G(x)$ have singular exponent $3/2$.

Accordingly we have the estimates

(2.1) $b_n \sim b n^{-7/2} R^{-n} n!, \qquad c_n \sim c n^{-7/2} \rho^{-n} n!, \qquad g_n \sim g n^{-7/2} \rho^{-n} n!$
(2.2) $b_n \sim b n^{-7/2} R^{-n} n!, \qquad c_n \sim c n^{-5/2} \rho^{-n} n!, \qquad g_n \sim g n^{-5/2} \rho^{-n} n!$
(2.3) $b_n \sim b n^{-5/2} R^{-n} n!, \qquad c_n \sim c n^{-5/2} \rho^{-n} n!, \qquad g_n \sim g n^{-5/2} \rho^{-n} n!$

(3) If $T_z(x,z)$ has singular exponent $\alpha = 3/2$, and in addition a critical condition is satisfied, then a singular exponent $5/3$ appears and one of the following holds:

(3.1) $b_n \sim b n^{-8/3} R^{-n} n!, \qquad c_n \sim c n^{-5/2} \rho^{-n} n!, \qquad g_n \sim g n^{-5/2} \rho^{-n} n!$
(3.2) $b_n \sim b n^{-7/2} R^{-n} n!, \qquad c_n \sim c n^{-8/3} \rho^{-n} n!, \qquad g_n \sim g n^{-8/3} \rho^{-n} n!$

In all cases b, c, g, R, ρ are computable positive constants and $\rho < R$.

Notice that in all cases the estimates for c_n and g_n are of the same type. This is because the relation $G(x) = \exp C(x)$ does not change the singular structure.

The cases covered by the previous theorem are those encountered in planar graphs, series-parallel graphs and related classes [39]. It would be very interesting to find examples in which $T(x, z)$ has other singular exponents. This is the case for maps in surfaces of higher genus, and likely also for *graphs* in surfaces.

We remark that not everything falls under the scheme of classes defined in terms of 3-connected components. For instance, the class of cubic planar graphs is not closed and needs a different analysis [14]. This is also the case for outerplanar graphs, a class which is easy to analyse since the 2-connected members correspond to polygon dissections [12].

Now we come back to random graphs. The basic parameters of random graphs in $\mathcal{G}$ are not affected by the singular structure of $T(x, z)$. We obtain systematically a normal limit law for the number of edges, as in Theorem 6.2, and a Poisson limit law for the number of components, as in Theorem 6.4; only the values of the constants change. This is also true for more complex parameters, like the size of the largest component, and the distribution of vertex degrees.

However, the difference in the analytic structure implies a deep difference in the structure of random graphs from the corresponding class. There is a drastic change in an important parameter, namely the size of the largest block. In case (2.1) of Theorem 9.1, the largest block is linear in n, as for planar graphs (see Theorem 8.3). This is because we have a critical composition scheme with exponents 3/2 as in the previous section. In the remaining cases the composition is not critical, and it can be shown that the size of the core converges to a discrete limit law with exponential decay; as a consequence, the size of the largest block is sublinear. The same analysis in both cases applies to the largest 3-connected component.

The picture that emerges from the former discussion is the following. A random connected planar graph G has a unique block of linear size, and within this block there is also a unique 3-connected component of linear size. The remaining 2- and 3-connected components are *small*, meaning of order $O(n^{2/3})$. Thus G can be seen as a large 3-connected map M in which the following operations are performed: first, edges of M are substituted by small networks, giving rise to the largest block L; then small connected graphs are attached to some of the vertices of L, which become cut vertices.

In contrast, in a random connected series-parallel graph G there is no block of linear size, and the size of the blocks tends to a discrete limit law; that is, the probability that a block has size k converges to a definite constant p_k when $n \to \infty$. We may say that G resembles a tree of small blocks.

A very interesting open question is whether there are other parameters besides the size of the largest block (or largest 3-connected component) for which planar graphs and series-parallel graphs differ in a qualitative way. More generally, the same question can be asked depending on which case of Theorem 9.1 applies.

10 Surfaces and minors

Planar graphs are those that can be embedded in the sphere. The situation for higher surfaces has been analysed by McDiarmid [44]. Let S be any surface, orientable or not, and let $g_n(S)$ be the number of labelled simple graphs with n vertices that can be embedded in S. The first result from [44] is that the growth constant is the same as in the planar case, that is,

$$\lim(g_n(S)/n!)^{1/n} \to \gamma, \qquad (10.1)$$

where γ is the planar constant from Theorem 5.1.

The analogous result was already known for *maps* (unlabelled, loops and multiple edges allowed) counted according to the number edges [2, 31]. If $M_n(S)$ denotes the number of maps with n edges in an orientable surface S of genus g, then

$$M_n(S) \sim t_g \cdot n^{5(g-1)/2} 12^n,$$

where t_g is a constant depending only on the genus (a similar estimate holds for nonorientable surfaces). The proof relies on an explicit expression for the generating function $\sum M_n(S)x^n$. In view of the above and the estimate (5.2), it seems natural to conjecture that for an orientable surface S of genus g we have

$$g_n(S) \sim c_g \cdot n^{5(g-1)/2-1} \gamma^n n!,$$

for some constant c_g depending on g. However there are fundamental differences with the planar case, starting with the fact that 3-connected graphs do not have a unique embedding.

The proof of (10.1) is by induction on the genus, using the fact that for any graph G embedded in S there is a noncontractible cycle C meeting G in at most $O(\sqrt{n})$ vertices. Cutting the surface along C and duplicating the vertices in C, one gets a new graph G' of smaller genus and applies induction. Now G' has only $O(\sqrt{n})$ additional vertices, and this can be accommodated into (10.1) without changing the limit.

Furthermore, significant results are proved in [44] for random graphs embeddable in a surface S, like the analogues of Theorem 6.1 and the bound (8.1). Further results, like the analogue of Theorem 6.4 with the same constants, are proved assuming the following technical condition: if g_n is the number of graphs embeddable in S, then $\lim_{n\to\infty} ng_{n-1}/g_n = \gamma$. That this condition holds for every surface has been proved in [3].

Now we turn to graphs defined in terms of minors. We recall that a graph H is a minor of G, written $H < G$, if H can be obtained from a subgraph of G by contracting edges. A class of graphs $\mathcal{G}$ is minor-closed if, whenever a graph is in $\mathcal{G}$, all its minors are also in $\mathcal{G}$. The basic example of minor-closed classes are graphs embeddable in a fixed surface, in particular planar graphs. Given a minor-closed class $\mathcal{G}$, a graph H is an excluded minor for $\mathcal{G}$ if H is not in $\mathcal{G}$ but every proper minor is in $\mathcal{G}$. It is an easy fact that a graph is in $\mathcal{G}$ if and only if it does not contain as a minor any of the excluded minors from $\mathcal{G}$. The fundamental result on graph minors is the theorem of Robertson and Seymour: for every minor-closed class, the number of excluded minors is finite [58]. We write $\mathcal{G} = \text{Ex}(H_1, \cdots, H_k)$ if $H_1, \ldots, H_k$ are the excluded minors of $\mathcal{G}$.

We are interested in the number of graphs in a minor-closed family $\mathcal{G}$. Let us start with K_5 and $K_{3,3}$, the excluded minors for planar graphs according to Kuratowski's theorem. Wagner [66] characterized graphs in $\text{Ex}(K_{3,3})$: the maximal ones are those obtained by gluing planar triangulations and copies of K_5 along edges. It follows that the 3-connected graphs in $\text{Ex}(K_{3,3})$ are either planar or the exceptional graph K_5, and the associated GF is obtained by adding the monomial $z^{10}x^5/5!$ to the known planar GF $T(x,z)$. This is shown in [35], and the estimate

$$g_n \sim c \cdot n^{-7/2}\theta^n n!,$$

where $\theta \approx 27.22935$ is slightly larger than the planar growth constant.

Wagner also characterized maximal graphs in $\text{Ex}(K_5)$, but the result in this case involves gluing planar triangulations along *triangles* (plus copies of an exceptional graph on 8 vertices glued along edges) and we cannot describe well enough the 3-connected graphs in the class. In fact the enumeration of graphs in $\text{Ex}(K_5)$ remains a challenging open problem; we do not even know the growth constant with any acceptable accuracy.

There are additional results, mainly by Halin and Wagner, giving characterizations of graphs in $\text{Ex}(H)$ for several H; they are reviewed in [22, Chapter X]. In several cases, from the characterization we can obtain the full list of 3-connected graphs in the class, from which we can compute the associated generating function, and the machinery of the previous section applies. In the following table we show several examples of excluded minors and the corresponding family of 3-connected graphs, where W_n is the wheel with n spokes, C_n is the cycle on n vertices, and $G \times H$ is the ordinary product. In addition, G^- denotes graph G after removing one edge, and G^+ after adding one edge.

Excluded minors	3-connected graphs in the associated class
K_4	Empty
W_4	K_4
K_5^-	$K_{3,3}, K_3 \times K_2, W_n$ for $n \geq 3$
$K_{3,3}$	Planar 3-connected, K_5
$K_{3,3}^+$	Planar 3-connected, K_5, $K_{3,3}$

Table 2: Excluded minors and the corresponding 3-connected graphs.

The full list of 3-connected graphs is also known when excluding W_5 and $K_3 \times K_2$, among others, but it takes longer to describe and is not shown in the table. As a rule, if the corresponding characterization involves gluing only along edges, the GF of 3-connected graphs in the class is not difficult to compute. Some cases are even simpler and we can obtain directly the 2-connected graphs in the class, as for instance in $\text{Ex}(C_4)$ or $\text{Ex}(K_4 - e)$.

Given a minor-closed class $\mathcal{G}$, let g_n be the number of graphs in $\mathcal{G}$ with n vertices. If $\mathcal{G}$ is not the class of all graphs then g_n cannot be too large. It is proved in [53] that there exists a constant c depending on $\mathcal{G}$ such that

$$g_n \leq c^n n!.$$

If follows that $\limsup(g_n/n!)^{1/n}$ is finite. If $\gamma = \lim(g_n/n!)^{1/n}$ exists, it is called the growth constant of $\mathcal{G}$. The question of which real numbers can be growth constants of minor-closed classes of graphs is studied in [9], where the following is proved. The number $\xi \approx 1.76$ is the inverse of the unique positive root of $x \exp(x) = 1$; it is the growth constant of the class of forests in which each component is a caterpillar, and a caterpillar is a path to which we attach vertices of degree one.

Theorem 10.1 *Let Γ be the set of real numbers which are growth constants of minor-closed classes of graphs.*

(1) *The values 0, 1, ξ and e are in Γ.*
(2) *If $\gamma \in \Gamma$ then $2\gamma \in \Gamma$.*
(3) *The only growth constants between 0 and 2 are $0, 1, \xi$ and 2.*

Further results, including an infinity of gaps inside the interval $(2, 2.25)$, have been obtained more recently by the authors of [9].

The proof of (2) uses the *apex* construction. For a class of graphs $\mathcal{G}$, let $\mathcal{AG}$ be the class of graphs G having a vertex v such that $G - v$ is in $\mathcal{G}$; we say that v is an apex of G. It is easy to check that if $\mathcal{G}$ is minor-closed, so is $\mathcal{AG}$, and that the growth constant of $\mathcal{AG}$ is twice that of $\mathcal{G}$. In particular this gives the growth constant of apex graphs (where the base family is the set of planar graphs), which appeared first in the study of Hadwiger's conjecture [59]. Even more, the number of apex graphs is asymptotically $(g/2\gamma)n^{-7/2}(2\gamma)^n n!$, where g and γ are as in Theorem 5.1 [44].

The proof of (3) works as follows. Let $\mathcal{G}$ be a minor-closed class with growth constant less than 1. Since the family consisting of all paths has growth constant 1 (because a path with n vertices can be labelled in $n!/2$ ways), there is some path P_k not in $\mathcal{G}$. This implies that graphs in $\mathcal{G}$ do not contain paths of length greater than k, and from this one can show that the growth constant of $\mathcal{G}$ is 0. The proof for the remaining gaps is similar but uses obstructions more complicated than paths.

We remark that, since the number of excluded minors is finite, the set Γ in the previous theorem is countable. This is not so for growth constants of arbitrary classes of graphs (as shown in Theorem 11.1 for planar graphs with given edge density) or for different combinatorial objects, in particular for permutations defined in terms of forbidden patterns [65, 41].

To conclude this section, we show in Table 3 approximate values of growth constants for several classes closed under minors. Some of the entries are computed using the results in Table 2.

11 Graphs with given edge density

The edge density of a graph is the number of edges divided by the number of vertices, which is the same as half the average degree. We are interested in an estimate for the number of planar graphs with fixed edge density.

Following the notation of previous sections, $g_{n, \lfloor \mu n \rfloor}$ is the number of planar graphs with n vertices and $\lfloor \mu n \rfloor$ edges. For $\mu \in (1, 3)$, this can estimated as follows. We choose a value $y_0 > 0$ depending on μ such that, if we give weight y_0^k to a graph with k edges, then only graphs with n vertices and μn edges have non-negligible weight. If $\rho(y)$ is the radius of convergence of $G(x, y)$, the right choice is the unique positive solution y_0 of

$$-y\rho'(y)/\rho(y) = \mu, \qquad (11.1)$$

Counting planar graphs

Class of graphs	Growth constant	Reference
$\mathrm{Ex}(P_k)$	0	[9]
Path forests $= \mathrm{Ex}(K_3, K_{1,3})$	1	Standard
Caterpillar forests	$\xi \approx 1.76$	[9]
Apex of path forests	2	[9]
Forests $= \mathrm{Ex}(K_3)$	$e \approx 2.72$	Standard
$\mathrm{Ex}(C_4)$	3.63	[39]
$\mathrm{Ex}(K_4^-)$	4.18	[39]
$\mathrm{Ex}(C_5)$	4.60	[39]
Outerplanar $= \mathrm{Ex}(K_4, K_{2,3})$	7.320	[12]
$\mathrm{Ex}(K_{2,3})$	7.327	[12]
Series parallel $= \mathrm{Ex}(K_4)$	9.07	[12]
$\mathrm{Ex}(W_4)$	11.54	[39]
$\mathrm{Ex}(W_5)$	14.67	[39]
$\mathrm{Ex}(K_5^-)$	15.65	[39]
$\mathrm{Ex}(K_3 \times K_2)$	16.24	[39]
Planar	27.226	[38]
Embeddable in a fixed surface	27.226	[44]
$\mathrm{Ex}(K_{3,3})$	27.2293	[35]
$\mathrm{Ex}(K_{3,3}^+)$	27.2295	[35]

Table 3: A table of growth constants. Approximate values are displayed, but all constants are well-defined analytically.

which is in fact a saddle point equation. As a consequence, $[x^n]G(x, y_0)$ can be used to estimate $g_{n, \lfloor \mu n \rfloor}$, and this can be obtained applying the transfer theorem to $G(x, y_0)$. The following is a simplified statement of a result from [38]. The proof uses a version of the local limit theorem [28, Thm. IX.14].

Theorem 11.1 *For μ in the open interval $(1, 3)$, the number of planar graphs with n vertices and $\lfloor \mu n \rfloor$ edges is asymptotically*

$$g_{n, \lfloor \mu n \rfloor} \sim c(\mu) \, n^{-4} h(\mu)^n n!, \qquad (11.2)$$

where $c(\mu)$ and $h(\mu)$ are analytic functions of μ.

The term n^{-4} must be read as $n^{-7/2}/\sqrt{n}$, where $1/\sqrt{n}$ comes from the application of the local limit theorem.

The function $h(\mu)$, which may be called the growth ratio of planar graphs with given edge density or average degree, can be computed explicitly; a plot is shown in Figure 11. It can be shown that the limit of $h(\mu)$ as $\mu \to 1^+$ is equal to e, which is the growth ratio of labelled trees: the limit as $\mu \to 3^-$ is equal to $256/27$, which is the growth ratio of triangulations [62]. The maximum is located precisely at the constant κ from Theorem 6.2, and $H(\kappa) = \gamma$. In the range $\mu \in (0, 1)$ we have $h(\mu) = 0$, as shown in [37] using different tools.

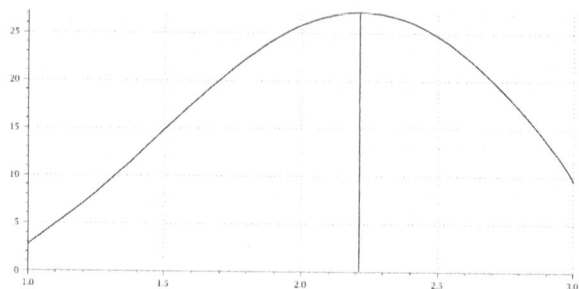

Figure 11: The growth ratio of planar graphs with a given edge density.

An important point is that we can analyse the basic parameters of random planar graphs if we fix the number of edges μn. Here is a succinct explanation. Suppose $G(x, y, w)$ is the GF associated to a given parameter, for instance the number of vertices of degree k, and w marks the parameter. We let $y_0 = y_0(\mu)$ be the solution of (11.1), and work with the generating function $G(x, y_0, w)$. If we know the singular expansion of G, for a given value of w we can estimate $g_n(y_0, w) = [x^n]G(x, y_0, w)$ and have access to the distribution of the parameter in graphs with μn edges.

This is illustrated for planar graphs and several parameters in Figures 12, 13 and 14. In all of them the abscissa corresponds to μ, and the ordinate is the value of the parameter. The abscissa $\kappa \approx 2.21$ is highlighted; the ordinate at κ is the value of the parameter for all planar graphs. The reason is that the number of edges in random planar graphs is strongly concentrated at κn (see Theorem 6.2).

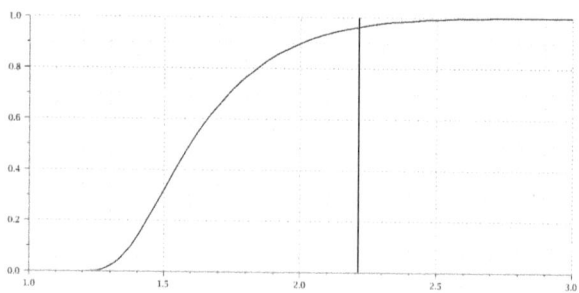

Figure 12: Probability of connectedness for planar graphs with μn edges, $\mu \in (1, 3)$ The value at κ is approximately 0.9632 as in Theorem 6.4.

We have seen in Theorem 11.1 that the estimates for the number of planar graphs with μn edges have the same shape for all values $\mu \in (1, 3)$. This is also the case for series-parallel graphs, where $\mu \in (1, 2)$ since maximal graphs in this class have only $2n - 3$ edges. It is natural to ask if there are classes in which there is a *critical phenomenon*, that is, a different behaviour depending on the edge density.

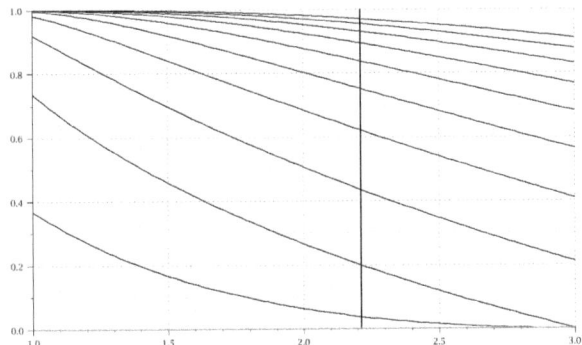

Figure 13: Cumulative degree distribution for random planar graphs with μn edges, $\mu \in (1,3)$. Each curve gives the probability that a vertex has degree at most k. The bottom curve corresponds to $k = 1$ and the top curve to $k \leq 10$. The values at κ agree with those obtained by accumulating the values in the first row of Table 1.

We have not found such phenomenon for 'natural' classes of graphs, in particular those defined in terms of forbidden minors. But we have been able to construct examples of critical phenomena by a suitable choice of the family $\mathcal{T}$ of 3-connected graphs, as we now explain.

The key point is that Theorem 9.1 can be generalized by taking into account the value of variable y, which as we have seen corresponds to fixing the number of edges (we do not treat here the critical case).

Theorem 11.2 *Let $\mathcal{T}$ be a family of 3-connected graphs with generating function $T(x,z)$, and let $\mathcal{G}$ be the associated closed family. Fix $y = y_0$, which corresponds to a given edge density.*

(1) If either $T(x,z)$ is analytic or has singular exponent $e < 2$ then $B(x,y_0)$, $C(x,y_0)$ and $G(x,y_0)$ have singular exponent $3/2$.

(2) If $T(x,z)$ has singular exponent $5/2$, then one of the following holds:

  (2.1) $B(x,y_0), C(x,y_0)$ and $G(x,y_0)$ have singular exponent $5/2$;

  (2.2) $B(x,y_0)$ has singular exponent $5/2$, and $C(x,y_0)$ and $G(x,y_0)$ have singular exponent $3/2$;

  (2.3) $B(x,y_0), C(x,y_0)$ and $G(x,y_0)$ have singular exponent $3/2$.

Now we have two sources for the main singularity of $B(x,y)$ for a given value of y: either (a) it comes from the singularities of $T(x,z)$; or (b) it comes from a branch point of the equation defining $D(x,y)$. For planar graphs the singularity always comes from case (a), and for series-parallel graphs always from case (b). If there is a value y_0 for which the two sources coalesce, then we get a different singular exponent depending on whether $y < y_0$ or $y > y_0$. The most important consequence in this situation is that there is a critical edge density μ_0, such that below μ_0 the largest 3-connected component has linear size, and above μ_0 it has sublinear size, or

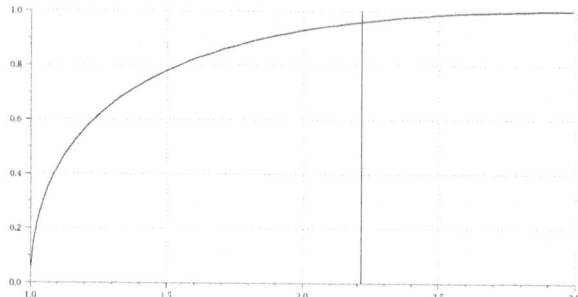

Figure 14: Size of the largest block for planar graphs with μn edges, $\mu \in (1,3)$. The ordinate gives the value $\alpha(\mu)$ such that the largest block has size $\sim \alpha(\mu)n$. The value at κ is approximately 0.9598 as in Theorem 8.3.

conversely. Similar considerations apply for the singularity of $C(x,y)$ and the size of the largest block in connected graphs; see the discussion below.

The following examples show that a variety of situations can occur.

1. If $\mathcal{T}$ is the family of 3-connected cubic planar graphs, then $B(x,y)$ has singular exponent 5/2 when $y < y_0 \approx 0.07422$, and 3/2 when $y > y_0$. The corresponding critical value for the number of edges is $\mu_0 \approx 1.3172$.

2. If $\mathcal{T}$ is the family of planar triangulations (maximal planar graphs), then $B(x,y)$ has singular exponent 3/2 when $y < y_0 \approx 0.4468$, and 5/2 when $y > y_0$. The corresponding critical value for the number of edges is $\mu_0 \approx 1.8755$.

3. This example shows that more than one critical value may occur. Let $\mathcal{T}$ be the family of planar triangulations plus the exceptional graph K_6. Then there are two critical values $y_0 \approx 0.4469$ and $y_1 \approx 108.88$, and the corresponding critical edge densities are $\mu_0 \approx 1.8756$ and $\mu_1 \approx 3.4921$. This last value is close to 7/2; this is the maximal edge density, which is approached by taking many copies of K_6 glued along a common edge. It turns out that $B(x,y)$ has singular exponent 3/2 when $y < y_0$, 5/2 when $y_0 < y < y_1$, and again 3/2 for $y_1 < y$.

The critical values in the former examples correspond to the density of 2-connected graphs in the class and come from the critical behaviour of $B(x,y)$ at y_0. The same phenomenon may arise for *connected* graphs in the class, with a corresponding critical density. In particular, in Example 1, there is a critical density for connected graphs $\mu_0^c \approx 1.18441$. Graphs with density less that μ_0^c share the characteristics of planar graphs and have a linear size block, while graphs with density above μ_0^c have only small blocks, as for series-parallel graphs. This is illustrated in Figure 15; observe that when the density approaches the critical value μ_0^c, the parameter α of the largest block goes to 0.

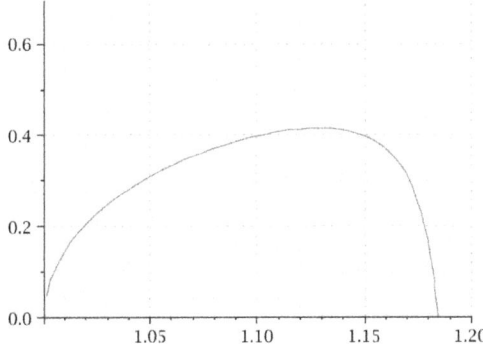

Figure 15: Size of the largest block for the class whose 3-connected members are the cubic 3-connected planar graphs. The edge density is in the interval $\mu \in (1, \mu_0^c \approx 1.18)$, below the critical value. The ordinate gives the value $\alpha(\mu)$ such that the largest block has size $\sim \alpha(\mu) n$.

12 Uniform sampling

The first algorithm for sampling planar graphs uniformly at random was a Markov chain algorithm proposed in [21] and used for experimental results. It converges to the uniform distribution but unfortunately the mixing rate has not been determined.

Two other algorithms have been proposed based on counting. The first one is based on the *recursive method*, which uses combinatorial decompositions and recurrence relations (see [52, 29]). The second one is the so-called method of *Boltzmann samplers*, developed in [26], which requires a combinatorial decomposition in terms of generating functions.

We explain both approaches, taking as a basic example the class $\mathcal{A}$ of rooted binary trees, which satisfies the combinatorial decomposition $\mathcal{A} = \{\epsilon\} \cup \{r\} \times \mathcal{A} \times \mathcal{A}$ (a rooted binary tree is either empty or a tree with a root plus left and right subtrees). The corresponding generating function $A(x)$ satisfies the equation

$$A(x) = 1 + xA(x)^2, \tag{12.1}$$

or, equivalently, the number a_n of binary trees with n nodes satisfies

$$a_n = \sum_{i=0}^{n-1} a_i a_{n-1-i} \quad n > 0, \qquad a_0 = 1.$$

In the recursive method, one uses the combinatorial decomposition associated with a recurrence relation to sample recursively an object from smaller objects of the same class. In our example, a random binary tree of size n has a left subtree of size i with probability $p_i = a_i a_{n-1-i}/a_n$. First we sample an integer i with distribution $\mathbf{P}(X = i) = p_i$ and determine the sizes of of the left and right subtrees; then we sample recursively using the same method, two trees of sizes i and $n - 1 - i$.

This method usually requires a preprocessing step in which the actual numbers a_n are computed. Optionally, one can also pre-compute the probability distributions used during the sampling. The main drawback of this method is that long and complex combinatorial decompositions give rise to recurrences with multi-index coefficients, so that the computations both for the preprocessing step and the actual sampling can become very costly. This method was used in [13] for labelled planar graphs with a time complexity of $O(n^{13/2})$ per sample. The result is mainly of theoretical interest, as it allows one to generate graphs of at most 100 vertices.

The method of Boltzmann samplers, as we are going to see, is better suited for sampling complex objects like planar graphs. Given a class $\mathcal{A}$ of combinatorial objects, define the generating function $A(x) = \sum_{\alpha \in \mathcal{A}} x^{|\alpha|}$, where $|\alpha|$ is the size of α. The Boltzmann distribution of parameter $x > 0$ associated to $\mathcal{A}$ assigns probability $x^{|\alpha|}/A(x)$ to $\alpha \in \mathcal{A}$. Clearly, x must be such that $A(x) < \infty$. This distribution has two key properties. First it assigns the same probability to all objects of the same size, thus objects sampled from it are indeed uniformly chosen. The second property is that to sample an object from a cartesian product of combinatorial classes $\mathcal{B} \times \mathcal{C}$, it is enough to sample *independently* one object from $\mathcal{B}$ and one from $\mathcal{C}$. This is because

$$\mathbf{P}(X_{\mathcal{B}\times\mathcal{C}} = (\beta,\gamma)) = \frac{x^{|(\beta,\gamma)|}}{B(x)C(x)} = \frac{x^{|\beta|}}{B(x)} \frac{x^{|\gamma|}}{C(x)} = \mathbf{P}(X_{\mathcal{B}} = \beta)\mathbf{P}(X_{\mathcal{C}} = \gamma), \quad (12.2)$$

where $X_{\mathcal{Z}}$ stands for a random variable with the Boltzmann distribution of $\mathcal{Z}$ with (implicit) parameter x.

As an illustration, let us describe an algorithm that samples a binary tree according to the Boltzmann distribution of parameter x, for some $x \leq 1/4$, the radius of convergence of the associated $A(x)$. Looking at Equation (12.1), we observe that the empty tree has probability $1/A(x)$ of being sampled. The algorithm first draws a Bernouilli sample with probability $1/A(x)$ to determine if the sampled tree is the empty one; if not, we sample a non-empty binary tree, an object enumerated by $xA(x)^2$. By Equation (12.2), it is enough to sample two independent binary trees according to the same Boltzmann distribution, and use them as the left and right subtrees. Since the resulting tree has been drawn according to a Boltzmann distribution, we know that it is chosen with uniform probability among all binary trees of the same size; a remarkable property for such a simple algorithm.

The same idea can be applied to other combinatorial constructions. For instance, to sample a sequence of elements of $\mathcal{A}$ with the Boltzmann distribution, we first notice that the GF associated to sequences is

$$\frac{1}{1-A(x)} = 1 + A(x) + A(x)^2 + A(x)^3 + \cdots.$$

Thus we first sample a geometric random variable X of parameter $A(x)$, and then we produce as many independent samples of $\mathcal{A}$ as required by the value of X. To sample an element from a composition $\mathcal{A}(\mathcal{B})$ of combinatorial classes, we note that

$$A(B(x)) = a_0 + a_1 B(x) + a_2 B(x)^2 + a_3 B(x)^3 + \cdots.$$

Thus we start by sampling an element α of $\mathcal{A}$ with Boltzmann parameter $B(x)$, and then we sample $|\alpha|$ independent copies of $\mathcal{B}$ with parameter x.

It can be shown that most combinatorial constructions can be translated into sampling algorithms as those described above, and that the resulting samplers run in linear time in the size n of the sampled object. Note that this is valid irrespectively of the complexity of the combinatorial construction, which only affects the multiplicative constant of the running time.

To sample objects of a desired size n, one typically tunes the parameter x, moving it towards the singularity of $A(x)$ (this increases the expected size of the sampled objects) or towards 0 (decreasing the expected size). It is shown in [26] that, for a combinatorial class with a generating function that is Δ-singular with singular exponent $e < 0$, we can tune the parameter $x = x(n)$ to sample objects of size belonging to $((1-\epsilon)n, (1+\epsilon)n)$ in expected constant number of samplings, and to sample objects of exact size n in expected linear number of samplings. If the singular exponent s is greater than 0, as for rooted binary trees ($s = 1/2$) or labelled planar graphs ($s = 5/2$), such a sampler can still be obtained by generating objects with extra roots, a process that decreases the singular exponent of the combinatorial class, at the cost of increasing the complexity of the combinatorial decomposition of the family.

Fusy [30] shows, solving delicate technical issues due to the complexity of the combinatorial decomposition, that the method of Boltzmann samplers can be applied to labelled planar graphs. It uses the generating functions described in Section 2 and gives an $\mathcal{O}(n^2)$ exact size sampler, and an $\mathcal{O}(n)$ approximate size sampler. The implementation of the algorithm in Java samples graphs of about 100000 vertices in a few seconds.

As a final remark, we mention that Boltzmann samplers have been used also for proving theoretical results on random graphs [10].

13 Unlabelled graphs

We discuss briefly the problem of counting unlabelled graphs, that is, isomorphism classes of labelled graphs. Almost invariably, the enumeration of unlabelled objects is harder than their labelled counterparts. This is very much the case for planar graphs, where the problem is wide open.

Let u_n be the number of unlabelled planar graphs. It is known [21] that $\gamma_u = \lim(u_n)^{1/n}$, the unlabelled growth constant, exists but we do not know its precise value. The fact that a.a.s. a labelled planar graph has an exponential number of automorphisms (see the discussion after Theorem 6.1) implies that $\gamma_u > \gamma$, where γ is the labelled growth constant from previous sections [47].

On the other hand, an unlabelled planar graph on n vertices can be encoded with at most αn bits for some constant α. If this is the case then clearly $u_n \le 2^{\alpha n}$. The first such result was proved in [61] with the value $\alpha = 12$. This has been improved over the years and presently the best result is $\alpha \approx 4.91$, obtained in [16]. Since $2^{4.91} \approx 30.06$, we have $u_n < 30.06^n$; and since $\gamma > 27.22$ we obtain

$$27.22 < \gamma_u < 30.06.$$

For graphs embeddable in a fixed surface it is also the case that the unlabelled growth constant is the same as γ_u [44]. On the other hand, if g_n is the number of unlabelled graphs in a proper minor-closed class, then $g_n < c^n$ for some constant

$c > 0$. This follows from the fact that graphs in a minor-closed class have bounded book thickness [45], and implies the bound for labelled classes from [53].

For simpler families of graphs much more is known. A classical result [56] gives the estimate $c \cdot n^{-5/2} \beta^m$ for the number of unlabelled trees, with $\beta \approx 2.95577$. The proof involves an equation relating the GFs of rooted and unrooted trees, an idea that has been formalized in [8] as the 'dissymmetry theorem'. A more complex example is the estimate from [11]

$$c \cdot n^{-5/2} \lambda^n,$$

for the number of unlabelled outerplanar graphs, where $\lambda \approx 7.50360$. The proof uses Pólya's cycle index sums and the fact that 2-connected outerplanar graphs are equivalent to polygon dissections, whose symmetries are easy to describe. The next natural step might be to count unlabelled series-parallel graphs, a more complex class containing outerplanar graphs.

Acknowledgements

We are grateful to Anna de Mier, Andrew Goodall, Dominic Welsh, and the referee for helpful comments. This research has been supported by project MTM2005-08618C02-01.

References

[1] C. Banderier, P. Flajolet, G. Schaeffer & M. Soria, Random maps, coalescing saddles, singularity analysis, and airy phenomena, *Random Structures Algorithms* **19** (2001), 194–246.

[2] E. A. Bender & E. R. Canfield, The asymptotic number of maps on a surface, *J. Combin. Theory Ser. A* **43** (1986), 244–257.

[3] E. A. Bender, E. R. Canfield & L. B. Richmond, Coefficients of functional compositions often grow smoothly, *Electron. J. Combin.* **15** (2008), R21.

[4] E. A. Bender, Z. Gao & N. C. Wormald, The number of 2-connected labelled planar graphs, *Electron. J. Combin.* **9** (2002), R43.

[5] E. A. Bender, Z. Gao & L. B. Richmond, Submaps of maps. I. General 0-1 laws, *J. Combin. Theory Ser. B* **55** (1992), 104–117.

[6] E. A. Bender & L. B. Richmond, Central and local limit theorems applied to asymptotic enumeration II: Multivariate generating functions, *J. Combin. Theory Ser. A* **34** (1983), 255–265.

[7] E. A. Bender, L. B. Richmond & N. C. Wormald, Largest 4-connected components of 3-connected planar triangulations, *Random Structures Algorithms* **7** (1995), 273–285.

[8] F. Bergeron, G. Labelle & P. Leroux, *Combinatorial species and tree-like structures*, Cambridge University Press, Cambridge (1998).

[9] O. Bernardi, M. Noy & D. Welsh, On the growth of minor-closed families of graphs, preprint, arXiv:0710.2995v1.

[10] N. Bernasconi, K. Panagiotou & A. Steger, On properties of random dissections and triangulations, in *Proc. Symp. Discrete Algorithms*, SIAM, San Francisco (2008), pp. 132–141.

[11] M. Bodirsky, E. Fusy, M. Kang & S. Vigerske, Enumeration and asymptotic properties of unlabeled outerplanar graphs, *Electron. J. Combin.* **14** (2007), R66.

[12] M. Bodirsky, O. Giménez, M. Kang & M. Noy, Enumeration and limit laws for series-parallel graphs, *European J. Combin.* **28** (2007), 2091–2105.

[13] M. Bodirsky, C. Gröpl & M. Kang, Generating labeled planar graphs uniformly at random, *Theoret. Comput. Sci.* **379** (2007), 377–386.

[14] M. Bodirsky, M. Kang, M. Löffler & C. McDiarmid, Random cubic planar graphs, *Random Structures Algorithms* **30** (2007), 78–94.

[15] B. Bollobás, *Random Graphs*, Academic Press, London (1985).

[16] N. Bonichon, C. Gavoille, N. Hanusse, D. Poulalhon & G. Schaeffer, Planar graphs, via well-orderly maps and trees, *Graphs Combin.* **22** (2006), 185–202.

[17] M. Bousquet-Mélou, Rational and algebraic series in combinatorial enumeration, in *International Congress of Mathematicians, Vol. III*, Eur. Math. Soc., Zürich (2006) pp. 789–826.

[18] W. G. Brown & W. T. Tutte, On the enumeration of rooted non-separable planar maps, *Canad. J. Math.* **16** (1964), 572–577.

[19] G. Chapuy, E. Fusy, M. Kang & B. Shoilekova, A complete grammar for decomposing a family of graphs into 3-connected components, *Electron. J. Combin.* **15** (2008), R148.

[20] P. Chassaing & G. Schaeffer, Random planar lattices and integrated super Brownian excursion, *Probab. Theory Relat. Fields* **128** (2004), 161–212.

[21] A. Denise, M. Vasconcellos & D. J. A. Welsh, The random planar graph, *Congr. Numer.* **113** (1996), 61–79.

[22] R. Diestel, *Graph decompositions. A study in infinite graph theory*, Oxford University Press, New York (1990).

[23] M. Drmota, Systems of functional equations, *Random Structures Algorithms* **10** (1997), 103–124.

[24] M. Drmota, O. Giménez & M. Noy, Degree distribution in random planar graphs, submitted.

[25] M. Drmota, O. Giménez & M. Noy, Vertices of given degree in series-parallel graphs, *Random Structures Algorithms*, in press.

[26] P. Duchon, P. Flajolet, G. Louchard & G. Schaeffer, Boltzmann samplers for the random generation of combinatorial structures, *Combin. Probab. Comput.* **13** (2004), 577–625.

[27] P. Flajolet & A. Odlyzko, Singularity analysis of generating functions, *SIAM J. Discrete Math.* **3** (1990), 216–240.

[28] P. Flajolet & R. Sedgewick, *Analytic Combinatorics*, Cambridge University Press, Cambridge (2009).

[29] P. Flajolet, P. Zimmerman & B. Van Cutsem, A calculus for the random generation of labelled combinatorial structures, *Theoret. Comput. Sci.* **132** (1994), 1–35.

[30] E. Fusy, A linear approximate-size random sampler for labelled planar graphs, extended abstract in *Discrete Math. Theor. Comput. Sci. Proc.* **AD** (2005), 125–138. See arXiv:0705.1287v1.

[31] Z. Gao, A pattern for the asymptotic number of rooted maps on surfaces, *J. Combin. Theory Ser. A* **64** (1993), 246–264.

[32] Z. Gao & N. C. Wormald, The size of the largest components in random planar maps, *SIAM J. Discrete Math.* **12** (1999), 217–228.

[33] Z. Gao & N. C. Wormald, The distribution of the maximum vertex degree in random planar maps, *J. Combin. Theory Ser. A* **89** (2000), 201–230.

[34] Z. Gao & N. C. Wormald, Asymptotic normality determined by high moments, and submap counts of random maps, *Probab. Theory Relat. Fields* **130** (2004), 368–376.

[35] S. Gerke, O. Giménez, M. Noy & A. Weissl, On the number of $K_{3,3}$-minor-free and maximal $K_{3,3}$-minor-free graphs, *Electron. J. Combin.* **15** (2008), R114.

[36] S. Gerke & C. McDiarmid, On the number of edges in random planar graphs, *Combin. Probab. Comput.* **13** (2004), 165–183.

[37] S. Gerke, C. McDiarmid, A. Steger & A. Weissl, Random planar graphs with given average degree, in *Combinatorics, Complexity, and Chance*, Oxford Lecture Ser. Math. Appl., 34, Oxford University Press, Oxford (2007), pp. 83–102.

[38] O. Giménez & M. Noy, Asymptotic enumeration and limit laws of planar graphs, *J. Amer. Math. Soc.*, in press, arXiv:math/0512435v1.

[39] O. Giménez, M. Noy & J. Rué, Graph classes with given 3-connected components: asymptotic counting, random properties, and critical phenomena, in preparation. Extended abstract in *Electron. Notes Discrete Math.* **29** (2007), 521–529.

[40] F. Harary & E. M. Palmer, *Graphical enumeration*, Academic Press, New York–London (1973).

[41] M. Klazar, On growth rates of permutations, set partitions, ordered graphs and other objects, *Electron. J. Combin.* **15** (2008), R75.

[42] S. K. Lando & A. K. Zvonkin, *Graphs on surfaces and their applications*, Springer-Verlag, Berlin (2004).

[43] V. A. Liskovets, A pattern of asymptotic vertex valency distributions in planar maps, *J. Combin. Theory Ser. B* **75** (1999), 116–133.

[44] C. McDiarmid, Random graphs on surfaces, *J. Combin. Theory Ser. B* **98** (2008), 778–797.

[45] C. McDiarmid, personal communication.

[46] C. McDiarmid & B. Reed, On the maximum degree of a random planar graph, *Combin. Probab. Comput.* **17** (2008), 591–601.

[47] C. McDiarmid, A. Steger & D. J. A. Welsh, Random planar graphs, *J. Combin. Theory Ser. B* **93** (2005), 187–205.

[48] C. McDiarmid, A. Steger & D. J. A. Welsh, Random graphs from planar and other addable classes, in *Topics in discrete mathematics*, Algorithms Combin., 26, Springer, Berlin (2006), pp. 231–246.

[49] B. Mohar & C. Thomassen, *Graphs on surfaces*, Johns Hopkins University Press, Baltimore, MD (2001).

[50] J. W. Moon, *Counting labelled trees*, Canadian Math. Monographs, 1, Canadian Mathematical Congress, Montreal, Que. (1970).

[51] R. C. Mullin & P. J. Schellenberg, The enumeration of c-nets via quadrangulations, *J. Combin. Theory* **4** (1968), 259–276.

[52] A. Nijenhuis & H. S. Wilf, *Combinatorial algorithms*, Academic Press, New York–London (1979).

[53] S. Norine, P. Seymour, R. Thomas & P. Wollan, Proper minor-closed families are small, *J. Combin. Theory Ser. B* **96** (2006), 754–757.

[54] M. Noy, Random planar graphs and the number of planar graphs, in *Combinatorics, Complexity, and Chance*, Oxford Lecture Ser. Math. Appl., 34, Oxford University Press, Oxford (2007), pp. 213–233.

[55] D. Osthus, H. J. Prömel & A. Taraz, On random planar graphs, the number of planar graphs and their triangulations, *J. Combin. Theory Ser. B* **88** (2003), 119–134.

[56] R. Otter, The number of trees, *Ann. Math.* **49** (1948), 583–599.

[57] G. Pólya & R. C. Read, *Combinatorial enumeration of groups, graphs, and chemical compounds*, Springer-Verlag, New York (1987).

[58] N. Robertson & P. Seymour, Graph minors **I-XX**, *J. Combin. Theory Ser. B* (1983–2004).

[59] N. Robertson, P. Seymour & R. Thomas, Hadwiger's conjecture for K_6-free graphs, *Combinatorica* **13** (1993), 279–361.

[60] B. A. Trakhtenbrot, Towards a theory of non-repeating contact schemes (in Russian), *Trudi Mat. Inst. Akad. Nauk SSSR* **51** (1958), 226–269.

[61] G. Turán, On the succinct representation of graphs, *Discrete Appl. Math.* **8** (1984), 289–294.

[62] W. T. Tutte, A census of planar triangulations, *Canad. J. Math.* **14** (1962), 21–38.

[63] W. T. Tutte, A census of planar maps, *Canad. J. Math.* **15** (1963), 249–271.

[64] W. T. Tutte, *Connectivity in graphs*, University of Toronto Press, Toronto (1966).

[65] V. Vatter, Permutation classes of every growth rate (a.k.a. Stanley-Wilf limit) above 2.48187, preprint, arXiv:0807.2815.

[66] K. Wagner, Über eine Erweiterung des Satzes von Kuratowski, *Deutsche Math.* **2** (1937), 280–285.

[67] T. R. S. Walsh, Counting labelled three-connected and homeomorphically irreducible two-connected graphs, *J. Combin. Theory Ser. B* **32** (1982), 1–11.

[68] H. Whitney, Congruent graphs and the connectivity of graphs, *Amer. J. Math.* **54** (1932), 150–168.

[69] N. C. Wormald, Models of random regular graphs, in *Surveys in Combinatorics, 1999*, Lond. Math. Soc. Lecture Note Ser., 267, Cambridge University Press, Cambridge (1999), pp. 239–298.

Departament de Llenguatges i Sistemes Informàtics
Universitat Politècnica de Catalunya
Jordi Girona 1-3
08034 Barcelona, Spain
omer.gimenez@gmail.com

Departament de Matemàtica Aplicada II
Universitat Politècnica de Catalunya
Jordi Girona 1-3
08034 Barcelona, Spain
marc.noy@upc.edu

Metrics for sparse graphs

Béla Bollobás and Oliver Riordan

Abstract

Recently, Bollobás, Janson and Riordan introduced a very general family of random graph models, producing inhomogeneous random graphs with $\Theta(n)$ edges. Roughly speaking, there is one model for each *kernel*, i.e. each symmetric measurable function from $[0,1]^2$ to the non-negative reals, although the details are much more complicated, to ensure the exact inclusion of many of the recent models for large-scale real-world networks.

A different connection between kernels and random graphs arises in the recent work of Borgs, Chayes, Lovász, Sós, Szegedy and Vesztergombi. They introduced several natural metrics on dense graphs (graphs with n vertices and $\Theta(n^2)$ edges), showed that these metrics are equivalent, and gave a description of the completion of the space of all graphs with respect to any of these metrics in terms of *graphons*, which are essentially bounded kernels. One of the most appealing aspects of this work is the message that sequences of inhomogeneous quasi-random graphs are in a sense completely general: any sequence of dense graphs contains such a subsequence. Alternatively, their results show that certain natural models of dense inhomogeneous random graphs (one for each graphon) cover the space of dense graphs: there is one model for each point of the completion, producing graphs that converge to this point.

Our aim here is to briefly survey these results, and then to investigate to what extent they can be generalized to graphs with $o(n^2)$ edges. Although many of the definitions extend in a simple way, the connections between the various metrics, and between the metrics and random graph models, turn out to be much more complicated than in the dense case. We shall prove many partial results, and state even more conjectures and open problems, whose resolution would greatly enhance the currently rather unsatisfactory theory of metrics on sparse graphs. This paper deals mainly with graphs with $o(n^2)$ but $\omega(n)$ edges: a companion paper will discuss the (more problematic still) case of *extremely sparse* graphs, with $O(n)$ edges.

Contents

1 **Introduction**   212

2 **Dense graphs**   214
    2.1 The subgraph distance 215
    2.2 The cut distance 217
    2.3 Kernels and (quasi-)random graphs 220
    2.4 Equivalent kernels 221

3 **Subgraph counts for sparse graphs**   227
    3.1 Bounded and unbounded kernels 229
    3.2 Non-uniform random graphs 231
    3.3 Subgraph counts in the uniform case 233
    3.4 Partial results in the almost dense, uniform case 238
    3.5 Extensions to lower densities 242

4 Szemerédi's Lemma and the cut metric — 243
4.1 Weakly regular partitions 245
4.2 Strongly regular partitions 248
4.3 Szemerédi's lemma and convergence in the cut norm 250

5 Comparison between cut and count convergence — 254
5.1 Admissible subgraphs and their counts 254
5.2 Conjectured equivalence between cut and count convergence 258
5.3 Partial results: embedding lemmas 261
5.4 Embeddings or homomorphisms? 272

6 The partition metric — 278
6.1 Partition matrices and the partition metric 278
6.2 The relationship between the cut and partition metrics 281

7 Discussion and closing remarks — 282
7.1 Models and metrics 282
7.2 Closing Remarks 284

1 Introduction

In recent years, much work has been done constructing and analyzing mathematical models of real-world networks. The random graphs in these models are inhomogeneous – in fact, many of them have degree sequences with power law distributions. In [8], Bollobás, Janson and Riordan defined a very general model of an n-vertex random graph $G(n, \kappa)$ with conditional independence between the edges which includes as special cases many of the models of real-world networks that have been studied, and proved numerous results about the random graphs generated by this model, including results about their component structure and the point and nature of the phase transition in them. Here the *kernel* κ is a symmetric measurable function from $[0,1]^2$ to $[0,\infty)$ satisfying some mild conditions. (Some of these conditions arise due to the very general nature of other parts of the model, and can be weakened in other contexts; see [9] and [10] for a discussion of this.) Just like the real-world graphs that motivated the construction of the BJR model, the random graphs $G(n, \kappa)$ are sparse in the sense that the expected number of edges is $O(n)$ (in fact, $(c + o(1))n$ for some constant c). In [8] the kernel κ was used to define a multi-type branching process $\mathcal{X}_\kappa$ whose survival probability is closely related to the component structure of $G(n, \kappa)$.

In order to decide how well our random graph $G(n, \kappa)$ approximates a given real-world graph G_n, it would be desirable to establish a *distance* between a random graph model and a graph, so that the approximation is judged to be better and better as the distance tends to 0. Putting it slightly differently, we should like to define a metric on the set of sparse finite graphs so that a Cauchy sequence consists of graphs that are in some sense 'similar', and the limit of such a (not eventually constant) sequence is naturally identified with a suitable random graph model. For *dense* graphs, graphs with n vertices and at least cn^2 edges, such a program has been carried out very successfully in a series of papers by (various subsets of) Borgs, Chayes, Lovász, Sós, Szegedy and Vesztergombi (see [13, 14, 15, 16, 33, 34] and the references therein). In particular, they introduced several metrics on the space of dense finite (weighted) graphs and showed them to be equivalent. The limiting objects, i.e. the additional points in the completion, turn out to be *graphons*, that is, bounded symmetric measurable functions from $[0,1]^2$ to $\mathbf{R}$. The corresponding random graph models, called *W-random graphs* in [34], are the natural dense version of $G(n,\kappa)$; see Subsection 2.3.

The only difference between kernels and graphons is that the latter are bounded, while the former must be allowed to be unbounded in order to model, for example, highly inhomogeneous real-world networks. In many fundamental questions (for example those concerning the phase transition), this difference is substantial. The appearance of graphons or kernels in the two different contexts described above suggests the existence of interesting connections between these areas. One such connection is described by Bollobás, Borgs, Chayes and Riordan [7], who study (sparse) random subgraphs of arbitrary dense graphs; this has recently been extended by Bollobás, Janson and Riordan [10].

We have several aims in this paper. First, we shall review some of the results of Borgs, Chayes, Lovász, Sós, Szegedy and Vesztergombi mentioned above. Our main aim is then to take the first tentative steps towards a general theory of metrics on sparse graphs; in particular, we shall investigate to what extent these ideas can be carried over to the sparse setting, and what can be said about the connection between the metrics and the ideas of Bollobás, Janson and Riordan. As we shall see, the difficulties that arise are considerably greater than in the dense case; in fact, the difficulties increase as the graphs get sparser. The *almost dense* case $e(G_n) = n^{2-o(1)}$ is already rather different from the dense case; the *extremely sparse* case $e(G_n) = \Theta(n)$, which will be studied in a companion paper [11], is *very* different indeed, having many novel features. We shall prove numerous results, but the picture we obtain is much less complete than that obtained by Borgs et al in the dense case. In fact, perhaps our most important aim is to identify some of the main problems and conjectures whose resolution would enhance the theory of metrics on sparse graphs.

An important tool in the study of metrics on spaces of dense graphs is Szemerédi's Regularity Lemma. While there is a version of Szemerédi's Lemma for sparse graphs (with $o(n^2)$ but $\omega(n)$ edges) satisfying a mild additional condition, there is no satisfactory counting/embedding lemma for counting (or even finding) small subgraphs using regular partitions. This is one of the reasons why sparse graphs are much more difficult to handle than dense ones. One of our main aims is to prove such a counting lemma for certain subgraphs, greatly extending a result of Chung and Graham [17].

The rest of the paper is organized as follows. The next section is about dense graphs and kernels; we start by briefly recalling some of the definitions and results of Borgs, Chayes, Lovász, Sós, Szegedy and Vesztergombi whose generalization we shall discuss, focussing in particular on the cut metric. Then, in Subsection 2.4, we show that these results are closely connected to the question of when two kernels are 'equivalent'; we shall need this notion of equivalence when we come to sparse graphs.

The rest of the paper concerns sparse graphs, i.e. graphs with n vertices and $o(n^2)$ edges: in Section 3 we consider subgraph counts in sparse (but mostly not too sparse) graphs, stating a conjecture that generalizes the main result of Lovász and Szegedy [34], and proving various partial results, concentrating especially on the *uniform* case, i.e. on sparse quasi-random graphs. In Section 4 we turn to Szemerédi's Lemma for sparse graphs satisfying an appropriate 'bounded density' assumption, and the consequences for questions of convergence in the cut metric.

Section 5 is the longest and most important section of the paper. In it we discuss the relationship between the cut metric and the count metric (to be defined) in the sparse case. As well as proposing various conjectures extending the results of Borgs, Chayes, Lovász, Sós and Vesztergombi, we prove several partial results, amounting to 'sparse counting lemmas' with various assumptions; these results, Theorem 5.14 and its variants Theorems 5.15 and 5.17, are the most substantial results in the paper.

In Section 6 we briefly discuss another metric considered by Borgs, Chayes, Lovász, Sós and Vesztergombi, showing that for graphs that are sparse, but not too sparse, it is equivalent to the cut metric. In the *extremely sparse* case, considering graphs with bounded average degree, the partition metric turns out to be much more useful than the cut metric. This and a discussion of the many problems and interesting open questions concerning metrics on extremely sparse graphs will be the topic of a companion paper [11].

In Section 7 we return briefly to the relationship between metrics and random graph models, and close with some final remarks summarizing our main results and conjectures.

Throughout the paper we use standard graph theoretic notation as in [4]. For example, $|G|$ and $e(G)$ denote respectively the number of vertices and number of edges of a graph G.

2 Dense graphs

There are many natural definitions of what it means for two graphs to be 'close', and corresponding metrics and notions of Cauchy/fundamental sequences. These tend to be particularly natural for 'dense' graphs, with $\Theta(n^2)$ edges. Several of these metrics have been studied by Borgs, Chayes, Lovász, Sós and Vesztergombi [15, 16], who showed that they are equivalent, and that there is a natural completion of the space of graphs under any of these metrics. In this section we briefly recall some of these definitions and results; we are not aiming to give a comprehensive survey of the results of these papers, discussing only those that will be relevant for us here. Although most of the results mentioned in Subsections 2.1–2.3 will be from Lovász and Szegedy [34] and [15, 16], we shall not always adopt their notation or terminology, or indeed follow their definitions exactly.

Borgs, Chayes, Lovász, Sós and Vesztergombi [15, 16] consider *weighted graphs*, with weights on the edges and on the vertices. For the results we shall describe, this makes essentially no difference. In what follows, we consider only unweighted graphs; while much of what we shall say presumably carries over to suitably weighted graphs, the definitions for weighted graphs are not as natural in the sparse case, and are likely to introduce more additional complications than new insights.

2.1 The subgraph distance

The basic starting point is to consider, for each fixed graph F, the number of copies of F in a large graph G, i.e. the number $X_F(G)$ of subgraphs of G isomorphic to F. Recall that a *homomorphism* from a graph F to a graph G is a function $\phi : V(F) \to V(G)$ such that $\phi(x)\phi(y) \in E(G)$ whenever $xy \in E(F)$. Although $X_F(G)$ (for example, the number of triangles in G) is the most natural basic notion in this context, it turns out to be cleaner to work with $\mathrm{emb}(F,G)$, the number of injective homomorphisms or *embeddings* of F into G. Note that

$$\mathrm{emb}(F,G) = \mathrm{aut}(F) X_F(G),$$

so $X_F(G)$ and $\mathrm{emb}(F,G)$ contain the same information. Working with the latter avoids constant factors $\mathrm{aut}(F)$ in many formulae.

If F has k vertices, then for $n \geq k$ we have $\mathrm{emb}(F, K_n) = n_{(k)} = n(n-1)\cdots(n-k+1)$, so the natural normalization is to work with

$$s(F,G) = \frac{\mathrm{emb}(F,G)}{n_{(k)}} = \frac{X_F(G)}{X_F(K_n)} \in [0,1],$$

where, as usual, $n = |G|$ is the number of vertices of G. If $|F| > |G|$ then the above ratio is not defined, and we set $s(F,G) = 0$.

Let $\mathcal{F}$ denote the set of isomorphism classes of finite graphs; sometimes it will be convenient to enumerate $\mathcal{F}$ in an arbitrary way, writing $\mathcal{F} = \{F_1, F_2, \ldots\}$. (More formally, we shall take each F_i to be a representative of an isomorphism class.) The graph parameters $s(F, \cdot)$, $F \in \mathcal{F}$, define a natural family of equivalent metrics on $\mathcal{F}$, by mapping $\mathcal{F}$ into $[0,1]^\infty$ (or into $[0,1]^{\mathcal{F}}$). Indeed, for any finite graph G, set

$$s(G) = (s_i(G))_{i=1}^\infty \in [0,1]^\infty,$$

where $s_i(G) = s(F_i, G)$. Let d be any metric on $X = [0,1]^\infty$ which gives the product topology, for example $d(s,t) = \sum_{i=1}^\infty 2^{-i}|s_i - t_i|$. We may define the *subgraph distance* of two graphs G_1, G_2 as

$$d_{\mathrm{sub}}(G_1, G_2) = d(s(G_1), s(G_2)).$$

It is easy to see that this defines a metric on $\mathcal{F}$: indeed, given $G \in \mathcal{F}$, among graphs F with $s(F,G) > 0$, there is a unique graph with $|F| + e(F)$ maximal, namely G. Thus the map $G \mapsto s(G)$ is injective. Furthermore, considering $s(E_{n+1}, G)$, where E_{n+1} is the empty graph with $n+1$ vertices, we see that the distance between any graph G with n vertices and the set of graphs with more than n vertices is positive. It follows that the metric space $(\mathcal{F}, d_{\mathrm{sub}})$ is discrete.

A sequence (G_n) of graphs is Cauchy with respect to d_{sub} if and only if, for each $F \in \mathcal{F}$, the sequence $s(F,G)$ converges. Such sequences are sometimes called 'convergent', although they do not converge in the metric space $(\mathcal{F}, d_{\mathrm{sub}})$. Note that if (G_n) is Cauchy then, since $(\mathcal{F}, d_{\mathrm{sub}})$ is discrete, either (G_n) is eventually constant, or $|G_n| \to \infty$.

Many minor variations on the definition of d_{sub} are possible. For example, instead of considering the number of embeddings of F into G, one can consider the number

hom(F,G) of homomorphisms from F to G. If $|F| = k$ and $|G| = n$, then the number of non-injective homomorphisms from F to G is at most $\binom{k}{2} n^{k-1} = O(n^{k-1})$, so setting
$$t(F,G) = \text{hom}(F,G)/n^k$$
we have
$$t(F,G) = s(F,G) + O(n^{-1}) \qquad (2.1)$$
for each F. Hence, in this dense case, the parameters $s(F,\cdot)$ and $t(F,\cdot)$ are essentially equivalent. [There is a minor difference that, working with homomorphisms, one ends up with a pseudo-metric: if G is any graph and $G^{(r)}$ is the *blow-up* of G obtained by making r copies of each vertex, joined to all copies of its neighbours, then $t(F, G^{(r)}) = t(F,G)$ for all $F \in \mathcal{F}$ and $r \geq 1$.] Also, one can pass easily back and forth between subgraph counts and counts of induced subgraphs using inclusion–exclusion.

One of the key properties of the metric d_{sub} is that there is a natural description of the (clearly compact) completion of $(\mathcal{F}, d_{\text{sub}})$, in terms of *standard kernels* (also called *graphons*). Here a *kernel* is a symmetric measurable function from $[0,1]^2$ to $[0,\infty)$; a standard kernel is one taking values in $[0,1]$. In other contexts, one considers more general bounded kernels, taking values in $[0,M]$ or $[-M,M]$, $M > 0$, or general *signed kernels* taking values in $\mathbf{R}$. One can extend the definition of $s(F,G)$ (or of $t(F,G)$) to kernels in a natural way: given a finite graph F with vertex set $\{1, 2, \ldots, k\}$, let

$$s(F,\kappa) = \int_{[0,1]^k} \prod_{ij \in E(F)} \kappa(x_i, x_j) \prod_{i=1}^{k} dx_i. \qquad (2.2)$$

(Some authors use the notation $t(F,\kappa)$ for the same quantity.) This formula has a natural interpretation as the normalized 'number' of embeddings of F into a weighted graph with the uncountable vertex set $[0,1]$, with edge weights given by κ. Of course, in this context there is no difference between embeddings and homomorphisms.

Lovász and Szegedy [34] proved (essentially) the following result.

Theorem 2.1 *Let (G_n) be a Cauchy sequence in $(\mathcal{F}, d_{\text{sub}})$. Then either (G_n) is eventually constant, or there is a standard kernel κ such that $s(F, G_n) \to s(F, \kappa)$.* □

Let us remark that the result proved in [34] concerns t rather than s, which makes no difference, except that a separate case for eventually constant sequences is then not needed. Here, the distinction is informative: considering the parameters $s(E_k, G_n)$ for each k shows that in the second case above we have $|G_n| \to \infty$.

Of course, (2.2) allows one to extend the metric d_{sub} to standard kernels, obtaining in the first instance a pseudo-metric on the set of standard kernels. There is a natural notion of equivalence for kernels, which one can think of as a two dimensional version of the equivalence relation on random variables given by $X \sim Y$ if X and Y have the same distribution; the details are somewhat technical, and not essential for understanding the metrics discussed here, so we postpone them to Subsection 2.4. We write $\sim$ for this relation, and $\mathcal{K}$ for the set of equivalence classes of standard kernels under $\sim$. Borgs, Chayes and Lovász [12] have shown that $\kappa_1 \sim \kappa_2$ if and only if $d_{\text{sub}}(\kappa_1, \kappa_2) = 0$ (see also Theorem 2.8), so d_{sub} induces a metric on $\mathcal{K}$.

The metric space $(\mathcal{K}, d_{\text{sub}})$ is complete (the result about Cauchy sequences of graphs above applies just as well to standard kernels). Hence, the completion of $(\mathcal{F}, d_{\text{sub}})$ is obtained by adding to $\mathcal{F}$ the set $\mathcal{K}$ of all equivalence classes of standard kernels, and using the map $s: \mathcal{F} \cup \mathcal{K} \to [0,1]^\infty$ to extend d_{sub} to $\mathcal{F} \cup \mathcal{K}$.

There is a natural way to associate a standard kernel κ_G to a graph G with n vertices: divide $[0,1]$ into n intervals $I_1, \ldots, I_n$ of equal length (we may and shall ignore the question of which endpoints are included), and set κ_G to be 1 on $I_i \times I_j$ if $ij \in E(G)$, and 0 otherwise. One slight advantage of using t rather than s is that

$$t(F, G) = s(F, \kappa_G)$$

for all graphs F and G. However, the metric obtained using t is only a pseudo-metric, since graphs on different numbers of vertices may correspond to the same kernel, for example if one is a blow-up of the other.

We say that a kernel κ is of *finite type* if there is a partition of $[0,1]$ into measurable sets $A_1, \ldots, A_k$ so that κ is constant on each of the rectangles $A_i \times A_j$. Note that κ_G is always of finite type.

2.2 The cut distance

Borgs, Chayes, Lovász, Sós and Vesztergombi [15] considered another natural metric on graphs or kernels, namely, the *cut metric*, based on a norm used by Frieze and Kannan [23]. For any integrable function $\kappa: [0,1]^2 \to \mathbf{R}$, its *cut norm* $\|\kappa\|_{\text{cut}}$ is defined by

$$\|\kappa\|_{\text{cut}} = \sup_{S, T \subset [0,1]} \left| \int_{S \times T} \kappa(x, y) \, dx \, dy \right|, \quad (2.3)$$

where the supremum is over all pairs of measurable subsets of $[0,1]$. It is easily seen that this defines a norm on $L^\infty([0,1]^2)$. In fact, there are several variations of this definition: one can take

$$\|\kappa\|_{\text{cut}} = \sup_{S \subset [0,1]} \left| \int_{S \times S^c} \kappa(x, y) \, dx \, dy \right|, \quad (2.4)$$

where $S^c = [0,1] \setminus S$, or one can take the supremum in (2.3) only over sets S, T with $S \cap T = \emptyset$. It is easy to check that these variations only affect the norm up to an (irrelevant) constant factor (see [15]), so we shall feel free to use whichever definition is most convenient in any given context.

There is yet another definition of $\|\kappa\|_{\text{cut}}$ that is more natural from the point of view of functional analysis, namely

$$\|\kappa\|_{\text{cut}} = \sup_{\|f\|_\infty, \|g\|_\infty \le 1} \int_{[0,1]^2} \kappa(x, y) f(x) g(y) \, dx \, dy,$$

where the supremum is taken over all pairs of measurable functions from $[0,1]$ to $[-1, +1]$. Since the integral above is linear with respect to each of f and g, the supremum is attained at some functions taking values in $\{-1, +1\}$, and it follows immediately that this version of the cut norm is again within a constant factor of that defined by (2.3). As noted in [10], for example, this last definition is the most natural from the point of view of functional analysis: it is the dual of the projective tensor product norm in $L^\infty \hat{\otimes} L^\infty$, and is thus the injective tensor product norm in $L^1 \check{\otimes} L^1$. Equivalently, this is just the norm of the integral operator with kernel κ, treated as a map from L^∞ to L^1.

Before turning to the cut metric we need one further definition. Given a kernel κ and a measure-preserving map $\tau : [0,1] \to [0,1]$, let $\kappa^{(\tau)}$ be the kernel defined by

$$\kappa^{(\tau)}(x,y) = \kappa(\tau(x), \tau(y)). \tag{2.5}$$

If τ is a bijection, then we call τ a *rearrangement* of $[0,1]$, and $\kappa^{(\tau)}$ a *rearrangement* of κ. (It is perhaps more natural to consider measure-preserving bijections between two subsets of $[0,1]$ with measure 1; this makes no difference.) Two kernels κ_1 and κ_2 are *naively equivalent* if one is a rearrangement of the other, more precisely, if there is a rearrangement τ of $[0,1]$ such that

$$\kappa_1(x,y) = \kappa_2^{(\tau)}(x,y) \quad \text{for a.e. } (x,y) \in [0,1]^2. \tag{2.6}$$

In this case we write $\kappa_1 \approx \kappa_2$, noting that $\approx$ is an equivalence relation.

The *cut metric* d_{cut} on the set of standard kernels may be defined as follows:

$$d_{\mathrm{cut}}(\kappa_1, \kappa_2) = \inf_{\kappa_2' \approx \kappa_2} \|\kappa_1 - \kappa_2'\|_{\mathrm{cut}}. \tag{2.7}$$

Clearly, this defines a pseudo-metric on standard kernels; in particular, if $\kappa_1 \approx \kappa_2$, then $d_{\mathrm{cut}}(\kappa_1, \kappa_2) = 0$. The reverse implication does not hold; in fact, $d_{\mathrm{cut}}(\kappa_1, \kappa_2) = 0$ if and only if $\kappa_1 \sim \kappa_2$, where $\sim$ is the equivalence relation to be defined in Subsection 2.4. Hence, d_{cut} induces a metric on the set $\mathcal{K}$ of equivalence classes of standard kernels under the relation $\sim$.

As noted above, there is a standard kernel κ_G naturally associated to each graph G, although the map $G \mapsto \kappa_G$ from $\mathcal{F}$ to $\mathcal{K}$ is not injective. One extends the cut metric to a pseudo-metric on graphs by setting

$$d_{\mathrm{cut}}(G_1, G_2) = d_{\mathrm{cut}}(\kappa_{G_1}, \kappa_{G_2}), \tag{2.8}$$

and to $\mathcal{F} \cup \mathcal{K}$ similarly.

For graphs G_1, G_2 on n vertices, there is a much more natural variant of their cut distance: let $\widehat{d}_{\mathrm{cut}}(G_1, G_2)$ be the smallest ε for which we can identify the vertices of G_1 with those of G_2 such that for any bipartition of the vertex set, the corresponding cuts in G_1 and G_2 have sizes within εn^2. In terms of kernels,

$$\widehat{d}_{\mathrm{cut}}(G_1, G_2) = \min_{\kappa \approx_n \kappa_{G_2}} \|\kappa_{G_1} - \kappa\|_{\mathrm{cut}}, \tag{2.9}$$

where $\kappa_1 \approx_n \kappa_2$ if (2.6) holds for some map τ that simply permutes the intervals I_n corresponding to the vertices, and we take (2.4) as the definition of the cut norm. Note that the supremum implied by (2.4) in the definition (2.9) is over all bipartitions of $[0,1]$, not just those corresponding to bipartitions of the vertices; it is very easy to see that this makes no difference: the supremum is attained at a vertex bipartition.

Comparing (2.8) and (2.9), since the infimum in the former is taken over a larger set, one trivially has $d_{\mathrm{cut}}(G_1, G_2) \le \widehat{d}_{\mathrm{cut}}(G_1, G_2)$. Borgs, Chayes, Lovász, Sós and

Vesztergombi [15] noted that strict inequality is possible. For example, taking (2.4) as the definition of the cut norm, let G_1 be a triangle, and let G_2 be the graph with 3 vertices and one edge. For any pairing of the vertices of G_1 with those of G_2, the 'worst' cut is the one in which the isolated vertex of G_2 is placed into one part and the other two vertices into the other part. This cut has 2 edges in G_1 but no edges in G_2, so $\widehat{d}_{\text{cut}}(G_1, G_2) = 2/9$. On the other hand, consider the blow-ups $G_1^{(2)}$, a complete tripartite graph with two vertices in each class, and $G_2^{(2)}$, a C_4 with two isolated vertices added. Pairing the vertices of $G_1^{(2)}$ and $G_2^{(2)}$ by placing two opposite vertices of the C_4 in one class of $G_1^{(2)}$, and the other vertices in different classes, we realize $G_2^{(2)}$ as a subgraph of $G_1^{(2)}$ in such a way that the 8 edges of $G_1^{(2)}$ not present in $G_2^{(2)}$ form a non-bipartite graph, so *every* cut cuts at most 7 of these extra edges. It follows that $\widehat{d}_{\text{cut}}(G_1^{(2)}, G_2^{(2)}) \leq 7/6^2$. In fact, one can check that with the vertices paired in this way the maximum difference between the sizes of corresponding cuts in $G_1^{(2)}$ and $G_2^{(2)}$ is 6, so

$$d_{\text{cut}}(G_1, G_2) \leq \widehat{d}_{\text{cut}}(G_1^{(2)}, G_2^{(2)}) \leq 6/6^2 = 1/6 < 2/9 = \widehat{d}_{\text{cut}}(G_1, G_2),$$

showing that d_{cut} and $\widehat{d}_{\text{cut}}$ do not always agree. For questions of convergence, however, the two metrics are equivalent: as shown in [15],

$$d_{\text{cut}}(G_1, G_2) \leq \widehat{d}_{\text{cut}}(G_1, G_2) \leq 32 d_{\text{cut}}(G_1, G_2)^{1/67}.$$

At first sight it is not clear why the cut metric should be interesting: after all, what is the significance of two graphs having almost the same number of edges in all corresponding cuts? One very important consequence of this property is that their subgraph counts are close, as shown by the following simple lemma from Borgs, Chayes, Lovász, Sós and Vesztergombi [15].

Lemma 2.2 *Let κ and κ' be two standard kernels. Then for every graph F we have*

$$|s(F, \kappa) - s(F, \kappa')| \leq e(F) \|\kappa - \kappa'\|_{\text{cut}}.$$

Proof Before we embark on the proof, we extend the definition of $s(F, \kappa)$ slightly. Fix the graph F, taking its vertex set to be $[k] = \{1, 2, \ldots, k\}$, as usual, and list the edges of F as $\{i_1 j_1, \ldots, i_m j_m\}$. Given a sequence $(\kappa_1, \ldots, \kappa_m)$ of standard kernels, set

$$s(F; \kappa_1, \ldots, \kappa_m) = \int_{[0,1]^k} \prod_{r=1}^{m} \kappa_r(x_{i_r}, x_{j_r}) \prod_{i=1}^{k} dx_i.$$

Thus $s(F, \kappa) = s(F; \kappa, \ldots, \kappa)$. We claim that for any graph F with m edges and any standard kernels $\kappa_1, \kappa_2, \ldots, \kappa_m$ and κ_1', we have

$$|s(F; \kappa_1, \kappa_2, \ldots, \kappa_m) - s(F; \kappa_1', \kappa_2, \ldots, \kappa_m)| \leq \|\kappa_1 - \kappa_1'\|_{\text{cut}}. \qquad (2.10)$$

Applying this $m = e(F)$ times, changing one kernel from κ to κ' each time, the lemma follows.

It remains to prove (2.10), which is easy. Suppose without loss of generality that the first edge is 12, so $i_1 = 1$ and $j_1 = 2$. Our task is to bound

$$\Delta = \int_{[0,1]^k} (\kappa_1(x_1,x_2) - \kappa_1'(x_1,x_2)) \prod_{r=2}^{m} \kappa_r(x_{i_r}, x_{j_r}) \prod_{i=1}^{k} dx_i$$

Collecting the terms in the product that involve x_1 or x_2, we may write this product as $f_0(\mathbf{x})f_1(x_1,\mathbf{x})f_2(x_2,\mathbf{x})$, where $\mathbf{x} = (x_3, \ldots, x_k)$ and each f_i (being a product of standard kernels evaluated at certain places) takes values in $[0,1]$. Now from (2.3), it is immediate that if f and g take values in $[0,1]$, then $\left|\int \kappa(x,y) f(x) g(y) \, dx \, dy\right| \leq \|\kappa\|_{\mathrm{cut}}$. Applying this with $\mathbf{x}$ fixed, and then integrating over $\mathbf{x}$, it follows that $|\Delta| \leq \|\kappa_1 - \kappa_1'\|_{\mathrm{cut}}$, as required. □

Corollary 2.3 *Let (G_n) be a sequence of graphs with $|G_n| \to \infty$, and let κ be a standard kernel. If $d_{\mathrm{cut}}(G_n, \kappa) \to 0$ then $d_{\mathrm{sub}}(G_n, \kappa) \to 0$.*

Proof Let $\kappa_n = \kappa_{G_n}$, so by definition $d_{\mathrm{cut}}(G_n, \kappa) = d_{\mathrm{cut}}(\kappa_n, \kappa)$. By Lemma 2.2, for every F we have $s(F, \kappa_n) \to s(F, \kappa)$. But $s(F, \kappa_n) = t(F, G_n)$, while from (2.1) we have $s(F, G_n) = t(F, G_n) + o(1)$. Thus $s(F, G_n) \to s(F, \kappa)$ for each F, i.e. $d_{\mathrm{sub}}(G_n, \kappa) \to 0$. □

We have just seen that convergence in d_{cut} implies convergence in d_{sub}; one of the main results of Borgs, Chayes, Lovász, Sós and Vesztergombi, namely Theorem 2.6 in [15], gives a converse of this. This result states that the metrics d_{sub} (defined using t rather than s) and d_{cut} are equivalent, in the sense that (G_n) is a Cauchy sequence for d_{sub} if and only if it is a Cauchy sequence for d_{cut}. In the light of the various other results of Lovász and Szegedy [34] and Borgs, Chayes, Lovász, Sós and Vesztergombi [15], this statement may be reformulated in our notation as follows.

Theorem 2.4 *Let (G_n) be a sequence of graphs or standard kernels with $|G_n| \to \infty$, where we take $|G_n| = \infty$ if G_n is a kernel, and let κ be a standard kernel. Then $d_{\mathrm{sub}}(G_n, \kappa) \to 0$ if and only if $d_{\mathrm{cut}}(G_n, \kappa) \to 0$.* □

An immediate consequence of this result is the following, Corollary 3.10 in [15].

Corollary 2.5 *Let κ and κ' be two bounded kernels. Then $s(F, \kappa) = s(F, \kappa')$ for every F if and only if $d_{\mathrm{cut}}(\kappa, \kappa') = 0$.* □

We shall return to a discussion of kernels at cut distance 0 shortly.

2.3 Kernels and (quasi-)random graphs

As well as going from graphs to kernels, one can go from kernels to *random graphs* in a very natural way, as in Section 2.6 of Lovász and Szegedy [34], or as in Bollobás, Janson and Riordan [8] for the sparse case. Indeed, given a standard kernel κ and an $n \geq 1$, let $G(n, \kappa)$ be the random graph on $[n]$ defined as follows: first let $x_1, \ldots, x_n$ be iid with the uniform distribution on $[0,1]$. Given the x_i, join each pair of vertices independently, joining i and j with probability $\kappa(x_i, x_j)$. The resulting graph is called a *κ-random graph* by Lovász and Szegedy [34], although they use W as their default symbol for a kernel. It is easy to check, for example by the second moment method, that for each F, the random variable $s(F, G(n, \kappa))$ converges (in probability and in fact almost surely) to $s(F, \kappa)$ as $n \to \infty$. Thus the sequence $G(n, \kappa)$ converges almost surely to κ in the metric d_{sub} or d_{cut}. Note that if κ is constant and takes the value p, then we recover the usual Erdős–Rényi model $G(n, p)$: no confusion should arise between the notation for the two models. (In fact, it was Gilbert [25] who introduced $G(n, p)$, while Erdős and Rényi [21] introduced a model, $G(n, m)$, that is essentially equivalent for many purposes. Since it was they who founded the theory of random graphs, both models are often referred to as Erdős–Rényi models.)

It is natural to view a sequence (G_n) converging to κ in d_{sub} as a sequence of 'inhomogeneous quasi-random graphs': when κ is constant, the convergence condition is equivalent to the standard notion of quasi-randomness, introduced by Thomason [37] in 1987 (although he called it pseudo-randomness) and studied in great detail by Chung, Graham and Wilson [18] and many others. The convergence of $G(n, \kappa)$ to κ in d_{sub} establishes that sequences generated by the natural inhomogeneous random model are also quasi-random, as one would hope. One of the most pleasing features of this whole subject area is the interpretation that inhomogeneous quasi-random graphs are completely general: any sequence of (dense) graphs has such a subsequence.

To take an alternative viewpoint, we may think of standard kernels as uncountable infinite graphs, and a 'typical' random graph $G(n, \kappa)$ as a good finite approximation to κ. Then the completion of $\mathcal{F}$ is obtained by adding these infinite graphs, and the approximations $G(n, \kappa)$ (n large) are examples of finite graphs close to a given infinite graph. Taking this viewpoint it is natural *not* to identify a finite graph with a kernel. For another, slightly different, point of view, see Diaconis and Janson [19], where connections to certain infinite random graphs are described.

2.4 Equivalent kernels

In the light of Corollary 2.5, it is clearly important to understand which pairs of kernels have $d_{\text{cut}}(\kappa_1, \kappa_2) = 0$; this is also important for understanding d_{cut} itself. Fortunately, it turns out that there is a natural notion of equivalence for kernels which gives the answer. Since this topic is only touched on in passing in Borgs, Chayes, Lovász, Sós and Vesztergombi [15], we shall go into some detail here.

Roughly speaking, we would like to say that two kernels are equivalent if one is obtained from the other simply by relabelling the 'types' in $[0, 1]$. It would seem that the notion $\approx$ of naive equivalence defined in (2.6) is thus the right one, but a little thought shows that this is not the case; for this, the random viewpoint is very helpful.

So far, as in [15], we defined kernels only on $[0, 1]^2$. In view of the connection to random graphs discussed in the previous subsection, it is *a priori* more natural to work with a general probability space $(\Omega, \mathcal{F}, \mu)$ rather than $[0, 1]$ with Lebesgue measure, defining a standard kernel as a symmetric measurable function from the square of a probability space to $[0, 1]$. (This is the approach taken in the sparse case by Bollobás, Janson and Riordan [8].) However, almost all the time, we shall consider only kernels on $[0, 1]$; there are two reasons for doing so. Firstly, graphs with

n vertices correspond to kernels on the discrete space with n equiprobable elements, and $[0,1]$ is the natural limit of these spaces. Secondly, all probability spaces that one would ever wish to work with (all so-called 'standard' probability spaces) are isomorphic to Lebesgue measure on an interval, combined with (possibly) a finite or countable number of atoms. When studying kernels, the presence of atoms makes no difference: for example, a kernel on a finite measure space corresponds in a natural way to a piecewise constant kernel on $[0,1]$. Hence it makes very good sense to consider only kernels on $[0,1]$. For a formal reduction to the case of kernels on $[0,1]$ in the context of random graphs, see Janson [27].

We may think of kernels as two-dimensional versions of random variables (not to be confused with vector-valued random variables). Two random variables are equivalent if they have the same distribution. Equivalently, they are equivalent if they may be coupled so as to agree with probability 1. This is the definition we shall use for kernels.

Working, for the moment, on general (standard) probability spaces, and suppressing the σ-field of measurable sets in the notation, let (Ω_1, μ_1) and (Ω_2, μ_2) be two probability spaces. A *coupling* of (Ω_1, μ_1) and (Ω_2, μ_2) is simply a probability space (Ω, μ) together with measure-preserving maps $\sigma_i : \Omega \to \Omega_i$, $i = 1, 2$. Thus, if X is a uniformly random point of (Ω, μ), then $\sigma_1(X)$ and $\sigma_2(X)$ are uniform on (Ω_1, μ_1) and (Ω_2, μ_2), respectively. Let κ_i be a kernel on (Ω_i, μ_i), $i = 1, 2$. Then κ_1 and κ_2 are *equivalent* if there is a coupling (Ω, μ) of the underlying probability spaces such that

$$\kappa_1(\sigma_1(x), \sigma_1(y)) = \kappa_2(\sigma_2(x), \sigma_2(y)) \text{ for } (\mu \times \mu)\text{-a.e. } (x,y) \in \Omega^2.$$

In other words, extending the notation in (2.5) to arbitrary spaces, we require $\kappa_1^{(\sigma_1)} = \kappa_2^{(\sigma_2)}$ almost everywhere; we write $\sim$ for the corresponding relation. Although this definition may seem a little complicated, as explained above it is in fact very natural.

Note that $\kappa_1 \approx \kappa_2$ implies $\kappa_1 \sim \kappa_2$: if $\kappa_1 = \kappa_2^{(\tau)}$, then one couples $x \in [0,1] = \Omega_1$ with $\tau(x) \in \Omega_2$. (More formally, we may take $\Omega = \Omega_1$, with σ_1 the identity and $\sigma_2 = \tau$.) It is easy to see that the reverse implication does not hold: for example, consider the random variables Λ_1, Λ_2 on $[0,1]$ given by $\Lambda_1(x) = x$ and $\Lambda_2(x) = 2x - \lfloor 2x \rfloor$; these both have the uniform distribution, but since one is 1-to-1 and the other 2-to-1, there is no measure-preserving bijection from one ground space to the other transforming one into the other. Setting $\kappa_i(x,y) = \Lambda_i(x)\Lambda_i(y)$, one obtains kernels with $\kappa_1 \sim \kappa_2$ but $\kappa_1 \not\approx \kappa_2$. (Recently, Borgs, Chayes and Lovász [12] have shown that if one excludes this phenomenon of 'twins', then $\sim$ and $\approx$ are equivalent; we refer the reader there for a precise statement.)

Returning to the special case of kernels on $[0,1]$, essentially equivalent to the general case, couplings have a very simple description. All that matters is that, for a uniform point X of (Ω, μ), the distribution of $(\sigma_1(X), \sigma_2(X))$ should have uniform marginals. Thus, couplings correspond to *doubly stochastic measures*, i.e. Borel measures μ on $[0,1]^2$ with both marginals Lebesgue measure. In other words, we have $\kappa_1 \sim \kappa_2$ if and only if there is a doubly stochastic measure μ such that

$$\kappa_1(x,y) = \kappa_2(u,v) \text{ for } (\mu \times \mu)\text{-a.e. } (x,u,y,v) \in [0,1]^4. \qquad (2.11)$$

At first sight, $[0,1]^2$ is the most natural space to use to couple two kernels on $[0,1]$, but there is another natural choice. Since $[0,1]^2$ is isomorphic as a probability space to $[0,1]$, we may construct the coupling on $[0,1]$! Hence, $\kappa_1 \sim \kappa_2$ if and only if there are measure-preserving maps $\sigma_1, \sigma_2 : [0,1] \to [0,1]$ such that $\kappa_1^{(\sigma_1)} = \kappa_2^{(\sigma_2)}$ for (Lebesgue) almost every $(x,y) \in [0,1]^2$. Putting this a little more symmetrically, we see that $\kappa_1 \sim \kappa_2$ if and only if

$$\exists \kappa, \sigma_1, \sigma_2 \text{ such that } \kappa = \kappa_1^{(\sigma_1)} \text{ a.e. and } \kappa = \kappa_2^{(\sigma_2)} \text{ a.e.,} \qquad (2.12)$$

where κ is a kernel on $[0,1]$ and σ_1 and σ_2 are measure-preserving maps from $[0,1]$ to itself. Note that $\kappa \sim \kappa^{(\sigma)}$ for any kernel κ on $[0,1]$ and any measure-preserving map from $[0,1]$ to itself.

Since couplings rather than rearrangements give the proper notion of equivalence for two kernels, it is natural to use couplings rather than rearrangements in the definition of the cut metric. Indeed, Borgs, Chayes, Lovász, Sós and Vesztergombi [15] define the cut metric on standard (or simply bounded) kernels as follows:

$$d_{\text{cut}}(\kappa_1, \kappa_2) = \inf_{\mu \in \mathcal{M}} \sup_{S,T} \left| \int_{S \times T} (\kappa_1(x,y) - \kappa_2(u,v))\, d\mu(x,u)\, d\mu(y,v) \right|, \qquad (2.13)$$

where $\mathcal{M}$ is the set of doubly stochastic measures on $[0,1]^2$, S and T run over measurable subsets of $[0,1]^2$, and the integral is over $(x,u) \in S$ and $(y,v) \in T$. As shown in [15], the definitions (2.7) and (2.13) coincide. (This is not hard to see – either formula defines a function that is continuous, indeed Lipschitz with constant 1, with respect to the cut norm, and hence continuous with respect to the L^1 norm. Since the finite-type kernels are dense in L^1, it suffices to check the equality of the two definitions for finite-type kernels, which is straightforward. For the details, see [15].) Since (2.7) is much easier to work with than (2.13), we shall take the former as our definition of d_{cut}.

Although (2.7) is more convenient, there is a sense in which (2.13) is the 'right' definition. For example, as we shall now show, the infimum in (2.13) is always attained, unlike that in (2.7). This is not discussed in [15], where it is of no particular significance. Here, and in the bulk of the paper, unless otherwise specified, we assume without loss of generality that the kernels we consider are are kernels on $[0,1]$, i.e. symmetric Lebesgue-measurable functions from $[0,1]^2 \to [0,\infty)$. Recall that we call a kernel *standard* if it takes values in $[0,1]$.

Lemma 2.6 *Let κ_1 and κ_2 be two standard kernels. Then there is a doubly stochastic measure μ achieving the infimum in (2.13).*

Proof For $\mu \in \mathcal{M}$ set

$$d_\mu(\kappa_1, \kappa_2) = \sup_{S,T} \left| \int_{S \times T} (\kappa_1(x,y) - \kappa_2(u,v))\, d\mu(x,u)\, d\mu(y,v) \right|, \qquad (2.14)$$

so our aim is to show that $\inf_{\mu \in \mathcal{M}} d_\mu(\kappa_1, \kappa_2)$ is attained. Before doing so, let us note that in the supremum one may restrict the sets S and T in (2.14) to 'nice' sets. Let $\mathcal{D}$ denote the set of finite unions of products of (half-open) intervals. Since μ is a finite Borel measure, for any measurable $S, T \subset [0,1]^2$ and any $\varepsilon > 0$, there are sets $S', T' \in \mathcal{D}$ with $\mu(S \Delta S'), \mu(T \Delta T') < \varepsilon$. Since $|\kappa_1 - \kappa_2| \leq 1$, replacing S, T by S'

and T' changes the value of the integral by at most 2ε. It follows that the supremum in (2.14) may be taken over $S, T \in \mathcal{D}$ without changing its value, as claimed.

It is well known that $\mathcal{M}$ is (sequentially) compact in the topology in which $\mu_n \to \mu$ if and only if $\mu_n(A) \to \mu(A)$ for every set $A \in \mathcal{D}$. Indeed, writing $\mathcal{D}_0$ for the set of products of intervals with rational endpoints, since $\mathcal{D}_0$ is countable any sequence in $\mathcal{M}$ has a subsequence (μ_n) such that $(\mu_n(A))$ converges for all $A \in \mathcal{D}_0$. Using the doubly stochastic property to bound the measure of a rectangle with one or more short sides, convergence for all $A \in \mathcal{D}$ follows easily, and one can check that the limiting values do define a measure μ. Note that one cannot require $\mu_n(A) \to \mu(A)$ for every measurable A: it is easy to construct sequences where μ is concentrated on, for example, the diagonal $S = \{(x,x)\}$, with $\mu_n(S) = 0$ for every n.

Let (μ_n) be a sequence of doubly stochastic measures for which $d_{\mu_n}(\kappa_1, \kappa_2) \to d_{\text{cut}}(\kappa_1, \kappa_2)$; such a sequence exists by the definition (2.13) of $d_{\text{cut}}(\kappa_1, \kappa_2)$. From the remark above, (μ_n) has a subsequence converging to some $\mu \in \mathcal{M}$ in the appropriate topology. Restricting to this subsequence, we may assume that $\mu_n(A) \to \mu(A)$ for every $A \in \mathcal{D}$.

Let $S = S_1 \times S_2$ and $T = T_1 \times T_2$, where S_1, S_2, T_1 and T_2 are all intervals in $[0,1]$. We claim that

$$\int_{S \times T} \kappa(x,y)\, d\mu_n(x,u)\, d\mu_n(y,v) \to \int_{S \times T} \kappa(x,y)\, d\mu(x,u)\, d\mu(y,v) \qquad (2.15)$$

as $n \to \infty$, for any standard kernel κ. Before proving this, let us show that the lemma follows.

For any $\nu \in \mathcal{M}$, let

$$f(S,T,\nu) = \int_{S \times T} \big(\kappa_1(x,y) - \kappa_2(u,v)\big)\, d\nu(x,u)\, d\nu(y,v),$$

so $d_\nu(\kappa_1, \kappa_2) = \sup_{S,T} |f(S,T,\nu)|$. Applying (2.15) with $\kappa = \kappa_1$ and $\kappa = \kappa_2$, we see that $f(S,T,\mu_n) \to f(S,T,\mu)$ holds whenever S and T are products of intervals. By additivity, it thus holds whenever S and T are in $\mathcal{D}$. Since $d_{\mu_n}(\kappa_1, \kappa_2) = \sup_{S,T} |f(S,T,\mu_n)|$, for $S, T \in \mathcal{D}$ we thus have

$$f(S,T,\mu) = \liminf f(S,T,\mu_n) \le \liminf d_{\mu_n}(\kappa_1, \kappa_2) = d_{\text{cut}}(\kappa_1, \kappa_2).$$

As noted earlier, when defining $d_\mu(\kappa_1, \kappa_2) = \sup_{S,T} |f(S,T,\mu)|$, we may take the supremum instead over $S, T \in \mathcal{D}$, so it follows that $d_\mu(\kappa_1, \kappa_2) \le d_{\text{cut}}(\kappa_1, \kappa_2)$. Since $d_{\text{cut}}(\kappa_1, \kappa_2) = \inf_{\mu' \in \mathcal{M}} d_{\mu'}(\kappa_1, \kappa_2)$, this infimum is attained (at μ), as claimed.

It remains to prove (2.15). But this is easy: for any interval $I \subset [0,1]$, let μ_n^I be the measure on $[0,1]$ defined by

$$\mu_n^I(A) = \mu_n(A \times I),$$

and define μ^I from μ similarly. Recall that $\mu_n \to \mu$ on products of intervals. Thus $\mu_n^I(A) \to \mu^I(A)$ whenever A is an interval, and hence whenever A is a finite union of intervals. Since $\mu_n, \mu \in \mathcal{M}$, we have that $\mu_n(A)$ and $\mu(A)$ are both at most the Lebesgue measure of A. It follows that $\mu_n^I(A) \to \mu^I(A)$ for any measurable $A \subset [0,1]$, since for any ε we can approximate A by a finite union of intervals A'

whose symmetric difference from A has Lebesgue measure at most ε. It also follows that if I and J are two intervals, and $A \subset [0,1]^2$ is Lebesgue measurable, then

$$(\mu_n^I \times \mu_n^J)(A) \to (\mu^I \times \mu^J)(A).$$

Indeed, this follows by approximating A by a finite union of products of intervals. Considering level sets, we see that

$$\int f(x,y)\, d\mu_n^I(x)\, d\mu_n^J(y) \to \int f(x,y)\, d\mu^I(x)\, d\mu^J(y)$$

for any bounded measurable function f. Taking $f = \kappa(x,y) 1_{x \in S_1} 1_{y \in T_1}$, $I = S_2$ and $J = T_2$, this is exactly (2.15), completing the proof. □

The special case of Lemma 2.6 where the distance is 0 is of particular interest.

Corollary 2.7 *Let κ_1 and κ_2 be two standard kernels. Then $d_{\mathrm{cut}}(\kappa_1, \kappa_2) = 0$ if and only if $\kappa_1 \sim \kappa_2$.*

Proof Using (2.13) as the definition of d_{cut}, if $\kappa_1 \sim \kappa_2$ then we certainly have $d_{\mathrm{cut}}(\kappa_1, \kappa_2) = 0$; see (2.11).

Suppose then than $d_{\mathrm{cut}}(\kappa_1, \kappa_2) = 0$. From Lemma 2.6, there is a $\mu \in \mathcal{M}$ such that $d_\mu(\kappa_1, \kappa_2) = 0$. Let ν be the signed measure on $[0,1]^4$ defined by

$$d\nu(x,u,y,v) = \bigl(\kappa_1(x,y) - \kappa_2(u,v)\bigr)\, d\mu(x,u)\, d\mu(y,v).$$

Then $d_\mu(\kappa_1, \kappa_2) = 0$ says exactly that $\nu(S \times T) = 0$ for all measurable $S, T \subset [0,1]^2$. Since ν is a signed Borel measure, it follows immediately that ν is the zero measure. Equivalently, $\kappa_1(x,y) - \kappa_2(u,v) = 0$ for $(\mu \times \mu)$-almost every points (x,u,y,v). Referring to (2.11) again, we see that $\kappa_1 \sim \kappa_2$. □

As we have seen, Corollary 2.7 is a simple exercise in measure theory. Using this corollary, and the equivalence of d_{cut} and d_{sub} proved by Borgs, Chayes, Lovász, Sós and Vesztergombi [15], one obtains the following characterization of equivalent (standard) kernels.

Theorem 2.8 *Let κ_1 and κ_2 be two standard kernels. Then $s(F, \kappa_1) = s(F, \kappa_2)$ holds for every finite graph F if and only if $\kappa_1 \sim \kappa_2$.*

Proof Immediate from Corollaries 2.5 and 2.7. □

The analogue of Theorem 2.8 for general (i.e. unbounded) kernels is false, even for 'rank 1' kernels with all counts $s(F, \kappa)$ finite. Indeed, if $\kappa(x,y) = f(x)f(y)$ for some $f: [0,1] \to [0,\infty)$, then the quantities $s(F, \kappa)$ are easily seen to be products of moments of f, viewed as a random variable. As is well known, there are non-negative random variables with the same finite moments but different distributions; using two such random variables, one can construct non-equivalent unbounded kernels κ_1, κ_2 with $s(F, \kappa_1) = s(F, \kappa_2) < \infty$ for all F.

We have shown that it is not hard to deduce Theorem 2.8 from Theorem 2.4. In fact, these results are equivalent! The reverse implication is actually much easier.

Proof [Theorem 2.8 $\implies$ Theorem 2.4] We write out the argument for a sequence of graphs; the treatment for kernels is essentially the same. Let (G_n) be a sequence of graphs with $|G_n| \to \infty$, and let κ be a standard kernel. From Corollary 2.3, if $d_{\mathrm{cut}}(G_n, \kappa) \to 0$, then $d_{\mathrm{sub}}(G_n, \kappa) \to 0$; it remains to prove the reverse implication.

As shown by Lovász and Szegedy [34] (see their Lemmas 5.1 and 5.2), repeatedly applying even the weak Frieze–Kannan [23] form of Szemerédi's Lemma, it is easy to prove that any sequence (G_n) with $|G_n| \to \infty$ has a subsequence converging in d_{cut} to some standard kernel κ'. We shall not give the details of this argument here as we shall prove a corresponding statement in a more general setting in Corollary 4.7.

Suppose then that $d_{\mathrm{sub}}(G_n, \kappa) \to 0$. Then by the observation above there is a subsequence (G_{n_k}) that converges in d_{cut} to some standard kernel κ'. But then, by Corollary 2.3, we have $d_{\mathrm{sub}}(G_{n_k}, \kappa') \to 0$. Since $d_{\mathrm{sub}}(G_n, \kappa) \to 0$ we must have $d_{\mathrm{sub}}(\kappa, \kappa') = 0$, i.e. $s(F, \kappa) = s(F, \kappa')$ for all F. Thus, by Theorem 2.8, we have $\kappa \sim \kappa'$, so $d_{\mathrm{cut}}(\kappa, \kappa') = 0$. Thus $d_{\mathrm{cut}}(G_{n_k}, \kappa) \to 0$.

We have shown that (G_n) has a subsequence converging to κ in d_{cut}. This argument applies equally well to any subsequence of (G_n), and it follows immediately that the whole sequence converges, i.e. $d_{\mathrm{cut}}(G_n, \kappa) \to 0$, as required. $\square$

As we have just seen, Theorem 2.4, one of the main results of Borgs, Chayes, Lovász, Sós and Vesztergombi [15], is equivalent to Theorem 2.8. As far as we are aware, this observation is new. Now Theorem 2.8 is a fundamental analytic fact about bounded kernels: it says that a bounded kernel is characterized up to equivalence by the quantities $s(F, \kappa)$, which are the natural analogues for a kernel of the moments of a random variable. When the first version of this paper was written, we thus had the following rather unsatisfactory situation: the only known proof of the analytic fact Theorem 2.8 was that given above, relying on the hard results of Borgs, Chayes, Lovász, Sós and Vesztergombi [15] about sequences of graphs. Fortunately, this situation has now been resolved: Borgs, Chayes and Lovász [12] have given a very clever direct proof of Theorem 2.8. In fact, they proved a little more.

Recall from (2.12) that $\kappa_1 \sim \kappa_2$ means that

$$\exists \kappa, \sigma_1, \sigma_2 \text{ such that } \kappa = \kappa_1^{(\sigma_1)} \text{ a.e. and } \kappa = \kappa_2^{(\sigma_2)} \text{ a.e.},$$

where κ is a kernel on $[0,1]$ and σ_1 and σ_2 are measure-preserving maps from $[0,1]$ to itself. Turning this 'upside-down', let us write $\kappa_1 \sim' \kappa_2$ if

$$\exists \kappa, \sigma_1, \sigma_2 \text{ such that } \kappa_1 = \kappa^{(\sigma_1)} \text{ a.e. and } \kappa_2 = \kappa^{(\sigma_2)} \text{ a.e. .} \qquad (2.16)$$

In (2.16), we require κ to be a kernel on $[0,1]$; it makes no difference if we allow κ to be a kernel on an arbitrary standard probability space. Note that if $\kappa_1 \sim' \kappa_2$, then using the observation that $\kappa \sim \kappa^{(\sigma)}$ twice, we have $\kappa_1 \sim \kappa_2$.

Borgs, Chayes and Lovász [12] proved the following result.

Theorem 2.9 *For two bounded kernels κ_1, κ_2, the following are equivalent. (a) $s(F, \kappa_1) = s(F, \kappa_2)$ for every finite graph F, (b) $\kappa_1 \sim \kappa_2$ and (c) $\kappa_1 \sim' \kappa_2$.*

The important implication is that if $s(F, \kappa_1) = s(F, \kappa_2)$ for all F, then $\kappa_1 \sim' \kappa_2$. As noted above, this trivially implies $\kappa_1 \sim \kappa_2$, which in turn easily implies $s(F, \kappa_1) = s(F, \kappa_2)$. The proof in [12] is direct, but somewhat technical.

As shown above, Theorem 2.9, which trivially implies Theorem 2.8, implies Theorem 2.4. This gives a proof of Theorem 2.4 that is very different from that given by Borgs, Chayes, Lovász, Sós and Vesztergombi [15].

Our aim in the rest of this paper is to investigate the extent to which the various results and observations above carry over to sparse graphs, graphs with n vertices and $o(n^2)$ edges. As we shall see, this gives rise to many difficult questions, so we shall present many more questions than answers.

3 Subgraph counts for sparse graphs

In this section we consider sparse graphs, where the number of edges is $o(n^2)$ as the number n of vertices goes to infinity. We shall assume throughout that we have at least $\omega(n)$ edges, i.e. that the average degree tends to infinity; often, we shall make much stronger assumptions. Given a function $p = p(n)$, one can adapt many of the notions of Section 2 to graphs with $\Theta(pn^2)$ edges. Indeed, let

$$s_p(F, G) = \frac{\mathrm{emb}(F, G)}{p^{e(F)} n_{(|F|)}} = \mathrm{aut}(F) \frac{X_F(G)}{p^{e(F)} X_F(K_n)},$$

noting that

$$s_p(F, G) = \frac{\mathrm{emb}(F, G)}{\mathbb{E}\big(\mathrm{emb}(F, G(n, p))\big)}.$$

Also, let

$$t_p(F, G) = \frac{\mathrm{hom}(F, G)}{p^{e(F)} n^{|F|}}.$$

If $p = 1$, then we recover the definitions in Section 2. Furthermore, if $0 < p < 1$ is constant, then we can define a map s as before, but now s maps $\mathcal{F}$ into the compact space $\prod_{F \in \mathcal{F}}[0, p^{-e(F)}]$, and everything proceeds as before. More generally, changing p by a constant factor will be irrelevant: just as we can use s_c for any c to study $G(n, 1/2)$, we may use s_p to study $G(n, p/2)$ or $G(n, 2p)$, say, for any $p = p(n)$.

From now on, we suppose that $p = p(n)$ is some given function of n, with $p(n) \to 0$ as $n \to \infty$. We wish to work in a compact space, so we shall assume that there are constants c_F, $F \in \mathcal{F}$, such that $s_p(F, G) \le c_F$ for all graphs G we consider. Enumerating $\mathcal{F}$ as $\{F_1, F_2, \ldots\}$, we may thus define a map

$$s_p : \mathcal{F} \to X = \prod_{i=1}^{\infty}[0, c_{F_i}], \qquad G \mapsto (s_p(F_i, G))_{i=1}^{\infty}, \qquad (3.1)$$

and, using any metric d on X giving the product topology, an associated metric

$$d_{\mathrm{sub}}(G_1, G_2) = d(s_p(G_1), s_p(G_2)). \qquad (3.2)$$

We suppress the dependence on p in our notation for the metric to avoid clutter. As in the dense case, we can extend d_{sub} to bounded kernels κ, setting

$$d_{\mathrm{sub}}(G, \kappa) = d(s_p(G), s(\kappa)) \quad \text{and} \quad d_{\mathrm{sub}}(s(\kappa_1), s(\kappa_2)) = d(s(\kappa_1), s(\kappa_2))$$

for a graph G and bounded kernels κ, κ_1 and κ_2. Here, for a kernel κ, $s(\kappa)$ is the vector with coordinates defined by (2.2).

Much of the time, we think of a sequence (G_n) of finite graphs. Throughout, we are only interested in sequences with $|G_n| \to \infty$. For notational convenience we always assume that $|G_n| = n$; this make no difference to our conjectures and results. As usual, we need not assume that G_n is defined for every $n \in \mathbf{N}$, but only for an infinite subset of $\mathbf{N}$. In this setting, the assumption described above may be stated as follows.

Assumption 3.1 (bounded subgraph counts) For each fixed graph F, we have $\sup_n s_p(F, G_n) < \infty$.

In particular, if (G_n) satisfies Assumption 3.1 then, taking $F = K_2$, we see that $e(G_n) = O(pn^2)$, so our graphs are sparse. There is a stronger version of Assumption 3.1 that is perhaps even more natural:

Assumption 3.2 (exponentially bounded subgraph counts) There is a constant C such that, for each fixed F, we have $\limsup s_p(F, G_n) \leq C^{e(F)}$ as $n \to \infty$.

In this case, changing p by a constant factor, we may take $C = 1$ if we like. This is not always the most natural normalization, however. There is a reason for writing lim sup in Assumption 3.2: for any graph G_n with $|G_n| = n$ and n large, there will be some F with $s_p(F, G_n)$ very large. Indeed, G_n contains at least one embedding of itself, so $s_p(G_n, G_n) \geq 1/(n! p^{e(G_n)})$, which typically grows much faster than any constant to the power $e(G_n)$.

Turning to kernels, there is no longer any good reason to restrict our kernels to take values in $[0, 1]$: in the dense case, the maximum possible 'local density' of edges is 1. Here, if we normalize so that G_n has $pn^2/2$ edges, say, local densities larger than p are certainly possible. We shall thus consider general kernels, i.e. symmetric measurable functions from $[0, 1]^2$ to $[0, \infty)$, rather than only standard kernels. We define $s(F, \kappa)$ as before, using (2.2); in general, $s(F, \kappa)$ may be infinite, but we shall always assume it is finite for the graphs F and kernels κ we consider.

Although we allow unbounded kernels in general, it may be that they give rise to difficulties (as they do in the general (very) sparse inhomogeneous model of Bollobás, Janson and Riordan [8]). Assumption 3.2 corresponds to the limiting kernel (if it exists) being bounded, as shown by Lemma 3.5 below.

Our main conjecture states that, if p is large enough, then, under Assumption 3.2, the equivalent of Theorem 2.1 holds.

Conjecture 3.3 Let $p = p(n) = n^{-o(1)}$, and let $C > 0$ be constant. Suppose that (G_n) is a sequence of graphs with $|G_n| = n$ such that, for every F, $s_p(F, G_n)$ converges to some constant $0 \leq c_F \leq C^{e(F)}$. Then there is a bounded kernel κ such that $c_F = s(F, \kappa)$ for every F.

As noted above, without loss of generality we may take $C = 1$. As we shall observe later, it is very easy to see that if $s_p(K_2, G_n) \to 0$ and $s_p(F, G_n)$ is bounded for every F, then $s_p(F, G_n) \to 0$ for every F. Thus we may assume that $s_p(K_2, G_n)$ is bounded away from zero, and we may normalize in a different way by assuming that $s_p(K_2, G_n) = 1$, i.e, that $e(G_n) = p\binom{n}{2}$.

Assumption 3.2 is trivially stronger than Assumption 3.1. Thus, if (G_n) satisfies Assumption 3.2, then the sequence $s_p(G_n)$ defined by (3.1) lives in a compact product space, and has a convergent subsequence. Hence there are real numbers $c_F \geq 0$, $F \in \mathcal{F}$, and a subsequence (G_{n_i}) with $s_p(F, G_{n_i}) \to c_F$ for every F, to which Conjecture 3.3 applies. Conjecture 3.3 is thus a statement about the possible limit points of the sequences $s_p(G_n)$.

It may well be that the restriction to bounded kernels is not necessary.

Conjecture 3.4 *Let $p = p(n) = n^{-o(1)}$, and let (G_n) be a sequence of graphs with $|G_n| = n$ such that, for every F, we have $s_p(F, G_n) \to c_F$ for some $0 \leq c_F < \infty$. Then there is a kernel κ with $c_F = s(F, \kappa)$ for every F.*

We have stated the above conjectures under the assumption that $p = n^{-o(1)}$; we shall call this the *almost dense* case. The reason for this assumption is discussed further below. Let us note that, in the almost dense case, for each fixed F with k vertices, the denominator in the formula $\mathrm{emb}(F, G_n)/(p^{e(F)} n_{(k)})$ for $s_p(F, G_n)$ is asymptotically $p^{e(F)} n^k$, which is $n^{k-o(1)}$. Since there are at most n^{k-1} non-injective homomorphisms from F to G_n, it follows that $t_p(F, G_n) \sim s_p(F, G_n)$ as $n \to \infty$, so it makes no difference whether we consider s_p or t_p. In general, this is not true: for example, considering homomorphisms which map all t vertices on one side of $K_{t,t}$ into a single vertex, we see that in any graph G_n with $pn^2/2$ edges there are at least $n(np)^t = (n^{2t} p^{t^2})/(np^t)^{t-1}$ non-injective embeddings of $K_{t,t}$. If np^t is bounded, then this is comparable to (or larger than) the denominator in the definition of $t_p(K_{t,t}, G_n)$, and it follows that $t_p(K_{t,t}, G_n) - s_p(K_{t,t}, G_n)$ is bounded away from zero. Thus, for $t_p(K_{t,t}, G_n) \sim s_p(K_{t,t}, G_n)$ to hold with both quantities bounded, we need $np^t \to \infty$. This condition holds for every t only in the almost dense case $p = n^{-o(1)}$.

3.1 Bounded and unbounded kernels

The following simple observation illuminates the relationship between Conjectures 3.3 and 3.4.

Lemma 3.5 *Let $\kappa : [0,1]^2 \to [0, \infty)$ be a kernel, and $C \geq 0$ a constant. Then we have $s(F, \kappa) \leq C^{e(F)}$ for every F if and only if $\kappa \leq C$ holds almost everywhere.*

Proof The result is trivial if $C = 0$. Otherwise, rescaling, we may assume that $C = 1$. If $\kappa \leq 1$ almost everywhere, then $s(F, \kappa) \leq s(F, 1) = 1$ for every F. We may thus suppose that $\kappa > 1$ on a set of positive measure. It follows that there is some $\eta > 0$ such that $\kappa > (1+\eta)^2$ on a set A of positive measure. Applying the Lebesgue Density Theorem to A, there is some $\varepsilon > 0$ and some rectangle $R = [a, a+\varepsilon] \times [b, b+\varepsilon] \subset [0,1]^2$ such that $\mu(A \cap R) \geq \mu(R)/(1+\eta)$. Thus, the average value of κ on the set R is at least $1 + \eta$. Let κ' be the kernel taking the value $1 + \eta$ on R and 0 elsewhere. Standard arguments from convexity show that, for each t,

$$s(K_{t,t}, \kappa) \geq s(K_{t,t}, \kappa') = \varepsilon^{2t}(1+\eta)^{t^2}.$$

Taking t large enough, we find an $F = K_{t,t}$ for which $s(F, \kappa) > 1$. □

Lemma 3.5 shows that a kernel κ is bounded if and only if the counts $s(F,\kappa)$ grow at most exponentially in $e(F)$. It also shows that, in Conjecture 3.3, we need only consider kernels $\kappa : [0,1]^2 \to [0,C]$.

Let us say that a kernel has *finite moments* if $s(F,\kappa) < \infty$ for all F. There are unbounded kernels with finite moments: the simplest way to construct such an example is to consider the 'rank 1' case, where $\kappa(x,y) = f(x)f(y)$ for some $f : [0,1] \to [0,\infty)$. Indeed, let f be any function from $[0,1]$ to $[0,\infty)$ with $\mathbb{E}(f^k) = \int_0^1 f(x)^k\, dx$ finite for every k; for example, let $f(x) = \log(1/x)$ for $x > 0$. Set $\kappa(x,y) = f(x)f(y)$. If F is a graph on $\{1,2,\ldots,k\}$ in which vertex i has degree d_i, then

$$s(F,\kappa) = \int_{[0,1]^k} \prod_{ij \in E(F)} f(x_i)f(x_j) \prod_{i=1}^{k} dx_i$$

$$= \int_{[0,1]^k} \prod_{i=1}^{k} f(x_i)^{d_i} \prod_{i=1}^{k} dx_i = \prod_{i=1}^{k} \mathbb{E}(f^{d_i}) < \infty.$$

The calculation above shows that a rank one kernel $\kappa(x,y) = f(x)f(y)$ has finite moments if and only if $\|f\|_p < \infty$ for every $p > 1$, and hence if and only if $\|\kappa\|_p < \infty$ for every $p > 1$. It is tempting to think that this holds in general. In one direction, for any kernel κ and any graph F on $\{1,2,\ldots,k\}$, we may write

$$s(F,\kappa) = \int_{[0,1]^k} \prod_{ij \in E(F)} \kappa_{ij}(x_1,\ldots,x_d) \prod_{i=1}^{k} dx_i,$$

where $\kappa_{ij}(x_1,\ldots,x_d) = \kappa(x_i,x_j)$. Thus, by Hölder's inequality,

$$s(F,\kappa) = \int \prod_{ij \in E(F)} \kappa_{ij} \le \prod_{ij \in E(F)} \|\kappa_{ij}\|_{e(F)} = \prod_{ij \in E(F)} \|\kappa\|_{e(F)}.$$

Hence, if $\|\kappa\|_p < \infty$ for every $p > 1$, then $s(F,\kappa) < \infty$ for every F. The reverse implication does not hold, however, as shown by the following example.

Example 3.6 A kernel with finite moments but infinite 2-norm. Let us define a sequence of independent random kernels $\kappa_0, \kappa_1, \kappa_2, \ldots$, as follows. For $r \ge 0$, let $\mathcal{P}_r$ be the partition of $[0,1]$ into 2^{2^r} equal intervals, and let $\mathcal{P}_r^2$ be the corresponding partition of $[0,1]^2$: divide $[0,1]^2$ into 4^{2^r} squares in the obvious way, and take as one part of $\mathcal{P}_r^2$ the union of a square and its reflection in the line $x = y$ (which may be the same square). Our kernel κ_r will be constant on each element of $\mathcal{P}_r^2$, taking the value 2^r with probability 2^{-2r} and 0 otherwise, with the values on different parts independent. Note that κ_0 is simply the constant kernel with value 1.

Let $\kappa(x,y) = \sum_{r=0}^{\infty} \kappa_r(x,y)$. It is easy to see that with probability 1 the sum converges almost everywhere (for example, recalling that μ denotes Lebesgue measure, use the fact that $\mathbb{E}\mu\{\kappa_r > 0\} = 2^{-2r}$ to deduce that, with probability 1, $\mu\{\exists s > r : \kappa_s > 0\}$ tends to 0 as $r \to \infty$). Also, for large r, $\|\kappa_r\|_2^2$ is concentrated around its mean of $(2^r)^2 2^{-2r} = 1$. Hence, with probability 1 we have $\|\kappa_r\|_2^2 \ge 0.99$ for infinitely many r. Using $(a+b)^2 \ge a^2 + b^2$ for $a, b \ge 0$, it follows that $\|\kappa\|_2^2$ is infinite with probability 1; in particular, κ does not have all p-norms finite.

Turning to the finite moments property, let F be any fixed graph with t vertices. Since $\kappa \geq \kappa_0 = 1$, we have $s(F,\kappa) \leq s(K_t,\kappa)$, so we may assume without loss of generality that $F = K_t$. Since κ is random, $s(K_t,\kappa)$ is a random variable. We may write its expectation as

$$\mathbb{E}_\kappa \mathbb{E}_\mathbf{x} \prod_{i<j} \kappa(x_i,x_j) = \mathbb{E}_\mathbf{x}\mathbb{E}_\kappa \prod_{i<j} \kappa(x_i,x_j),$$

where $\mathbb{E}_\kappa$ denotes expectation over the random choice of κ, and $\mathbb{E}_\mathbf{x}$ over the random choice of $(x_1,\ldots,x_t)$, a sequence of t iid uniform elements of $[0,1]$. Let us fix $\mathbf{x}$ for the moment, assuming as we may that $x_i \neq x_j$ for $i \neq j$. Let ℓ be the largest r such that some pair x_i, x_j lie in the same part of $\mathcal{P}_r$, so $0 \leq \ell < \infty$. Let $\sigma = \sum_{r \leq \ell} \kappa_r$ and $\tau = \sum_{r > \ell} \kappa_r$, so $\kappa = \sigma + \tau$. For $r > \ell$, the $\binom{t}{2}$ pairs (x_i,x_j), $i<j$, all lie in different parts of $\mathcal{P}_r^2$, so the values of κ_r on these pairs are independent. Since different κ_r are independent, it follows that the values of τ on the pairs are also independent. Now $\|\sigma\|_\infty \leq \sum_{r=0}^\ell \|\kappa_r\|_\infty = 2^{\ell+1} - 1$. Thus,

$$\mathbb{E}_\kappa \prod_{i<j} \kappa(x_i,x_j) \leq \mathbb{E}_\kappa \prod_{i<j}(2^{\ell+1} + \tau(x_i,x_j)) = \prod_{i<j}(2^{\ell+1} + \mathbb{E}_\kappa \tau(x_i,x_j)).$$

For any x and y we have $\mathbb{E}_\kappa \kappa_r(x,y) = 2^r 2^{-2r} = 2^{-r}$, from which it follows that $\mathbb{E}_\kappa \tau(x,y) \leq 2$, and hence, very crudely, that

$$\mathbb{E}_\kappa \prod_{i<j} \kappa(x_i,x_j) \leq \prod_{i<j}(2^{\ell+1} + 2) \leq 2^{2\ell t^2}.$$

It remains to take the expectation over $\mathbf{x}$. Since $\mathbb{P}(\ell = r) \leq \binom{t}{2} 2^{-2^r}$, we find that

$$\mathbb{E} s(F,\kappa) \leq \sum_{r=0}^\infty \binom{t}{2} 2^{-2^r} 2^{2rt^2} < \infty,$$

noting that for any fixed t the 2^{-2^r} term dominates. If follows that with probability 1 we have $s(F,\kappa) < \infty$ for every F, giving a kernel with finite moments but with $\|\kappa\|_2$ infinite. A simple modification, taking the probability that κ_r takes the value 2^r on a given square to be $2^{-(1+\varepsilon)r}$ rather than 2^{-2r} gives, for each $\varepsilon > 0$, an example with $\|\kappa\|_{1+\varepsilon}$ infinite.

3.2 Non-uniform random graphs

As in the dense case, there is a key connection between convergence of the counts $s_p(F,G_n)$ and random graphs. Given a kernel κ, let $G_p(n,\kappa)$ be the random graph on $[n]$ obtained as follows: first choose $x_1,\ldots,x_n$ independently and uniformly from $[0,1]$. Then, conditional on this choice, join each pair $\{i,j\}$ of vertices independently, with probability $\min\{p\kappa(x_i,x_j),1\}$. If $p\kappa$ is bounded by 1, then $G_p(n,\kappa)$ is simply $G(n,p\kappa)$; we write the parameter p as a subscript to emphasize that it is part of the overall normalization: we think of a sparse graph generated from the kernel κ, rather than a 'sparse kernel' $p\kappa$. If $p = 1/n$, then $G_p(n,\kappa)$ is a special case of the general sparse inhomogeneous model of Bollobás, Janson and Riordan [8].

Remark 3.7 In what follows, we shall consider many statements about the convergence of various sequences of random graphs. As usual in the theory of random graphs, the precise notion of convergence is not important: one thinks of 'a random graph' with certain asymptotic properties, although this makes no formal sense. Formally, it is most natural to work throughout with convergence in probability, but this would require us to consider 'in probability' versions of our various assumptions, for example the (exponentially) bounded counts assumptions 3.1 and 3.2. In fact, it is easy to check that in all cases considered here, the error probabilities decay fast enough to give almost sure convergence for any coupling of the relevant probability spaces. However, we shall not verify this explicitly, noting that one can in any case ensure almost sure convergence by passing to a suitable subsequence.

Lemma 3.8 *Let $p = p(n) = n^{-o(1)}$, and let κ be a kernel with $s(F, \kappa) < \infty$ for every F. Then $s_p(F, G_p(n, \kappa)) \xrightarrow{p} s(F, \kappa)$ for each fixed graph F, so $d_{\text{sub}}(G_p(n, \kappa), \kappa) \xrightarrow{p} 0$. In fact, the sequence $G_p(n, \kappa)$ converges almost surely to κ in the metric d_{sub}.*

Proof It is very easy to check that, for every F, $s_p(F, G_p(n, \kappa))$ is concentrated around its mean $s(F, \kappa)$: indeed, the second moment of the number of copies of F can be written as a sum of terms $(1 + o(1))n_{(|H|)}p^{e(H)}s_p(H, \kappa)$, and the dominant term is the unique one with the largest power of n, where H is the disjoint union of two copies of F. (The $1 + o(1)$ correction is only needed if κ is unbounded, and appears due to the $\max\{1, \cdot\}$ in the edge probabilities.) This proves the first part of the result. Convergence in probability in d_{sub} follows, since convergence in probability in a product topology is equivalent to convergence in probability of each coordinate. For the final statement, see Remark 3.7. □

Lemma 3.8 implies that if κ has finite moments, then the sequence $G_n = G_p(n, \kappa)$ has bounded subgraph counts (i.e. satisfies Assumption 3.1) with probability 1. If κ is bounded, then G_n has exponentially bounded subgraph counts with probability 1.

Using Lemma 3.8, it is easy to see that we must allow unbounded kernels in Conjecture 3.4. Indeed, set $\kappa(x, y) = \log(1/x)\log(1/y)$ for $0 < x, y \leq 1$, say, and let $p(n) = 1/\log n$. Then the random graphs $G_p(n, \kappa)$ satisfy Assumption 3.1 with probability 1, and

$$s_p(F, G_p(n, \kappa)) \to s(F, \kappa) < \infty$$

holds with probability 1 for every F. Since κ is unbounded, by Lemma 3.5 there is no C with $s(F, \kappa) \leq C^{e(F)}$ for every F, so there is no bounded κ' with $s_p(F, G_p(n, \kappa)) \to s(F, \kappa')$ for every F.

Note that if p decreases too fast with n, then $s_p(F, G_p(n, \kappa))$ is no longer concentrated around its mean: for example, this is the case if $\mathbb{E}\,\text{emb}(F, G_p(n, k))$ does not tend to infinity. This is the reason for the assumption $p = n^{-o(1)}$ in the various conjectures and results above: otherwise, there will be some F for which the expected number of embeddings does not tend to infinity. Note also that, for smaller p, when $s_p(F, \cdot)$ and $t_p(F, \cdot)$ are no longer asymptotically equal, the former is the more natural parameter: for a given F, the lower limit on p below which the corresponding parameter for $G_p(n, \kappa)$ is no longer close to $s(F, \kappa)$ is in general much smaller for $s_p(F, \cdot)$ than for $t_p(F, \cdot)$. It may well be, however, that the conjectures in this section (or perhaps just their proofs) fail when the relevant parameters $s_p(F, \cdot)$ and $t_p(F, \cdot)$ are no longer asymptotically equal.

3.3 Subgraph counts in the uniform case

Using convexity, it is very easy to check that the only possible kernel κ with $s(K_2,\kappa) = s(C_4,\kappa) = 1$ is the uniform kernel, with $\kappa = 1$ almost everywhere. The following conjecture is thus a very special case of Conjecture 3.4.

Conjecture 3.9 *Let $p = p(n) = n^{-o(1)}$, and let (G_n) be a sequence of graphs with $|G_n| = n$, $e(G_n) = p\binom{n}{2}$, $s_p(C_4, G_n) \to 1$, and $\sup_n s_p(F, G_n) < \infty$ for each F. Then $s_p(F, G_n) \to 1$ for every F.*

Of course, there is a variant of Conjecture 3.9 where we replace Assumption 3.1 by Assumption 3.2, i.e. we demand that $\limsup_n s_p(F, G_n) \le C^{e(F)}$ for some $C < \infty$. In this uniform context there is perhaps less reason to expect this to make a difference.

In the dense case, it is one of the basic results about quasi-random graphs that $s_p(K_2, G_n) \to 1$ and $s_p(C_4, G_n) \to 1$ imply $s_p(F, G_n) \to 1$ for every F, with no further assumptions; see Chung, Graham and Wilson [18]. In the sparse case, this result extends easily to certain graphs F; here it turns out to be simpler to work with $t_p(F, G_n)$ rather than $s_p(F, G_n)$.

Lemma 3.10 *Let $p = p(n)$ with $pn^{1/2} \to \infty$, and let (G_n) be a sequence of graphs with $|G_n| = n$ such that $t_p(K_2, G_n) \to 1$ and $t_p(C_4, G_n) \to 1$. Then $t_p(C_k, G_n) \to 1$ for each $k \ge 5$.*

Proof Suppressing the dependence on n, let A denote the adjacency matrix of G_n, and let $\lambda_1 \ge \lambda_2 \ge \cdots \ge \lambda_n$ be the eigenvalues of A. For $k \ge 3$ we have

$$\hom(C_k, G_n) = \sum_{v_1, v_2, \ldots, v_k \in V(G_n)} A_{v_1 v_2} A_{v_2 v_3} \cdots A_{v_k v_1} = \mathrm{tr}(A^k) = \sum_{i=1}^n \lambda_i^k,$$

so

$$t_p(C_k, G_n) = n^{-k} p^{-k} \sum_{i=1}^n \lambda_i^k = \sum_{i=1}^n \mu_i^k, \qquad (3.3)$$

where $\mu_i = \lambda_i/(np)$ is the ith normalized eigenvalue of G_n. In particular,

$$\sum_i \mu_i^4 \to 1. \qquad (3.4)$$

The maximum eigenvalue of the adjacency matrix of any graph is at least the average degree, so

$$\mu_1 = (np)^{-1} \lambda_1 \ge (np)^{-1}(1 + o(1))(n^2 p)/n = 1 + o(1).$$

From (3.4) it follows that $\mu_1 \sim 1$ and that $\sum_{i \ge 2} \mu_i^4 \to 0$. Hence $\mu_2 \le 1$ and $\mu_n \ge -1$ if n is large enough, and then for $k \ge 5$ we have

$$\sum \mu_i^k = \mu_1^k + \sum_{i \ge 2} \mu_i^k \le \mu_1^k + \max\{\mu_2^{k-4}, \mu_n^{k-4}\} \sum_{i \ge 2} \mu_i^4 \le \mu_1^k + \sum_{i \ge 2} \mu_i^4 = 1 + o(1).$$

Using (3.3) again, the result follows. □

Informally, when $pn^{1/2} \to \infty$, the parameters $s_p(C_k, G_n)$ and $t_p(C_k, G_n)$ are equivalent. More precisely, Lemma 3.10 implies the analogous statement with all occurrences of t_p replaced by s_p, but this requires a little work to show.

The restriction on p in Lemma 3.10 was not used in the proof. However, if G_n has average degree $\bar{d}$, then it contains at least $n\binom{\bar{d}}{2}$ pairs of adjacent edges. Thus, writing $N_{i,j}$ for the number of common neighbours of i and j, the sum of $N_{i,j}$ over ordered pairs $i \ne j$ is at least $2n\binom{\bar{d}}{2} = n\bar{d}(\bar{d}-1)$. Hence, the number of homomorphisms from C_4 to G_n with a given pair of opposite vertices mapped to distinct vertices is

$$\sum_{i \ne j} N_{i,j}^2 \ge \frac{1}{n(n-1)} \left(\sum N_{i,j} \right)^2 \ge \frac{n\bar{d}^2(\bar{d}-1)^2}{n-1}.$$

The number of homomorphisms with a given pair of opposite vertices mapped to the same vertex is simply the sum of the squares of the degrees in G_n, which is at least $n\bar{d}^2$. Thus,

$$\hom(C_4, G_n) \ge \frac{n\bar{d}^2(\bar{d}-1)^2}{n-1} + n\bar{d}^2 \qquad (3.5)$$

for any graph G_n with n vertices and average degree $\bar{d}$. With $\bar{d} \sim pn \to \infty$, this gives $\hom(C_4, G_n) \ge (1+o(1))(n^4p^4 + n^3p^2)$, i.e. $t_p(C_4, G_n) \ge (1+o(1))(1+n^{-1}p^{-2})$. Consequently, $t_p(C_4, G_n) \sim 1$ implies $pn^{1/2} \to \infty$. When $pn^{1/2} \to \infty$, (3.5) reduces to the well-known fact that, in this case, $e(G_n) \sim p\binom{n}{2}$ implies that

$$t_p(C_4, G_n), \; s_p(C_4, G_n) \ge 1 - o(1).$$

In the dense case, Lemma 3.10 extends to triangles. Indeed, $\mathrm{tr}(A^2)$ counts the number of walks of length 2 in G, which is just $2e(G)$. Thus

$$\sum \mu_i^2 = \frac{2e(G)}{n^2 p^2} \sim p^{-1}.$$

If p is bounded away from zero then it follows that $\sum_{i \ge 2} \mu_i^2$ is bounded as $n \to \infty$. Since $\sum_{i \ge 2} \mu_i^4 \to 0$, it follows by the Cauchy–Schwarz inequality that $\sum_{i \ge 2} \mu_i^3 \to 0$, and hence that $t_p(C_3, G_n) \to 1$. Since $\mathrm{emb}(C_3, G) = \hom(C_3, G)$ for any graph G, it follows that $s_p(C_3, G_n) \to 1$.

To obtain a result for triangles in the sparse case by this method, one needs stronger assumptions. Defining p by $e(G) = n^2 p/2$, if we assume that $t_p(C_4, G_n) = 1 + o(p)$, then arguing as above we find that $\sum_{i \ge 2} \mu_i^4 = o(p)$ and $\sum_{i \ge 2} \mu_i^4 \le p^{-1}$, so Cauchy–Schwarz does give $\sum_{i \ge 2} \mu_i^3 \to 0$. In general, many results for quasi-random graphs extend to the sparse case with similar modifications, where $o(1)$ error terms are replaced by suitable functions of p; see, for example, the results of Thomason [37, 38] on (p, α)-jumbled graphs. Our aim here is different; we wish to assume only convergence in the relevant metric, making no assumption about the rate of convergence.

When $p \to 0$, the conditions of Lemma 3.10 do not guarantee the 'right' number of triangles, as our next two examples will show.

Example 3.11 Very sparse graphs with too few triangles. Throughout this example we assume that $p_1(n)$ and $p_2(n)$ are functions of n satisfying

$$p_2 = (1 - p_1^2)^{n-2} \qquad (3.6)$$

and $p_1, p_2 = \Theta(\sqrt{\log n}/\sqrt{n})$. To be concrete, we may take $p_2 = \sqrt{\log n}/\sqrt{n}$, in which case the corresponding p_1 satisfies $p_1 \sim p_2/\sqrt{2}$. Suppressing the dependence on n, let G be the usual Erdős–Rényi random graph $G = G(n, p_1)$, and let H be the graph on the same vertex set $[n]$ in which vertices i and j are joined if and only if they do not have a common neighbour in G. From (3.6), each edge of H is present with probability p_2; note that the edges of H are *not* present independently of one another. For any set E of $r = O(1)$ possible edges of H, the edges of E are all present if and only if no vertex of G is joined to both ends of some edge in E. Considering each vertex of G separately, we see that the probability of this event is

$$(1 - rp_1^2 + O(p_1^3))^{n+O(1)} = e^{-rp_1^2 n + O(np_1^3)} \sim \left((1 - p_1^2)^{n-2}\right)^r = p_2^r,$$

where the $O(1)$ correction in the first exponent is to account for vertices that are endpoints of one or more edges in E. In other words, the probability that a bounded number of edges is present in H is asymptotically the corresponding probability for $G(n, p_2)$.

For $E_1, E_2 \subset E(K_n)$, the event $E_2 \subset E(H)$ is a down-set in terms of G (it says that certain pairs of edges of G are not present), so $E_2 \subset E(H)$ and $E_1 \subset E(G)$ are negatively correlated. Hence, if $|E_2| = O(1)$, we have

$$\mathbb{P}(\{E_1 \subset E(G)\} \cap \{E_2 \subset E(H)\}) \leq (1 + o(1))p_1^{|E_1|} p_2^{|E_2|}. \tag{3.7}$$

Considering all ways of splitting a set E, it follows that $\mathbb{P}(E \subset G \cup H) \leq (1 + o(1))(p_1 + p_2)^{|E|}$, and hence that

$$\mathbb{E}(s_p(F, G \cup H)) \leq 1 + o(1) \tag{3.8}$$

for any fixed graph F, where $p = p_1 + p_2$.

Since G and H overlap in very few edges, and the numbers of edges of G and of H are concentrated, we have $s_p(K_2, G \cup H) \to 1$ almost surely. It follows that $s_p(C_4, G \cup H) \geq 1 - o(1)$ almost surely. Hence, from (3.8), $s_p(C_4, G \cup H) \xrightarrow{p} 1$, and it is not hard to deduce that $t_p(C_4, G \cup H) \xrightarrow{p} 1$.

On the other hand, there are by definition no triangles with two edges in G and one in H. Hence, from (3.7), the expectation of $\operatorname{emb}(K_3, G \cup H)$ is at most

$$(1 + o(1))n^3(p_1^3 + 0 + 3p_1 p_2^2 + p_2^3),$$

so $\mathbb{E}(s_p(K_3, G \cup H)) \leq (p^3 - 3p_1^2 p_2)/p^3 + o(1)$. Since p_1, p_2 and p are all of the same order, this final fraction is strictly less than 1, and our construction gives almost surely a sequence $G_n = G \cup H$ with $s_p(K_2, G_n) \to 1$, $s_p(C_4, G_n) \to 1$ but $s_p(C_3, G_n) \not\to 1$. Since $\operatorname{emb}(C_3, G) = \hom(C_3, G)$ for any G, we have $t_p(C_3, G_n) \sim s_p(C_3, G_n) \not\to 1$. Choosing p_1 and p_2 satisfying (3.6) so that $p_2 \sim p_1/2$, we may achieve $s_p(C_3, G_n) \to 5/9$. Alternatively, choosing p_1 and p_2 suitably, we may find a sequence with $s_p(C_3, G_n) \not\to 1$ for any $p = p(n)$ satisfying $pn^{1/2} \to \infty$ and $p = O(\sqrt{\log n}/\sqrt{n})$.

Example 3.12 Very sparse graphs with no triangles. In the context of finding explicit constructions giving lower bounds on Ramsey numbers, Alon [1] constructed a sequence of graphs G_n defined for infinitely many values of n, with the following properties, where $d = d(n) \sim n^{2/3}/4$: the graph G_n is a d-regular Cayley graph, it is triangle free and (which is irrelevant here) the largest independent set has size $O(n^{2/3})$. In proving the last property, Alon shows that all eigenvalues other than $\lambda_1 = d$ are uniformly bounded by $O(n^{1/3})$. Setting $p = d/n$, so $t_p(K_2, G_n) = 1$, and writing μ_i for $\lambda_i/(np)$, as in the proof of Lemma 3.10, one thus has $\mu_1 = 1$ and $\mu_i = O(n^{-1/3})$ for $i \neq 2$, so from (3.3) it follows that $t_p(C_4, G_n) = 1 + O(n^{-1/3}) = 1 + o(1)$. This gives another example of a graph with almost the minimal number of C_4s but too few (in this case no) triangles.

Example 3.13 Denser graphs with too few triangles. Let $n = mk$ where $m \to \infty$, and let $p = \sqrt{\log m}/\sqrt{m}$. Example 3.11 gives us a graph G' of order m with $t_p(K_2, G'), t_p(C_4, G') \sim 1$ and $t_p(K_3, G') \leq 0.9$, say, for all large enough m. Let G be the blow-up of G' obtained by replacing each vertex by k vertices. Since $t_p(F, \cdot)$ is unchanged by blow-ups, we have $t_p(K_2, G), t_p(C_4, G) \sim 1$ but $t_p(K_3, G) \leq 0.9$, from which $s_p(K_2, G), s_p(C_4, G) \sim 1$ and (for n large) $s_p(K_3, G) \leq 0.91$ follow immediately.

Although p has not changed, the number of vertices has. Seen as a function of n, we may choose $p = \sqrt{\log m}/\sqrt{m}$ for any m dividing n with $m \to \infty$. Exact divisibility is not essential. Either by using this fact, or by restricting to a subsequence, we see that any given function $p(n)$ can be realized up to a factor of $(1+o(1))$, provided $p(n)/(\sqrt{\log n}/\sqrt{n}) \to \infty$ and $p(n) = o(1)$. Hence, we may construct graphs with the right number of C_4s but too few triangles for any such function $p(n)$.

At first sight Example 3.13 seems to contradict Conjecture 3.9, but this is not the case. Indeed, for the graph G' that we blow up, (3.8) tells us that we do not have too many embeddings of any fixed F. However, while $s_p \sim t_p$ for $p = n^{-o(1)}$, the final p we consider, and while blowing up preserves t_p, G' is a very sparse graph: although it has the same absolute density as the final graph G, this density is much smaller than $|G'|^{-o(1)}$, since G' has many fewer vertices than G. It follows that the homomorphism counts in G' are *not* well behaved. In particular, G' contains around $m^4 p^3$ non-injective homomorphisms from $K_{2,3}$, which turns out to be much larger than the number $m^5 p^6$ of embeddings. It follows that G contains too many homomorphisms from, and thus embeddings of, $K_{2,3}$, i.e. that $s_p(K_{2,3}, G) \to \infty$.

Remark 3.14 Let us note in passing that the blowing-up argument above shows that replacing the assumption $p = n^{-o(1)}$ in Conjecture 3.9 (or Conjecture 3.3) with a stronger assumption such as $p(n) \geq 1/\log \log \log n$, say, makes no difference. Indeed, if the conjecture fails, and (G_n) is a counterexample, then blowing up G_n as above by replacing each vertex by $f(n)$ vertices for some rapidly growing $f(n)$ gives a counterexample for a different density function, where now the density goes to zero extremely slowly as a function of the number of vertices.

One possible approach to producing a counterexample to Conjecture 3.9 would be to consider *circulant graphs*, i.e. graphs on the vertex set $[n]$ in which whether or not ij is an edge depends only on $i - j$ modulo n. There is one circulant graph for each subset A of the integers modulo n satisfying $0 \notin A$ and $a \in A$ if and only if $-a \in A$. All our conjectures thus imply corresponding conjectures for subsets of $\mathbf{Z}_n$, the integers modulo n, in which the symmetry condition is not likely to be relevant.

Metrics for sparse graphs

Most subgraph counts in the graph have a rather unnatural interpretation in terms of the corresponding sets; the exception is cycles, where the number of k-cycles in G corresponds to (n times the) number of k-tuples in A^k summing to 0. There is a result corresponding to Lemma 3.10 for subsets of $\mathbf{Z}_n$, proved in the same way but using Fourier coefficients instead of eigenvalues. Unfortunately, Examples 3.11 and 3.13 also carry over to the set context, in a fairly straightforward way: instead of blowing up the graph, we replace each element of A by a block of consecutive integers. This shows that any result of the kind we want about subsets of $\mathbf{Z}_n$ must involve conditions other than constraints on the number of tuples summing to 0.

In the sparse case, even when $p = n^{-o(1)}$, it is not true that $s_p(K_2, G_n) \to 1$ and $s_p(C_4, G_n) \to 1$ together imply $s_p(F, G_n) \to 1$ for every F. We have just seen one example, with $F = C_3$. There are also much simpler examples.

Example 3.15 Adding a dense part. Let $p = 1/\log n$, say, and let $m = m(n) = n/(\log n)^c$ where $c > 0$ is constant. (We ignore rounding to integers.) Let G' be any graph on $n - m$ vertices, and let G be the disjoint union of G' and a complete graph on m vertices. Since K_m contains roughly $m^{|F|}$ embeddings of any fixed F, we have

$$s_p(F, G) \sim s_p(F, G') + \frac{m^{|F|}}{p^{e(F)} n^{|F|}} = s_p(F, G') + (\log n)^{e(F) - c|F|}.$$

Taking $G' = G(n - m, p)$ and $c = 3/2$, say, we have $s_p(K_2, G) \sim s_p(K_2, G') \sim 1$, $s_p(C_4, G) \sim s_p(C_4, G') \sim 1$, but $s_p(K_4, G) \sim 1 + 1 = 2$. Note that $s_p(K_5, G) \to \infty$, so the assumptions of Conjecture 3.9 are not satisfied.

The above example is rather artificial: there are too many copies of K_4 (and of K_5), but these sit on a small number of vertices. However, the same effect can be achieved by taking the union on the same vertex set of $G(n, p)$ and a disjoint union of n/m copies of K_m. Also, we can use complete bipartite graphs instead of complete graphs.

Example 3.16 A blown-up random graph. Let $n = mk$, where $k = k(n)$ and $m = m(n)$ both tend to infinity. (As usual, we ignore divisibility issues, or consider a sequence $n_i \to \infty$.) Let G_1 be the random graph $G(m, p)$, where $p = p(n)$, and let $G = G_1^{(k)}$ be formed by replacing each vertex of G by an independent set of size k, and each edge by a k-by-k complete bipartite graph. The number of edges of G is $k^2 e(G_1)$, which is asymptotically $k^2 m^2 p/2 = n^2 p/2$, so $s_p(K_2, G) \to 1$ in probability and almost surely. Similarly, for any fixed graph F, each embedding of F into G_1 gives rise to $k^{|F|}$ embeddings into G; the expected number of embeddings arising in this way is essentially the expected number in $G(n, p)$, so whenever this expectation tends to infinity, such embeddings will contribute $1 + o(1)$ to $s_p(F, G)$.

There are other embeddings of F into G, however, where some distinct vertices of F are mapped to the same vertex in G_1. For C_4, we have roughly $m^2 p k^4$ such embeddings within our complete bipartite graphs, and roughly $2m^3 p^2 k^4$ from embeddings involving three vertices of G_1. Provided $mp^2 \to \infty$, we still have $s_p(C_4, G) \to 1$.

Fix an integer $t \geq 3$, and suppose now that $m = m(n)$ and $p = p(n)$ are chosen so that m and $k = n/m \to \infty$, and $mp^t \to c$ for some constant $0 < c < \infty$; for example, set $p = 1/\log n$, $m = c(\log n)^t$ and $k = c^{-1} n/(\log n)^t$. Note that $mp^2 \to \infty$. Then we have roughly $m^{2+t}k^{2+t}p^{2t}$ embeddings of $K_{2,t}$ into G coming from embeddings into G_1. But we also have roughly $m^{1+t}k^{2+t}p^t$ embeddings into G coming from maps from $K_{2,t}$ into G_1 sending the two vertices on one side to the same vertex. It is easy to check that these two are the dominant terms (mapping the two vertices on one side to the same place we gain t factors of $1/p$ and lose one factor of m; any other identifications gain fewer factors of $1/p$ per factor of m lost), and it follows that $s_p(K_{2,t}, G) \to 1 + 1/c$.

Taking a 'typical' sequence of random graphs constructed as above gives an example with $s_p(K_2, G_n) \to 1$, $s_p(C_4, G_n) \to 1$ (and indeed $s_p(K_{2,t'}, G_n) \to 1$ for $2 \le t' < t$), but $s_p(K_{2,t}, G_n) \to 1 + 1/c \ne 1$. Once again, the assumptions of Conjecture 3.9 are not satisfied, this time because $s_p(K_{2,t+1}, G_n) \to \infty$.

We have seen from the examples above that if $p(n) \to 0$, then $s_p(K_2, G_n)$ and $s_p(C_4, G_n) \to 1$ do not themselves imply that $s_p(F, G_n) \to 1$ for every F. However, attempted counterexamples to Conjecture 3.9 seem to be doomed to failure by the additional assumption that $s_p(F, G_n)$ is bounded for every F. In the next section we shall see that we can make some progress towards proving Conjecture 3.9.

3.4 Partial results in the almost dense, uniform case

In the examples in the previous subsection, each vertex is in about the same number of copies of any fixed graph F, but there are relatively few ($o(n^2)$) pairs that are in too many copies of $K_{2,t}$, for example. It is easy to see that, under the assumptions of Conjecture 3.9, this cannot happen. In fact, we can make a much more general statement. For this it is convenient to work with homomorphism counts and $t_p(F, G_n)$ rather than embeddings and $s_p(F, G_n)$. As noted earlier, in the almost dense case that we consider in this subsection, i.e. when $p = n^{-o(1)}$, the quantities $t_p(F, G_n)$ and $s_p(F, G_n)$ differ by $o(1)$.

Let F be a fixed graph, and F' a subgraph of F. Without loss of generality, suppose that $V(F') = [\ell] \subset [k] = V(F)$. Then any homomorphism $\phi_F : F \to G_n$ restricts to a homomorphism $\phi_{F'} : F' \to G_n$. With $e(G_n) \sim p\binom{n}{2}$, we expect a typical $\phi_{F'}$ to have around $n^{k-\ell}p^{e(F)-e(F')}$ extensions. For each n, let us define a random variable $Z_n(F', F)$ as follows: let $\phi_{F'}$ be chosen uniformly at random from among all homomorphisms from F' into G (if there are any), and let $Z_n(F', F)$ be the number of extensions of $\phi_{F'}$ divided by $n^{k-\ell}p^{e(F)-e(F')}$. (The reader may well prefer to picture copies of F' and F in G_n rather than homomorphisms. In fact, it is better to picture embeddings, i.e. labelled copies. There are essentially the same number of these as of homomorphisms.) Since $\hom(F, G_n)$ is the sum over $\phi_{F'}$ of the number of extensions, we have

$$\hom(F, G_n) = \hom(F', G_n)\mathbb{E}(Z_n(F', F))n^{k-\ell}p^{e(F)-e(F')},$$

and hence

$$t_p(F, G_n) = t_p(F', G_n)\mathbb{E}(Z_n(F', F)).$$

For $r \ge 2$, let rF/F' denote the graph formed by the union of r copies of F which all meet in the same subgraph F', so rF/F' has $|F'| + r(|F| - |F'|)$ vertices and $e(F') + r(e(F) - e(F'))$ edges. A homomorphism from rF/F' to G_n consists of a homomorphism ϕ from F' to G_n together with r extensions of ϕ to homomorphisms from F to G, which may or may not be distinct. (They almost always will be.) Since we have normalized by the right powers of n and p, it follows that

$$t_p(rF/F', G_n) = t_p(F', G_n)\mathbb{E}(Z_n(F', F)^r). \tag{3.9}$$

Let $\mu_F = \mu_F(n) = n^{|F|}p^{e(F)}$, which is asymptotically equal to the expected number of homomorphisms from F into $G(n,p)$. Then, under the assumptions of any of Conjectures 3.3, 3.4 and 3.9, it is easy to see that for $F' \subset F$, any $o(\mu_{F'})$ copies of F' meet $o(\mu_F)$ copies of F. (Here 'copies' may be subgraphs of G_n, embeddings, or homomorphisms; it makes no difference.) Otherwise $t_p(2F/F', G_n)$ would not remain bounded. This rules out any construction of a potential counterexample similar to those above; it also shows that if $t_p(K_2, G_n) \to 0$ and Assumption 3.1 holds (i.e. (G_n) has bounded subgraph counts), then $t_p(F, G_n) \to 0$ for every F.

Conjecture 3.9 states that infinitely many conclusions (one for each F) hold under the same assumptions. We have already proved some of these conclusions, with $F = C_k$, $k \geq 5$. Our next aim is to prove a corresponding result for a much wider class of graphs. In doing so, the following observation will be useful.

Lemma 3.17 *Let $X_n \geq 0$ be a sequence of random variables with $\sup_n \mathbb{E}(X_n^k) < \infty$ for every $k \geq 1$. Then $\mathbb{E}(X_n^k) \to 1$ for every k if and only if $X_n \xrightarrow{p} 1$.*

Proof For the forward implication we have $\mathbb{E}(X_n) \to 1$ and $\mathbb{E}(X_n^2) \to 1$; applying Chebyshev's inequality it follows that $X_n \xrightarrow{p} 1$. The reverse implication is not much harder. Suppose that $X_n \xrightarrow{p} 1$, but that $\mathbb{E}(X_n^k) \not\to 1$ for some k. For any M, the variables $X_n^k 1_{X_n \leq M}$ are uniformly bounded and converge in probability to 1, so $\mathbb{E}(X_n^k 1_{X_n \leq M}) \to 1$. It follows that there is some $M(n) \to \infty$ such that $\mathbb{E}(X_n^k 1_{X_n \leq M(n)}) \to 1$. But then $\mathbb{E}(X_n^k 1_{X_n > M(n)}) \not\to 0$, so

$$\mathbb{E}(X_n^{k+1}) \geq \mathbb{E}(X_n^{k+1} 1_{X_n > M(n)}) \geq M(n) \mathbb{E}(X_n^k 1_{X_n > M(n)})$$

is unbounded, contradicting our assumptions. □

Corollary 3.18 *Under the assumptions of Conjecture 3.9, if F' and F are fixed graphs with $F' \subset F$ and $t_p(F', G_n) \to 1$, then $Z_n(F', F) \xrightarrow{p} 1$ if and only if $t_p(rF/F', G_n) \to 1$ for every $r \geq 1$.*

Proof Apply Lemma 3.17 to the random variable $Z_n(F', F)$, using (3.9) to evaluate its moments. □

We shall say that the distribution of F is *flat* over that of F' in G_n, or simply that F is *flat over* F', if $Z_n(F', F) \xrightarrow{p} 1$.

Lemma 3.19 *Under the assumptions of Conjecture 3.9 we have $s_p(K_{s,t}, G_n) \to 1$ for all $s, t \geq 1$. Moreover, $K_{1,s}$ is flat over E_s, where E_s is the empty subgraph of $K_{1,s}$ induced by the vertices in the second part.*

Proof Let $d_1, \ldots, d_n$ denote the degrees of the vertices of G_n, and $\bar{d}$ the average degree. Fix $s \geq 1$. By convexity, we have

$$\hom(K_{1,s}, G_n) = \sum_{i=1}^{n} d_i^s \geq n\bar{d}^s,$$

which we can rewrite as $t_p(K_{1,s}, G_n) \geq t_p(K_2, G_n)^s$. Since $t_p(K_2, G_n) \to 1$ by assumption, this gives
$$t_p(K_{1,s}, G_n) \geq 1 + o(1). \tag{3.10}$$

Specializing to $s = 2$ for the moment, let $Z_n = Z_n(E_2, K_{1,2})$ be the random variable describing the distribution of the number of common neighbours of a random pair of vertices of G_n. For any empty graph E_k we have $t_p(E_k, G_n) = 1$. Hence, from (3.9) and (3.10),
$$\mathbb{E}(Z_n) = t_p(K_{1,2}, G_n) \geq 1 + o(1).$$
On the other hand, since $tK_{1,2}/E_2 = K_{2,t}$,
$$\mathbb{E}(Z_n^2) = t_p(K_{2,2}, G_n) = t_p(C_4, G_n) \to 1.$$

Since $\mathbb{E}(Z_n^2) \geq \mathbb{E}(Z_n)^2$, it follows that $\mathbb{E}(Z_n) \to 1$ and (by Lemma 3.17) that $Z_n \xrightarrow{P} 1$. In other words, $K_{1,2}$ is flat over pairs of vertices. By Corollary 3.18 it then follows that $t_p(K_{2,t}, G_n) \to 1$ for every t.

Returning to general s, let $W_n = Z_n(E_s, K_{1,s})$. From (3.10) we have $\mathbb{E}(W_n) = t_p(K_{1,s}, G_n) \geq 1 + o(1)$. But we have just shown that $\mathbb{E}(W_n^2) = t_p(K_{2,s}, G_n) \to 1$, so $W_n \xrightarrow{P} 1$, i.e. $K_{1,s}$ is flat over E_s. Applying Corollary 3.18 again we thus have $t_p(K_{s,t}, G_n) \to 1$ for every t, as required. □

Theorem 3.20 *Let F be any fixed graph with girth at least 4, and let $F' \neq F$ be any induced subgraph of F. Under the assumptions of Conjecture 3.9, F is flat over F'. Furthermore, $s_p(F, G_n), t_p(F, G_n) \to 1$ as $n \to \infty$.*

Proof Note first that the definition of $Z_n(F', F)$ makes perfect sense when F' is the empty 'graph' with no vertices; there is one homomorphism from F' to G_n, and $Z_n(F', F)$ is constant and takes the value $t_p(F, G_n)$. Hence, F is flat over the empty subgraph means exactly that $t_p(F, G_n) \to 1$. Since $p = n^{-o(1)}$, we have $s_p(F, G_n) \sim t_p(F, G_n)$, so it suffices to prove the first statement.

We prove the first statement of the theorem by induction on $|F|$. If $|F| = 1$, there is nothing to prove. Suppose then that F and F' are given, with $|F| \geq 2$, and that the result holds for all smaller F.

Suppose first that $F' = F - v$ for some vertex v of F. Let E_s denote the subgraph of F' induced by the neighbours of v, noting that E_s has no edges, as F is triangle free. Set $X_n = Z_n(E_s, F')$ and $Y_n = Z_n(E_s, K_{1,s})$. Note that these random variables are defined on the same probability space: the elements of this space are simply s-tuples of vertices of G_n. If $F' = E_s$, then F' is trivially flat over E_s. If not, then F' is flat over E_s by the induction hypothesis. Hence, in either case, $\mathbb{E}(X_n^k) \to 1$ for every k. By the last part of Lemma 3.19, $K_{1,s}$ is flat over E_s, so $\mathbb{E}(Y_n^k) \to 1$ for every k. It follows that $\mathbb{E}((X_n - 1)^k) \to 0$ and $\mathbb{E}((Y_n - 1)^k) \to 0$ for all $k \geq 1$. Hence, by the Cauchy–Schwarz inequality,
$$\mathbb{E}((X_n - 1)^k (Y_n - 1)^\ell) \leq \sqrt{\mathbb{E}((X_n - 1)^{2k})\mathbb{E}((Y_n - 1)^{2\ell})} \to 0$$
for all $k, \ell \geq 0$ with $k + \ell > 0$. Writing $\mathbb{E}(X_n^k Y_n^\ell) = \mathbb{E}((X_n - 1 + 1)^k (Y_n - 1 + 1)^\ell)$ as 1 plus a sum of terms $\mathbb{E}((X_n - 1)^{k'} (Y_n - 1)^{\ell'})$, $k', \ell' \geq 0$, $k' + \ell' > 0$, it follows that $\mathbb{E}(X_n^k Y_n^\ell) \to 1$ for any $k, \ell \geq 0$.

Any homomorphism $\phi_{F'}$ from F' into G_n is the extension of a unique homomorphism ϕ_{E_s} from E_s into G_n. Furthermore, to extend $\phi_{F'}$ to F we must choose for the image of v a common neighbour of the vertices in the image of ϕ_{E_s}. Hence, the value of $Z_n = Z_n(F', F)$ on $\phi_{F'}$ is simply the value of Y_n on ϕ_{E_s}. Choosing $\phi_{F'}$ uniformly at random, to obtain the correct distribution for Z_n, the probability of obtaining a particular restriction ϕ_{E_s} is proportional to the number of extensions of ϕ_{E_s} to F', i.e. to X_n. Thus the distribution of Z_n is that of Y_n 'size biased' by X_n. In particular,

$$\mathbb{E}(Z_n^k) = \frac{\mathbb{E}(X_n Y_n^k)}{\mathbb{E}(X_n)} \sim 1/1 = 1.$$

Taking $k = 1, 2$, it follows that $Z_n \xrightarrow{p} 1$, i.e. that F is flat over F', as required.

It remains to handle the case $|F| - |F'| \geq 2$. In this case, we can find an induced subgraph $F'' = F - v$ of F with $F' \subset F'' \subset F$. Note that $t_p(F', G_n), t_p(F'', G_n) \sim 1$ by induction, that F'' is flat over F' by induction, and that F is flat over F'' by the case treated above. In particular, we certainly have

$$t_p(F, G_n) = t_p(F'', G_n) \mathbb{E}(Z_n(F'', F)) \sim 1.$$

Fix $\varepsilon > 0$. Let us call a copy of F'' (more precisely, a homomorphism from F'' into G_n) bad if it has fewer than $(1 - \varepsilon)\mu_F/\mu_{F''}$ extensions to copies of F. Since F is flat over F'' and $t_p(F'', G_n) \sim 1$, there are fewer than $\varepsilon^2 \mu_{F''}$ bad copies of F'' if n is large enough. Since each copy of F'' extends a unique copy of F', it follows that at most $\varepsilon \mu_{F'}$ copies of F' have more than $\varepsilon \mu_{F''}/\mu_{F'}$ extensions to bad copies of F''.

Let $\mathcal{B}_1$ denote the set of copies of F' that have more than $\varepsilon \mu_{F''}/\mu_{F'}$ extensions to bad copies of F'', so $|\mathcal{B}_1| \leq \varepsilon \mu_{F'}$ if n is large. Let $\mathcal{B}_2$ denote the set of copies of F' that have fewer than $(1 - \varepsilon)\mu_{F''}/\mu_{F'}$ extensions to copies of F''. Since F'' is flat over F', we have $|\mathcal{B}_2| \leq \varepsilon \mu_{F'}$ if n is large enough, which we assume from now on. If ϕ is a copy of F' not in $\mathcal{B}_1 \cup \mathcal{B}_2$, then ϕ has at least $(1 - 2\varepsilon)\mu_{F''}/\mu_{F'}$ extensions to good copies of F'', which in turn have at least $(1 - \varepsilon)\mu_F/\mu_{F''}$ extensions to copies of F, so the value of $Z_n(F', F)$ on ϕ is at least $(1 - 2\varepsilon)(1 - \varepsilon)$. Since there are $(1 + o(1))\mu_{F'}$ copies of F' in total, the proportion of these copies in $\mathcal{B}_1 \cup \mathcal{B}_2$ is at most $\varepsilon + o(1)$. Since $\varepsilon > 0$ was arbitrary, it follows that the negative part of $Z_n(F', F) - 1$ tends to zero in probability. Since $\mathbb{E}(Z_n(F', F)) = t_p(F, G_n)/t_p(F', G_n) \to 1$, it follows that $Z_n(F', F) \xrightarrow{p} 1$, i.e. that F is flat over F'. □

The reader may find many of the arguments above familiar from the dense case; for example, the proof for $K_{2,t}$ is an absolutely standard convexity argument. The key point is that many arguments for the dense case do not carry over. In particular, we have shown that almost all, i.e. all but $o(n^2)$, pairs of vertices have about the right number of common neighbours. In the dense case, it follows immediately that almost all (all but $o(pn^2) = o(n^2)$) edges are in the right number of triangles, and hence that $t_p(K_3, G_n) \to 1$. Similarly, the proof above shows that any F is flat over all of its subgraphs in the dense case, without restriction to girth at least 4. In the sparse case, there are only $o(n^2)$ edges, and there seems to be no simple way to rule out the possibility that a large fraction, or even all, of the pairs of vertices corresponding to edges fall in the $o(n^2)$ set with too few common neighbours. Nevertheless, we conjecture that this cannot happen. The simplest graph for which we cannot prove the conclusion of Conjecture 3.9 is the triangle.

Conjecture 3.21 *Under the conditions of Conjecture 3.9 we have* $s_p(K_3, G_n) \to 1$.

In fact, we do not even have a proof that G_n must contain at least *one* triangle for n large enough!

3.5 Extensions to lower densities

Let us return to the study of general subgraphs F, rather than simply triangles. If true, the various conjectures above may extend to smaller values of p, but one must be careful. Firstly, s_p and t_p no longer coincide, as noted above. One should work with s_p, because these quantities behave in the right way for $G_p(n, \kappa)$, while t_p does not. A simple modification of the proof of Lemma 3.19, considering the distribution of the number of common neighbours of a set of s *distinct* vertices, shows that if $np^s \to \infty$, then $s_p(K_2, G_n) \to 1$, $s_p(C_4, G_n) \to 1$ and $s_p(K_{s,t+1}, G_n)$ bounded together imply $s_p(K_{s,t}, G_n) \to 1$. Taking $p = n^{-\alpha}$, with $0 < \alpha < 1/2$ constant, there is no corresponding result for t_p, even with $s = 2$. Indeed, if $t_p(K_2, G_n) = 1$, then there are at least $n^{t+1} p^t$ homomorphisms from $K_{2,t}$ into G_n mapping the two vertices in the smaller class to the same vertex. It follows that $t_p(K_{2,t}, G_n)$ will be unbounded for any $t > 1/\alpha$.

Secondly, even working with s_p rather than t_p, we cannot in general hope to conclude in the analogue of Conjecture 3.4 that $s_p(F, G_n) \to s(F, \kappa)$ for *all* fixed graphs F. For example, set $p = n^{-1/2}$ and consider the *polarity graphs* G_n of Erdős and Rényi [22], defined (for suitable n) by taking as vertices the points of the projective plane over $GF(q)$, q a prime power, and joining $x = (x_0, x_1, x_2)$ and $y = (y_0, y_1, y_2)$ if and only if $x_0 y_0 + x_1 y_1 + x_2 y_2 = 0$ in $GF(q)$. These graphs satisfy $e(G_n) \sim n^{3/2}/2 = pn^2/2$ but contain no C_4s, and thus satisfy $s_p(K_2, G_n) \to 1$ and $s_p(C_4, G_n) = 0$. Since $s(C_4, \kappa) \geq s(K_2, \kappa)^4$ for any κ, we cannot have $s_p(F, G_n) \to s(F, \kappa)$ for $F = K_2$ and for $F = C_4$ in this case. More generally, whenever $pn^{1/2} \not\to \infty$, then there are graphs G_n with pn^2 edges but too few C_4s, so we should only consider the counts $s_p(C_4, G_n)$ if $pn^{1/2} \to \infty$. This problem is not unique to C_4, so it seems that to extend our conjectures for $p = n^{-o(1)}$ to sparser graphs, we should modify them to refer only to a certain set of 'admissible' subgraphs F, depending on the function $p = p(n)$.

In fact, we should only consider subgraphs F for which the expected number $\mu_F \sim n^{|F|} p^{e(F)}$ of embeddings of F into $G(n,p)$ is much larger than the number $(1 + o(1))n^2 p/2$ of edges, at least if $pn^{1/2} \to \infty$. To see this, first suppose that $n^{|F|} p^{e(F)} \sim An^2 p$, for some constant $0 < A < \infty$. Form a graph G' from $G = G(n,p)$ by adding $\varepsilon n^2 p/(2e(F))$ copies $F_1, F_2, \ldots$ of F, chosen uniformly at random from all subgraphs of K_n isomorphic to F. After deleting the small number of duplicate edges, we have added around $\varepsilon n^2 p/2$ edges, so $s_p(K_2, G') \sim 1 + \varepsilon$. It is easy to check that the number of C_4s in G' containing two or more edges from one single F_i is negligible and thus, considering C_4s formed from all combinations of edges from $G(n,p)$ and from different F_i, that $s_p(C_4, G') \sim (1+\varepsilon)^4$ with high probability. Hence, the appropriate limiting kernel is the constant kernel $\kappa = 1 + \varepsilon$. Copies of F itself containing at most one edge from each F_i contribute $(1 + \varepsilon)^{e(F)}$ to $s_p(F, G')$, but there are $\Theta(n^{|F|} p^{e(F)})$ extra copies of F, namely the F_i themselves. It follows that $s_p(F, G_n) \not\to 1$. If $n^{|F|} p^{e(F)} = o(n^2 p)$, then the argument is much simpler:

adding a few copies of F to $G(n,p)$ does not change the number of edges or C_4s significantly, but does change the number of copies of F.

We can go somewhat further: the construction in Example 3.11 shows that for C_3 to be admissible, the expected number of C_3s per edge should be larger than $\log n$. A similar construction can be carried out for any fixed F, and shows that, at least for suitable balanced F, we should require $n^{|F|}p^{e(F)}/(np^2 \log n) \to \infty$ for F to be admissible. In general, for F to be admissible, we need all induced subgraphs F' of F to be admissible; otherwise, the distribution of copies of F over F' cannot be flat as we expect in the uniform case.

Returning to triangles, in the light of the comments above, perhaps the strongest conceivable extension of Conjecture 3.21 to smaller p would be that if $p = p(n) = \omega(\sqrt{\log n}/\sqrt{n})$, and $s_p(K_2, G_n) \to 1$, $s_p(C_4, G_n) \to 1$, and $\sup_n s_p(K_{2,t}, G_n) < \infty$ for each t, then $s_p(C_3, G_n) \to 1$. However, it may well be that the graphs constructed by Alon [1] mentioned in Example 3.12 have $s_p(K_{2,t}, G_n) \to 1$ for each t. (This may also be true of Kim's random construction [30] giving his famous lower bound on the Ramsey numbers $R(3,t)$.) If so, blowing these graphs up as in Example 3.13 would show that even in the almost dense case, controlling the $K_{2,t}$ counts is not enough, so one should control (at least) the $K_{s,t}$ counts for some larger s. Returning to much sparser graphs, we then have to limit ourselves to $p = p(n)$ for which $K_{s,t}$ is admissible, suggesting the following conjecture.

Conjecture 3.22 *There are constants $s \geq 2$ and $a > 0$ such that, if $p = p(n) = \omega((\log n)^a n^{-1/s})$ and G_n is a sequence of graphs with $|G_n| = n$, $s_p(K_2, G_n) \to 1$, $s_p(C_4, G_n) \to 1$, and $\sup_n s_p(K_{s,t}, G_n) < \infty$ for each t, then $s_p(C_3, G_n) \to 1$.*

It may be that if the conjecture holds for a given s, it holds with $a = 1/s$. It may also be that one needs to control the counts for $K_{s,t}$ and at the same time to consider p larger than n^{-b} for some $b < 1/s$.

There is a potential pitfall in handling subgraph counts when p is smaller than $n^{-o(1)}$: in proving that $s_p(F, G_n) \to 1$ for various graphs F above, we made use of the assumption that $s_p(F', G_n)$ is bounded for other graphs F'. In particular, with $F = K_{2,t}$, we used this assumption for $F' = K_{2,t+1}$. It may be that F' is admissible whenever F is (as is likely in this case: $K_{2,t}$ should be admissible as soon as C_4 is), but perhaps not. In the latter case we may be forced to work with a larger admissible set for which we impose the hypothesis of Conjecture 3.3 (or Conjecture 3.4), and a smaller set for which we obtain the conclusion. In any case, the (smaller) admissible set should have the following property: if $\mathcal{F}_\alpha$ denotes the set of admissible graphs when $p = n^{-\alpha}$, $\alpha > 0$, then the sets $\mathcal{F}_\alpha$ should increase as α decreases, and their union should contain all finite graphs. We shall return to this question in Section 5, in particular in Subsections 5.3 and 5.4, where we prove results that are steps towards (non-uniform) versions of the various conjectures in this section.

4 Szemerédi's Lemma and the cut metric

In the next section we shall discuss the relationship between the cut and count metrics. As in the dense case, a key tool in the study of the cut metric is some variant of Szemerédi's Lemma [36]: this will be discussed in this section. Unlike in the dense case, we need an assumption on the graphs we consider to make this useful; roughly speaking, our assumption is that no subgraph of G_n containing a constant fraction of the vertices has density more than a constant factor larger than it should have. Several of the usual proofs of Szemerédi's Lemma extend easily to the sparse case under this assumption; this was noted independently by Kohayakawa and Rödl; see [32]. (The much earlier Theorem 2 of Kohayakawa [31] is slightly different.)

Throughout this section, $p = p(n)$ with $p = o(1)$ and $np \to \infty$. (Often, $n^2 p \to \infty$ is enough in the proofs, but see Remark 4.4.) As before, (G_n) always denotes a sequence of graphs with $|G_n| = n$, which need not be defined for all n, but only for some infinite set.

For disjoint sets A, B of vertices of a graph $G = G_n$ with n vertices, we write $e_G(A, B)$ for the number of edges of G joining A to B, and

$$d_p(A, B) = \frac{e_G(A, B)}{p|A||B|} \tag{4.1}$$

for the normalized density of G between A and B. It is convenient to extend this definition to sets A and B that need not be disjoint: in this case, we write $e_G(A, B)$ for the number of ordered pairs (i, j) with $i \in A$, $j \in B$ and $ij \in E(G)$; we then define $d_p(A, B)$ as above. Note that $e_G(A, A) = 2e(G[A])$. We shall make the following assumption:

Assumption 4.1 (bounded density) There is a constant C and a function $n_0(\varepsilon)$ such that, for every $\varepsilon > 0$ and $n \geq n_0(\varepsilon)$, and any $A, B \subset V(G_n)$ with $|A|, |B| \geq \varepsilon n$, we have $d_p(A, B) \leq C + \varepsilon$.

It suffices to impose this assumption only when $A = B$, replacing C by $C/2$ and ε by $\varepsilon/2$. Indeed, if $|A|, |B| \geq \varepsilon n$, $n \geq n_0(\varepsilon)$, and $d_p(A, B) > C + \varepsilon$ then, by averaging, we may find $A' \subset A$ and $B' \subset B$ with $|A'| = |B'| = \lceil \varepsilon n \rceil$ such that $d_p(A', B') > C + \varepsilon$. Then $e_G(A' \cup B', A' \cup B') \geq 2e_G(A', B') > 2(C + \varepsilon)|A'|^2 \geq (C/2 + \varepsilon/2)|A' \cup B'|^2$.

The condition above may be written more compactly as follows:

$$\forall \varepsilon > 0: \limsup_{n \to \infty} \max\{d_p(A, B) : A, B \subset V(G_n), |A|, |B| \geq \varepsilon n\} \leq C. \tag{4.2}$$

Note that we shall often assume that (4.2) holds for a particular value of C: in this case, we say that (G_n) has *density bounded by* C. This is the reason for including the final $+\varepsilon$ in Assumption 4.1.

It will be convenient to phrase the proof of Szemerédi's Lemma in terms of kernels. In this sparse setting, the way in which we associate a kernel to a graph is different from in the dense case. Indeed, our aim is that the random graph $G(n, p)$ should approximate the constant kernel taking value 1. For this reason, to a graph G with n vertices $1, 2, \ldots, n$ we associate the kernel κ_G taking the value $1/p$ on each square $((i-1)/n, i/n] \times ((j-1)/n, j/n]$ whenever $ij \in E(G)$, and zero elsewhere. This association will often be implicit: for example, given a graph G and a kernel κ, we write $d_{\text{cut}}(G, \kappa)$ for $d_{\text{cut}}(\kappa_G, \kappa)$.

The following observation shows the importance of bounded density. In the proof, and throughout this section, given a subset A of the vertices of a graph G, we shall often abuse notation by also writing A for the corresponding subset of $[0, 1]$.

Lemma 4.2 *Let $p = p(n)$ be any function of n, let $\kappa : [0,1]^2 \to [0, C]$ be a kernel, and let (G_n) be a sequence of graphs with $|G_n| = n$ and $d_{\mathrm{cut}}(G_n, \kappa) \to 0$. Then (G_n) has density bounded by C.*

Proof Suppose that (G_n) does not have density bounded by C. Then there is an $\varepsilon > 0$ such that, for infinitely many n, there are sets $A_n, B_n \subset V(G_n)$ with $|A_n|, |B_n| \geq \varepsilon n$ and $d_p(A_n, B_n) \geq C + \varepsilon$. Identifying A_n and B_n with subsets of $[0, 1]$, and writing μ for Lebesgue measure, we have

$$\int_{A_n \times B_n} \kappa_{G_n} = d_p(A_n, B_n)\mu(A_n)\mu(B_n) \geq (C + \varepsilon)\mu(A_n)\mu(B_n).$$

Since κ is bounded by C, it follows that

$$\left| \int_{A_n \times B_n} \kappa_{G_n} - \kappa^{(\tau)} \right| \geq \varepsilon\mu(A_n)\mu(B_n) \geq \varepsilon^3$$

for any rearrangement $\kappa^{(\tau)}$ of κ, which contradicts $d_{\mathrm{cut}}(G_n, \kappa) \to 0$. □

4.1 Weakly regular partitions

If G is a graph with vertex set $\{1, 2, \ldots, n\}$, and $\Pi = (P_1, \ldots, P_k)$ is a partition of $V(G)$, then we write G/Π for the kernel on $[0, 1]^2$ taking the value $d_p(P_a, P_b)$ on the union of the squares $((i-1)/n, i/n] \times ((j-1)/n, j/n]$, $i \in P_a$, $j \in P_b$. We say that a partition Π of a graph G is *weakly (ε, p)-regular* if $\|\kappa_G - G/\Pi\|_{\mathrm{cut}} \leq \varepsilon$. Note that the normalizing function p comes in via the definition of the kernels κ_G and G/Π.

For a kernel κ, the definitions are similar: for $A, B \subset [0, 1]$ we write $\kappa(A, B)$ for the integral of κ over $A \times B$, and

$$d(A, B) = d_\kappa(A, B) = \frac{\kappa(A, B)}{\mu(A)\mu(B)}$$

for the average value of κ on $A \times B$. Then $d_p(A, B)$, defined using G, is exactly $d(A, B)$, defined using κ_G, so the kernel G/Π is obtained from κ_G by replacing the value at each point by the average over the relevant rectangle $P_a \times P_b$. For κ a kernel and Π a partition of $[0, 1]^2$, we define κ/Π similarly. The partition Π is *weakly ε-regular* with respect to κ if $\|\kappa - \kappa/\Pi\|_{\mathrm{cut}} \leq \varepsilon$.

The next lemma is a a sparse equivalent of (a version of) the Frieze–Kannan 'weak' form of Szemerédi's Lemma from [23]. As with many proofs of the various forms of Szemerédi's Lemma, the proof of the dense result is not hard to adapt to the sparse setting: the only additional complication is that one must make sure that the parts of the partition remain large enough so that we can make use of the bounded density assumption. In the following lemma, $p = p(n)$ is any normalizing function with $pn^2 \to \infty$. In principle, the various constants depend on the choice of p, but this is not the case if we impose an explicit lower bound on $p(n)$, such as the harmless bound $p \geq n^{-3/2}$.

Lemma 4.3 *Let $p = p(n)$ be any function with $0 < p \leq 1$ and $pn^2 \to \infty$. Let $\varepsilon > 0$, $C > 0$ and $k \geq 1$ be given. There exist constants n_0, K and $\eta > 0$, all depending on ε, C and k, such that, if G_n is any graph with $n \geq n_0$ vertices such that*

$$d_p(A, B) \leq C \text{ whenever } |A|, |B| \geq \eta n, \tag{4.3}$$

and Π is any partition of $V(G)$ into k parts $P_1, \ldots, P_k$ with sizes as equal as possible, then there is a weakly (ε, p)-regular partition Π' of $V(G_n)$ into K parts that refines Π.

Proof Reducing ε if necessary, we may assume that $\varepsilon \leq C$, say. We assume without comment that n is 'large enough' whenever this is needed.

Let $\Pi_0 = \Pi$. We shall inductively define a sequence Π_t of partitions of $V(G)$ into $k_t = 2^t k$ parts, stopping either when we reach some Π_t that is weakly $(\varepsilon/2, p)$-regular, or when $t \geq T = \lceil 16C^2/\varepsilon^2 \rceil + 1$. Every part of Π_t will have size at least $\gamma^t n/(2k)$, where $\gamma = \varepsilon/(100C) \leq 1/100$. Note that Π_0 satisfies this condition.

Set $\eta = \gamma^T/(2k)$, and let n_0 be a large constant to be chosen later. We shall write κ_t for the kernel G/Π_t, noting that, since all parts of Π_t have size at least ηn, the kernel κ_t is bounded by C.

Given Π_t as above, suppose that Π_t is not weakly $(\varepsilon/2, p)$-regular. Then there is a cut $[0, 1] = A \cup A^c$ exhibiting this, i.e. a set $A \subset [0, 1]$ for which $|\kappa_G(A, A^c) - \kappa_t(A, A^c)| \geq \varepsilon/2$. Since both κ_G and κ_t correspond to weighted graphs on $V(G) = \{1, 2, \ldots, n\}$, we may choose the cut A to correspond to a subset of $V(G)$: among all 'worst' cuts, there is a cut of this form.

Our aim is to modify A slightly to obtain a set B (which we may think of as a subset of $V(G)$ or as a subset of $[0, 1]$) and then take two parts $P_i \cap B$ and $P_i \cap B^c$ of Π_{t+1} for each part of Π_t; in doing so, we must ensure that neither of these parts is too small. We modify the set A to obtain B in k_t stages, one for each part P_i. At each stage, we move a set S of at most $\gamma|P_i| \geq \eta n$ vertices from A to A^c or vice versa, to ensure that both B and B^c meet P_i in at least $\gamma|P_i|$ vertices. Since κ_t is bounded by C, this changes the value of the cut $\kappa_t(A, A^c)$ by at most $2C\gamma|P_i|/n$.

From (4.3), the set S meets at most $Cpn\gamma|P_i|$ edges of G: to see this, apply (4.3) to S and $V(G)$ if $|S| \geq \eta n$, and to S' and $V(G)$ otherwise, for any $S' \supset S$ with $\lceil \eta n \rceil$ vertices. Hence, the value of the cut $\kappa_G(A, A^c)$ changes by at most $2C\gamma|P_i|/n$ when we move our set S from one side of the cut to the other. After all these changes, we have

$$|\kappa_t(A, A^c) - \kappa_t(B, B^c)|, |\kappa_G(A, A^c) - \kappa_G(B, B^c)| \leq 2C\gamma \leq \varepsilon/8.$$

It follows that

$$|\kappa_G(B, B^c) - \kappa_t(B, B^c)| \geq \varepsilon/4. \tag{4.4}$$

Let Π_{t+1} be the partition obtained by intersecting each part of Π_t with B and B^c, noting that Π_{t+1} has all the required properties. Set $\kappa_{t+1} = G/\Pi_{t+1}$, noting that $\kappa_{t+1}(B, B^c) = \kappa_G(B, B^c)$, since Π_{t+1} refines the partition (B, B^c). From (4.4) it thus follows that

$$\|\kappa_{t+1} - \kappa_t\|_1 \geq \|\kappa_{t+1} - \kappa_t\|_{\text{cut}} \geq \varepsilon/4,$$

with the final inequality witnessed by the cut (B, B^c). Hence, $\|\kappa_{t+1} - \kappa_t\|_2^2 \geq \|\kappa_{t+1} - \kappa_t\|_1^2 \geq \varepsilon^2/16$. Since κ_t may be obtained from κ_{t+1} by averaging over rectangles, κ_t and $\kappa_{t+1} - \kappa_t$ are orthogonal: for any two parts P_i, P_j of Π_t, the kernel κ_t is constant on $P_i \times P_j$. Also, $\int_{P_i \times P_j} \kappa_{t+1} = \int_{P_i \times P_j} \kappa_G = \int_{P_i \times P_j} \kappa_t$. Thus $\int_{P_i \times P_j} \kappa_t(\kappa_{t+1} - \kappa_t) = 0$. Summing over i and j it follows that $\int \kappa_t(\kappa_{t+1} - \kappa_t) = 0$. Thus,

$$\|\kappa_{t+1}\|_2^2 = \|\kappa_t\|_2^2 + \|\kappa_{t+1} - \kappa_t\|_2^2 \geq \|\kappa_t\|_2^2 + \varepsilon^2/16.$$

It follows by induction that $\|\kappa_t\|_2^2 \geq t\varepsilon^2/16$ as long as our construction continues. But, as noted above, κ_t is bounded by C, so our construction must stop after at most $16C^2/\varepsilon^2$ steps. Since this number is smaller than T, we must stop at a weakly $(\varepsilon/2, p)$-regular partition.

To complete the proof we modify the final partition Π_t slightly. Set $K = k\lceil\gamma^{-T}\rceil$, and note that, since $t \leq T-1$, each part of Π_t has size at least $\gamma^{-1}n/K$. First, adjust the parts slightly so that the size of each is of the form $a\lfloor n/K \rfloor + b\lceil n/K \rceil$, $a, b \in \mathbf{Z}^+$, replacing the kernel κ_t by a new kernel κ' corresponding to the altered partition Π'. Arguing as above, $\|\kappa_t - \kappa'\|_{\mathrm{cut}} \leq 2C\gamma \leq \varepsilon/4$, so, by the triangle inequality and weak $(\varepsilon/2, p)$-regularity of Π_t, we have

$$\|\kappa' - \kappa_G\|_{\mathrm{cut}} \leq \|\kappa_t - \kappa_G\|_{\mathrm{cut}} + \|\kappa_t - \kappa'\|_{\mathrm{cut}} \leq \varepsilon/2 + \varepsilon/4 = 3\varepsilon/4.$$

Finally, we split each part randomly into parts of sizes exactly $\lfloor n/K \rfloor$ and $\lceil n/K \rceil$, obtaining a partition Π'' into K parts whose sizes are as equal as possible. We write κ'' for the corresponding kernel. Since Π' has $O(1)$ parts, and we have $\Theta(pn^2)$ edges between any two parts with density at least $\varepsilon/100$, say, it follows from Chernoff's inequality that if n is large enough, which we enforce by choosing n_0 suitably, then with probability at least 0.99 the density $d_p(A, B)$ between every pair (A, B) of new parts A and B coming from parts P_i and P_j of Π' with $d_p(P_i, P_j) \geq \varepsilon/100$ is $d_p(P_i, P_j)(1 + o(1))$. Since the densities $d_p(P_i, P_j)$ are uniformly bounded by C, it follows that with probability at least 0.99 we have $\|\kappa'' - \kappa'\|_1 \leq \varepsilon/100$. But then

$$\|\kappa'' - \kappa_G\|_{\mathrm{cut}} \leq \|\kappa'' - \kappa'\|_{\mathrm{cut}} + \|\kappa' - \kappa_G\|_{\mathrm{cut}} \leq \|\kappa'' - \kappa'\|_1 + \|\kappa' - \kappa_G\|_{\mathrm{cut}} \leq \varepsilon/100 + \varepsilon/2,$$

so our final partition Π'' is indeed weakly (ε, p)-regular. $\square$

If for any reason we want a weakly (ε, p)-regular partition into a particular number K of parts (which must be a multiple of the number in the original partition if we are refining a given partition), the proof above gives such a partition for any large enough K, indeed, for any $K \geq k\lceil\gamma^{-T}\rceil$. Of course, n_0 then depends on K.

Remark 4.4 The proof of Lemma 4.3 works even if p is very small, say of order $1/n$. However, this is of no help – it is impossible for Assumption 4.1 (the sequence version of (4.3)) to be satisfied in this range, except in the trivial case where $e(G_n) = o(pn^2)$ (so p is not the appropriate normalizing function). Indeed, passing to a subsequence where $e(G_n)/(pn^2)$ is bounded away from zero, picking any εpn^2 edges of G_n, and putting one endpoint of each edge into A and the other into B, we find sets A, B with $|A|, |B| \leq \varepsilon pn^2$ but $e(A, B) \geq \varepsilon pn^2$, which gives $d_p(A, B) \geq 1/(\varepsilon p^2 n^2)$, which tends to infinity as $\varepsilon \to 0$.

4.2 Strongly regular partitions

Usually, when working with the cut metric, weak ε-regularity turns out to be just as good as the usual stronger ε-regularity. In the dense case, this is true also when considering subgraph counts. However, for the subgraph counts we consider in the next section, it turns out that we do in fact need the usual form of ε-regularity.

As usual, a pair (A, B) of (not necessarily disjoint) subsets of $V(G)$ is an (ε, p)-*regular pair* if $|d_p(A', B') - d_p(A, B)| \leq \varepsilon$ whenever $A' \subset A$ and $B' \subset B$ satisfy $|A'| \geq \varepsilon|A|$ and $|B'| \geq \varepsilon|B|$. A partition $\Pi = (P_1, \ldots, P_k)$ of $V(G)$ is (ε, p)-*regular* if the parts P_i each have size $\lceil n/k \rceil$ or $\lfloor n/k \rfloor$, and all but at most $\varepsilon \binom{k}{2}$ of the unordered pairs $\{P_i, P_j\}$, $i \neq j$, are (ε, p)-regular. The definition (now simply of ε-regularity) for a kernel is similar, although here one partitions the interval $[0, 1]$ into parts with measure exactly $1/k$.

The following is (essentially) the sparse version of Szemerédi's Lemma observed by Kohayakawa and Rödl; see [32], where a closely related result is proved. For a proof, see also Gerke and Steger [24]. We shall include a proof here as we state the result in a slightly different way (which makes no real difference), and the use of kernels allows one to phrase the proof a little more simply than in [32] or [24].

Lemma 4.5 *Let $p = p(n)$ be any function with $0 < p \leq 1$ and $pn^2 \to \infty$. Let $\varepsilon > 0$, $C > 0$ and $k \geq 1$ be given. There exist constants n_0, K and $\eta > 0$, all depending on ε, C and k, such that, if G_n is any graph with $n \geq n_0$ vertices such that*

$$d_p(A, B) \leq C \text{ whenever } |A|, |B| \geq \eta n, \tag{4.5}$$

and Π is any partition of $V(G)$ into k parts $P_1, \ldots, P_k$ with sizes as equal as possible, then there is an (ε, p)-regular partition Π' of $V(G_n)$ into at most K parts that refines Π.

Proof Reducing ε and/or increasing C if necessary, we may suppose for convenience that $\varepsilon \leq 1$ and $C \geq 1$.

Set $\gamma = \varepsilon^3/(100C)$. This time we inductively define a sequence Π_t of partitions of $V(G)$ into k_t parts, where $\Pi_0 = \Pi$, $k_0 = k$, and $k_{t+1} = k_t \lceil k_t 2^{k_t}/\gamma \rceil$, stopping either when we reach some Π_t that is (ε, p)-regular, or when $t \geq T = \lceil 20C^2/\varepsilon^5 \rceil + 1$. The parts of each Π_t will have sizes as equal as possible. Note that Π_0 satisfies this condition.

Set $\eta = 1/(2k_T)$, and let n_0 be a large constant to be chosen later. We assume throughout that $n \geq n_0$. As before, we write κ_t for the kernel G/Π_t, noting that, since all parts of Π_t have size at least ηn, the kernel κ_t is bounded by C.

The key (standard) observation is the following. Let A and B be parts of Π_t, so κ_t is by definition constant on $A \times B$, and let $A' \subset A$ and $B' \subset B$. Let Π' be any partition refining Π such that each of A' and B' is a union of parts of Π', and let $\kappa' = G/\Pi'$ be the corresponding kernel. Restricted to $A \times B$, the function κ' integrates to $d_p(A, B)\mu(A)\mu(B) = \int_{A \times B} \kappa_t$, since A and B are unions of parts of κ'. Hence, κ_t and $\kappa' - \kappa_t$ are orthogonal on this set. Using the fact that A' and B' are unions of parts of κ', we see that $\int_{A' \times B'} \kappa' = d_p(A', B')\mu(A')\mu(B')$, which differs from the integral of κ_t over the same set by $|d_p(A', B') - d_p(A, B)|\mu(A')\mu(B')$. It follows that $\|\kappa' - \kappa_t\|_2^2$ is at least $\left(d_p(A', B') - d_p(A, B)\right)^2 \mu(A')\mu(B')$, and hence, using orthogonality, that

$$\int_{A \times B} (\kappa')^2 \geq \int_{A \times B} \kappa_t^2 + \left(d_p(A', B') - d_p(A, B)\right)^2 \mu(A')\mu(B'). \tag{4.6}$$

Suppose then that Π_t is not (ε, p)-regular, and let $A_1, \ldots, A_{k_t}$ denote the parts of Π_t. Then there are at least $\varepsilon \binom{k_t}{2}$ pairs $\{A_i, A_j\}$ of parts of Π_t that are not (ε, p)-regular. For each, pick sets $A_{ij} \subset A_i$ and $A_{ji} \subset A_j$ witnessing this, i.e. with $|d_p(A_{ij}, A_{ji}) - d_p(A_i, A_j)| \geq \varepsilon$ and $|A_{ij}| \geq \varepsilon |A_i|$, $|A_{ji}| \geq \varepsilon |A_j|$. Let Π' be the partition whose parts are all atoms formed by the sets A_i and the sets A_{ij} taken together, so Π' refines Π_t, and each A_{ij} is a union of parts of Π'. We could estimate the L^2-norm of G/Π' using (4.6), but this will not be useful if some parts of Π' are too small, so we first adjust the part sizes.

Define Π_{t+1} by dividing each A_i into k_{t+1}/k_t parts whose sizes are as equal as possible, so that each part of Π' differs from a union of parts of Π_{t+1} in at most n/k_{t+1} vertices: to do this, keep taking for a part of Π_{t+1} a subset of some part of Π', until what is left of every part of Π' is too small. For each i, there are at most k_t sets A_{ij} inside A_i, so A_i is a union of at most 2^{k_t} parts of Π'. It follows that there is some union A'_{ij} of parts of Π_{t+1} with

$$|A_{ij} - A'_{ij}| \leq 2^{k_t} n/k_{t+1} \leq \gamma n/k_t^2.$$

Arguing as in the proof of Lemma 4.3, it follows from (4.5) that the symmetric difference S_{ij} of A_{ij} and A'_{ij} meets at most

$$Cpn|S_{ij}| \leq Cp\gamma n^2/k_t^2 \leq \varepsilon^3 p |A_i||A_j|/99$$

edges of G, if n is sufficiently large. Since $|S_{ij}| \leq \varepsilon^3 |A_i|/100$, say, while $|A_{ij}| \geq \varepsilon |A_i|$ and $|A_{ji}| \geq \varepsilon |A_j|$, it follows crudely that

$$|d_p(A'_{ij}, A'_{ji}) - d_p(A_{ij}, A_{ji})| \leq \varepsilon/2,$$

which implies that

$$|d_p(A'_{ij}, A'_{ji}) - d_p(A_i, A_j)| \geq \varepsilon/2.$$

Now A'_{ij} and A'_{ji} are unions of parts of Π_{t+1}, and these sets have size at least $\varepsilon n/(2k_t)$. Hence, from (4.6),

$$\int_{A_i \times A_j} \kappa_{t+1}^2 \geq \int_{A_i \times A_j} \kappa_t^2 + \varepsilon^4/(16k_t^2)$$

for each of the at least $\varepsilon \binom{k_t}{2}$ irregular pairs $\{A_i, A_j\}$. Since $\int_{A_i \times A_j} \kappa_{t+1}^2 \geq \int_{A_i \times A_j} \kappa_t^2$ always holds, it follows that $\|\kappa_{t+1}\|_2^2 \geq \|\kappa_t\|_2^2 + \varepsilon^5/20$.

If the construction above does not stop before step T, then by induction we have $\|\kappa_t\|_2 \geq t\varepsilon^5/20$ for $0 \leq t \leq T$. But each κ_t is bounded by C, so $\|\kappa_T\|_2^2 \leq C^2$, giving a contradiction. Hence the construction does stop before step T, giving an (ε, p)-regular partition with $k_t \leq k_T$ parts. $\square$

Note that Lemma 4.5 implies (essentially) Lemma 4.3: it is easy to check that an (ε, p)-regular partition is, say, weakly $(10(C+1)\varepsilon, p)$-regular, provided the parts are large enough for (4.5) to hold. However, one of course obtains much worse bounds on the number of parts using the stronger notion of regularity.

Remark 4.6 Let us illustrate once again the difference between the dense and sparse cases with a simple observation. Given a pair (A, B) of sets of vertices of a graph G, let $C_4(A, B)$ denote the number of homomorphisms from C_4 into the subgraph spanned by $A \cup B$ mapping a given pair of opposite vertices into A and the other pair into B. Standard convexity arguments show that $C_4(A, B) \geq d(A, B)^4 |A|^2 |B|^2$. The pair (A, B) is (ε, p)-C_4-minimal if $C_4(A, B) \leq (d(A, B)^4 + \varepsilon p^4)|A|^2|B|^2$. In the dense case (with $p = 1$) it is well known and very easy to check that ε-regularity and ε-C_4-minimality are essentially equivalent: ε-regularity implies $f(\varepsilon)$-C_4-minimality, and ε-C_4-minimality implies $g(\varepsilon)$-regularity, for some $f(\varepsilon), g(\varepsilon)$ with $f(\varepsilon), g(\varepsilon) \to 0$ as $\varepsilon \to 0$.

Let $\varepsilon > 0$ and M be given. By counting C_4s it is easy to see that there is a function $f(\varepsilon)$ with $f(\varepsilon) \to 0$ as $\varepsilon \to 0$ such that, if n is large enough and (A, B) is ε-regular with $|A| = |B| = n$, then we may partition A and B into sets $A_1, \ldots, A_M$ and $B_1, \ldots, B_M$ of almost equal sizes so that every pair (A_i, B_j) is $f(\varepsilon)$-regular. Indeed, a random partition has this property with probability tending to 1, since by standard concentration results (for example, the Hoeffding–Azuma inequality), the edge densities and 'C_4-densities' of the pairs (A_i, B_j) are highly concentrated about the corresponding densities for (A, B). It follows immediately that in the usual dense Szemerédi's Lemma [36], we may specify in advance the number of parts K we would like our partition to have, provided (as in the weak case) that K is large enough given ε, and n large enough given ε and K.

In the sparse case, the fact about random partitioning above is presumably true, but the simple proof using C_4-counts fails totally. It is still true that (ε, p)-C_4-minimality implies $(f(\varepsilon), p)$-regularity, but the reverse implication fails. Indeed, whenever $p = p(n) \to 0$, given any pair (A, B), we may add a small dense (say complete bipartite) subgraph with too few edges to disturb regularity, but containing many more than $p^4 |A||B|$ copies of C_4.

4.3 Szemerédi's lemma and convergence in the cut norm

We start with a consequence of Lemma 4.3 concerning the cut norm.

Corollary 4.7 *Let (G_n) be a sequence of graphs satisfying Assumption 4.1. Then there is a kernel $\kappa : [0,1]^2 \to [0, C]$ and a subsequence (G_{n_i}) of (G_n) such that $d_{\mathrm{cut}}(G_{n_i}, \kappa) \to 0$. Moreover, we may label the vertices of G_{n_i} with $1, 2, \ldots, n_i$ so that $\|\kappa_{G_{n_i}} - \kappa\|_{\mathrm{cut}} \to 0$.*

Proof We shall only sketch the proof as the argument is exactly the same as that of Lovász and Szegedy [34] for the dense case. Note that given any $\eta > 0$ and $\varepsilon > 0$, our graphs G_n satisfy the assumption (4.3) of Lemma 4.3 with $C + \varepsilon$ in place of C whenever n is large enough.

First, let us apply Lemma 4.3 with $k = 1$ and $\varepsilon = \varepsilon_1 = 1/2$, say, to obtain a weakly (ε_1, p)-regular partition $\Pi_{n,1}$ of G_n into $k_1 = K$ parts, for all large enough

n. We may relabel the vertices of each G_n so that the parts of $\Pi_{n,1}$ are all intervals. Each kernel $G_n/\Pi_{n,1}$ is characterized by a k_1-by-k_1 density matrix, whose entries all lie in $[0, C + \varepsilon_1]$. (Indeed, if k_1 happens to divide n, then the kernel is exactly the kernel obtained from the matrix in the obvious way.) Since these matrices live in a compact set, $[0, C + \varepsilon_1]^{k_1^2}$, they have a convergent subsequence. Passing to the corresponding subsequence of G_n, we then have $G_n/\Pi_{n,1} \to \kappa_1$ pointwise almost everywhere, and hence in L^1 and in the cut norm. Since the partitions $\Pi_{n,1}$ are weakly (ε_1, p)-regular, we have $\|\kappa_{G_n} - G_n/\Pi_{n,1}\|_{\mathrm{cut}} \leq \varepsilon_1$. Passing far enough along our subsequence, it follows that $\|\kappa_{G_n} - \kappa_1\|_{\mathrm{cut}} \leq 2\varepsilon_1$.

Working within the subsequence defined above, apply Lemma 4.3 again with $\varepsilon = \varepsilon_2 = 1/4$, say, and $k = k_1$. For each n we find a partition $\Pi_{n,2}$ refining $\Pi_{n,1}$, with $k_2 = K(\varepsilon_1, C, k_1)$ parts. Relabelling vertices, we may assume that each part of each $\Pi_{n,2}$ is an interval. (Note that we only reorder the vertices within parts of $\Pi_{n,1}$.) As before, on a subsequence we have $G_n/\Pi_{n,2} \to \kappa_2$, for some kernel κ_2 constant on squares of side-length $1/k_2$. Since $\Pi_{n,2}$ refines $\Pi_{n,1}$ for each n, it follows that the value of κ_1 on each $1/k_1$-by-$1/k_1$ square is exactly the average of κ_2 over this set; to see this, let $n \to \infty$.

Iterating, we find kernels $\kappa_1, \kappa_2, \ldots$, each of which can be obtained by averaging the next one, and graphs G_{n_i} with $\|\kappa_{G_{n_i}} - \kappa_i\|_{\mathrm{cut}} \leq 2\varepsilon_i = 2^{1-i}$, say. To complete the proof we simply observe that the sequence (κ_t) is a martingale on the state space $[0,1]^2$. Since each κ_t is bounded by $C + \varepsilon_t \leq C + 1$, by the Martingale Convergence Theorem there is a kernel $\kappa : [0,1]^2 \to [0, C]$ with $\kappa_t \to \kappa$ pointwise almost everywhere, and hence in L^1 and in the cut-norm. Then $\|\kappa_{G_{n_i}} - \kappa\|_{\mathrm{cut}} \to 0$ as required. $\square$

The corollary above says that any (suitable) sequence of graphs has a subsequence converging to a kernel, and is a simple consequence of Szemerédi's Lemma and the Martingale Convergence Theorem. Together with Lemma 4.2, it shows that Assumption 4.1 is the correct assumption to impose on sequences of graphs when we seek limits that are bounded kernels $\kappa : [0,1]^2 \to [0, C]$. Before turning to an application of Corollary 4.7, let us note an even simpler consequence of the Martingale Convergence Theorem.

Lemma 4.8 *Let κ be a bounded kernel, and for $k \geq 1$, let κ_k be the piecewise constant kernel obtained by dividing $[0,1]^2$ into 2^{2k} squares of side 2^{-k}, and replacing κ by its average over each square. Then $\kappa_k \to \kappa$ pointwise almost everywhere and also in L^p for any p.*

Proof The sequence κ_k is a bounded martingale on $[0,1]^2$, so pointwise convergence is given by the Martingale Convergence Theorem. Since the sequence κ_k is bounded by $\sup \kappa$, convergence in L^p follows by dominated convergence. $\square$

A consequence of Corollary 4.7 is that it allows us to compare the two different versions of the cut metric. Recall that for graphs G_1, G_2, we defined $d_{\mathrm{cut}}(G_1, G_2)$ by first passing to kernels taking the values 0 and $1/p$. If G_1 and G_2 have the same number of vertices, then there is a more natural definition of their cut-distance, $\widehat{d}_{\mathrm{cut}}(G_1, G_2)$, defined in the same way but only allowing rearrangements that 'map whole vertices to whole vertices'. As in the dense case, $d_{\mathrm{cut}}(G_1, G_2)$ and $\widehat{d}_{\mathrm{cut}}(G_1, G_2)$

are defined by (2.8) and (2.9), respectively; the difference between the sparse and dense cases is in the normalization of κ_{G_i}. Writing d^1_{cut} and $\widehat{d}^1_{\text{cut}}$ for the metrics defined using $p = 1$, Borgs, Chayes, Lovász, Sós and Vesztergombi [15, Theorem 2.3] showed that these metrics are equivalent, proving that

$$d^1_{\text{cut}}(G_1, G_2) \leq \widehat{d}^1_{\text{cut}}(G_1, G_2) \leq 32 d^1_{\text{cut}}(G_1, G_2)^{1/67}. \quad (4.7)$$

In fact, they proved (4.7) for edge-weighted graphs, as long as all edge weights lie in $[-1, 1]$. Unlike simple Lipschitz equivalence, which may also hold, this does not directly carry over to the sparse setting: we have $d_{\text{cut}} = p^{-1} d^1_{\text{cut}}$ and $\widehat{d}_{\text{cut}} = p^{-1} \widehat{d}^1_{\text{cut}}$, so (4.7) can be written as

$$d_{\text{cut}}(G_1, G_2) \leq \widehat{d}_{\text{cut}}(G_1, G_2) \leq 32 d_{\text{cut}}(G_1, G_2)^{1/67} p^{-66/67},$$

which is of little if any use here. However, the equivalence of the two metrics in the sparse case is not too hard to deduce from (4.7), using Corollary 4.7.

Lemma 4.9 *For $i = 1, 2$, let $(G^{(i)}_n)$ be a sequence of graphs satisfying the bounded density assumption 4.1. Then $d_{\text{cut}}(G^{(1)}_n, G^{(2)}_n) \to 0$ if and only if $\widehat{d}_{\text{cut}}(G^{(1)}_n, G^{(2)}_n) \to 0$.*

Proof If $\widehat{d}_{\text{cut}}(G^{(1)}_n, G^{(2)}_n) \to 0$ then, since $d_{\text{cut}} \leq \widehat{d}_{\text{cut}}$, it follows trivially that $d_{\text{cut}}(G^{(1)}_n, G^{(2)}_n) \to 0$.

Suppose now that $d_{\text{cut}}(G^{(1)}_n, G^{(2)}_n) \to 0$; our aim is to show that $\widehat{d}_{\text{cut}}(G^{(1)}_n, G^{(2)}_n) \to 0$, so we may suppose that this is not the case. Hence, passing to a subsequence, we may assume that $\widehat{d}_{\text{cut}}(G^{(1)}_n, G^{(2)}_n) \geq \delta$ for some positive δ and all n in our subsequence.

Applying Corollary 4.7 twice, the second time to a suitable subsequence, we find kernels $\kappa_1, \kappa_2 : [0, 1]^2 \to [0, C]$, and subsequences of the sequences $(G^{(i)}_n)$, defined for the same values of n, on which $\|\kappa_{G^{(i)}_n} - \kappa_i\|_{\text{cut}} \to 0$. Since $d_{\text{cut}}(G^{(1)}_n, G^{(2)}_n) \to 0$, it follows that $d_{\text{cut}}(\kappa_1, \kappa_2) = 0$.

For any $\varepsilon > 0$, by Lemma 4.8 we may find a K and kernels $\kappa'_1, \kappa'_2 : [0, 1]^2 \to [0, C]$ that are constant on squares of side $1/K$, with $\|\kappa'_i - \kappa_i\|_{\text{cut}} \leq \varepsilon$. Since the kernels κ'_i may be thought of as weighted graphs, it would appear that we have gone round in circles, but the point is that they are *dense* weighted graphs. Regarding the kernels κ'_1/C and κ'_2/C as weighted graphs with edge weights in $[0, 1]$, we have

$$d_{\text{cut}}(\kappa'_1/C, \kappa'_2/C) = d_{\text{cut}}(\kappa'_1, \kappa'_2)/C \leq 2\varepsilon/C,$$

so (4.7) gives

$$\widehat{d}_{\text{cut}}(\kappa'_1, \kappa'_2) = C \widehat{d}_{\text{cut}}(\kappa'_1/C, \kappa'_2/C) \leq 32 C (2\varepsilon/C)^{1/67} = O(\varepsilon^{1/67}).$$

Hence, there is a rearrangement of κ'_1 preserving intervals that is close to κ'_2 in the cut norm. Ignoring divisibility, adapting this rearrangement to the graph $G^{(1)}_n$, n much larger than K, and using $\|\kappa_{G^{(i)}_n} - \kappa'_i\|_{\text{cut}} \leq \varepsilon + o(1)$, it follows that that $\widehat{d}_{\text{cut}}(G^{(1)}_n, G^{(2)}_n) \leq O(\varepsilon) + O(\varepsilon^{1/67})$. Choosing ε small enough, the final bound is less than δ, contradicting our assumptions. □

Corollary 4.7 shows that one property of the cut metric carries over to the sparse setting: for every suitable sequence (G_n), i.e. any sequence satisfying the bounded density assumption 4.1, there is a kernel κ and a subsequence converging to κ in d_{cut}. In the other direction, as in the dense case, such a sequence is given by the natural random construction.

Lemma 4.10 *Let $p = p(n)$ satisfy $np \to \infty$, let $C > 0$ be constant, let $\kappa : [0, 1]^2 \to [0, C]$ be a bounded kernel, and let $G_n = G_p(n, \kappa)$. Then $d_{\text{cut}}(G_n, \kappa) \to 0$ almost surely. Also, the sequence (G_n) satisfies the bounded density assumption 4.1 with probability 1.*

Proof The second statement is essentially immediate from Chernoff's inequality, constructing G_n as a subgraph of the Erdős–Rényi random graph $G(n, Cp)$; it also follows from the first statement and Lemma 4.2.

We now turn to the proof that $d_{\text{cut}}(G_n, \kappa) \to 0$. Recall that κ is of finite type if $[0, 1]$ may be partitioned into sets $A_1, \ldots, A_k$ so that κ is constant on each rectangle $A_i \times A_j$. We first suppose that κ is of finite type. Rearranging κ, and ignoring parts with measure zero, we may assume that each A_i is an interval with positive measure. Recall that $G_n = G_p(n, \kappa)$ is constructed by first choosing the 'types' $x_1, \ldots, x_n$ of the vertices independently and uniformly at random from $[0, 1]$. Let n_i denote the number of vertices of type i, noting that we have $n_i \sim \mu(A_i)n$ almost surely. Let us adjust the intervals A_i slightly, replacing A_i by a set A'_i $(= A'_i(n))$ with measure n_i/n. Let $\kappa' = \kappa'(n)$ be the adjusted kernel, taking on $A'_i \times A'_j$ the value that κ takes on $A_i \times A_j$. Since, almost surely, we adjust the length of each A_i by $o(1)$, the kernels κ' and κ differ on a set of measure $o(1)$. Since each is bounded, it follows that

$$||\kappa - \kappa'||_{\text{cut}} \leq ||\kappa - \kappa'||_1 \to 0 \qquad (4.8)$$

almost surely, as $n \to \infty$.

Given $x_1, \ldots, x_n$, let G' be the weighted graph in which each edge is present and has weight $w_{ij} = p\kappa(x_i, x_j)$. Then, relabelling the vertices so that those with $x_i = k$ correspond to the set A'_k, we see that $\kappa_{G'} = \kappa'$. The graph G_n may be constructed from G' by simply selecting each edge ij independently, with probability equal to its weight in G'. As noted earlier, for a kernel corresponding to a (weighted) graph, the cut norm (defined by (2.3)) is realized by a cut corresponding to a partition of the vertex set, so

$$||\kappa_{G_n} - \kappa'||_{\text{cut}} = ||\kappa_{G_n} - \kappa_{G'}||_{\text{cut}} = \max_{S \subset V(G_n)} \left| \frac{e_{G_n}(S, S^c) - \sum_{i \in S, j \in S^c} w_{ij}}{n^2 p} \right|.$$

Having conditioned on $x_1, \ldots, x_n$, for each S the random variable $X = e_{G_n}(S, S^c)$ has mean exactly $\sum_{i \in S, j \in S^c} w_{ij}$. Furthermore, $\mathbb{E}(X) = O(n^2 p)$. Since X is a sum of independent indicator variables, it follows from (for example) the Chernoff bounds, that for any $\varepsilon > 0$ we have $\mathbb{P}(|X - \mathbb{E}(X)| \geq \varepsilon n^2 p) \leq \exp(-c_\varepsilon n^2 p)$ for some $c_\varepsilon > 0$. Since $n^2 p = \omega(n)$, this probability decays superexponentially. Since there are only 2^n sets S to consider, we see that $\mathbb{P}(||\kappa_{G_n} - \kappa'||_{\text{cut}} \geq \varepsilon)$ decays superexponentially as $n \to \infty$. Since $\varepsilon > 0$ was arbitrary, using (4.8) it follows that $d_{\text{cut}}(G_n, \kappa) \to 0$ almost surely.

So far we assumed that κ was of finite type. Given an arbitrary κ, for each $\varepsilon > 0$ we can find a finite type approximation κ_ε to κ with

$$||\kappa_\varepsilon - \kappa||_{\text{cut}} \leq ||\kappa_\varepsilon - \kappa||_1 \leq \varepsilon;$$

see, for example, Lemma 4.8. One can couple the random graphs $G_n = G_p(n,\kappa)$ and $G'_n = G_p(n,\kappa_\varepsilon)$ using the same vertex types $x_1,\ldots,x_n$ for each, in such a way that the symmetric difference $G_n \Delta G'_n$ has the distribution of $G_p(n,\Delta\kappa)$, where $\Delta\kappa(x,y) = |\kappa(x,y) - \kappa_\varepsilon(x,y)|$. The expected number of edges of $G_p(n,\Delta\kappa)$ is at most $n(n-1)p||\Delta\kappa||_1/2$ (with equality if $p\Delta\kappa \leq 1$), which is at most $n^2 p\varepsilon/2$. It is easy to check that the actual number is tightly concentrated about the mean, so

$$d_{\text{cut}}(G_n, G'_n) \leq ||\kappa_{G_n} - \kappa_{G'_n}||_1 = \frac{2e(G_n \Delta G'_n)}{n^2 p} \leq 2\varepsilon$$

holds with probability tending (rapidly) to 1 as $n \to \infty$. Using the finite-type case to show that $d_{\text{cut}}(G'_n, \kappa_\varepsilon) \to 0$ and the bound $d_{\text{cut}}(\kappa, \kappa_\varepsilon) \leq \varepsilon$, and recalling that $\varepsilon > 0$ was arbitrary, the result follows. $\square$

5 Comparison between cut and count convergence

Throughout this section, we fix a function $p = p(n)$, and consider sequences (G_n) of graphs with $|G_n| = n$. In the dense case, with $p(n) = 1$ for all n, Borgs, Chayes, Lovász, Sós and Vesztergombi [15] showed that such a sequence converges to a kernel κ in d_{cut} if and only if it converges to κ in d_{sub}; here we wish to investigate whether this result can be extended to the sparse case. To do this, we first have to make sense of the definitions. For d_{cut}, as in the previous section, we simply associate a kernel κ_n to G_n as before, with κ_n taking the values 0 and $1/p$. Then we use the usual definition of d_{cut} for (dense) kernels to define $d_{\text{cut}}(G_n, G_m)$ and $d_{\text{cut}}(G_n, \kappa)$. In the light of Lemma 4.9, for questions of convergence the metrics d_{cut} and $\widehat{d}_{\text{cut}}$ are equivalent; we shall use d_{cut} rather than $\widehat{d}_{\text{cut}}$ in this section.

5.1 Admissible subgraphs and their counts

If $p = n^{-o(1)}$, then we use (3.1) and (3.2) to define d_{sub}, so convergence in d_{sub} is equivalent to convergence of $s_p(F, G_n)$ for every graph F. For smaller p, as noted in Subsection 3.5, it makes sense only to consider graphs F in a certain set $\mathcal{A}$ of *admissible* graphs. It is not quite clear exactly which graphs should be admissible (see Subsection 3.5), so there are several variants of the definitions. To keep things simple, we shall work here with one particular choice for the set $\mathcal{A}$, depending on the function p. It may be that the various conjectures we shall make, if true, extend to larger sets $\mathcal{A}$.

Recall that we write $\mathcal{F}$ for the set of isomorphism classes of finite (simple) graphs. Given a loopless multi-graph F and an integer $t \geq 1$, let F_t denote the graph obtained by subdividing each edge of F exactly $t - 1$ times, so $e(F_t) = te(F)$ and $|F_t| = |F| + (t-1)e(F)$. Writing $\mathcal{F}^m$ for the set of isomorphism classes of finite loopless multi-graphs, for $t \geq 2$ let

$$\mathcal{F}_t = \{F_t : F \in \mathcal{F}^m\},$$

and set $\mathcal{F}_1 = \mathcal{F}$ (not $\mathcal{F}^m$). Thus, for $t \geq 2$, the family $\mathcal{F}_t$ is the set of simple graphs that may be obtained as follows: starting with a set of paths of length t, identify subsets of the endpoints of these paths in an arbitrary way, except that the two endpoints of the same path may not be identified. Note that any $F_t \in \mathcal{F}_t$ has girth at least $2t$.

Similarly, let $\mathcal{F}_{\geq t}$ be the set of simple graphs that may be obtained as above but starting with paths of length at least t. Thus $\mathcal{F}_{\geq 1} = \mathcal{F}$ and, for $t \geq 2$, $\mathcal{F}_{\geq t}$ is the set of graphs that may be obtained from some $F \in \mathcal{F}^m$ by subdividing each edge at least $t-1$ times. Note that $\mathcal{F} = \mathcal{F}_{\geq 1} \supset \mathcal{F}_{\geq 2} \supset \cdots$. Let $\mathcal{T}$ denote the set of (isomorphism classes of) finite trees.

Throughout this subsection and the next we suppose that there is some $\alpha > 0$ such that $np \geq n^\alpha$ for all large enough n. Equivalently, there is some integer $t \geq 1$ such that
$$n^{t-1}p^t \geq n^{-o(1)}. \tag{5.1}$$

We shall set
$$\mathcal{A} = \mathcal{T} \cup \mathcal{F}_{\geq t}$$
for the smallest such t, noting that if $p = n^{-o(1)}$ then $t = 1$, so all graphs are admissible. (An alternative that would work just as well is to let $\mathcal{A}$ be the set of all subgraphs of graphs in $\mathcal{F}_{\geq t}$, which includes $\mathcal{T}$.) A key observation is that if $F \in \mathcal{F}_{\geq t}$ then (considering the internal vertices on the paths making up F) we have $|F| > e(F)(t-1)/t$. This also holds if $F \in \mathcal{T}$, or indeed if F is a subgraph of some $F' \in \mathcal{A}$. It follows that if $F \subset F' \in \mathcal{A}$ then
$$\mathbb{E}\operatorname{emb}(F, G(n,p)) \sim n^{|F|} p^{e(F)} = n^{|F|-e(F)(t-1)/t}(n^{t-1}p^t)^{e(F)/t} = n^{\Theta(1)-o(1)} \to \infty. \tag{5.2}$$

On the one hand, $\mathcal{A} = \mathcal{T} \cup \mathcal{F}_{\geq t}$ is small enough to satisfy the requirements for admissibility discussed in Subsection 3.5, including (5.2). (There may be requirements we have missed, in which case $\mathcal{A} = \mathcal{T} \cup \mathcal{F}_{\geq t}$ for some larger t is likely to work.) On the other hand, as we shall now see, this set $\mathcal{A}$ is large enough to ensure that the counts for $F \in \mathcal{A}$ determine a kernel, up to the equivalence relation $\sim$ defined in Subsection 2.4.

Theorem 5.1 *Let κ_1 and κ_2 be two bounded kernels, and $t \geq 1$ an odd integer. Suppose that $s(F, \kappa_1) = s(F, \kappa_2)$ for every $F \in \mathcal{F}_t$. Then $\kappa_1 \sim \kappa_2$.*

Proof Given a kernel κ, let κ^t be the kernel defined by
$$\kappa^t(x, y) = \int_{[0,1]^{t-1}} \kappa(x, x_1)\kappa(x_1, x_2) \cdots \kappa(x_{t-1}, y)\, dx_1 \cdots dx_{t-1}. \tag{5.3}$$

In other words, roughly speaking, $\kappa^t(x, y)$ counts the number of paths from x to y in κ with length t. The key observation is that if F is a graph, κ a kernel, and $t \geq 1$, then
$$s(F_t, \kappa) = s(F, \kappa^t). \tag{5.4}$$

Indeed, $s(F_t, \kappa)$ is defined as an integral over one variable for each vertex of F_t. We may evaluate this integral by first fixing the variables corresponding to vertices of F, then using (5.3) once for each edge of F to integrate over the remaining variables. What remains is exactly the integral defining $s(F, \kappa^t)$.

By assumption, $s(F, \kappa_1) = s(F, \kappa_2)$ for every $F \in \mathcal{F}_t$. Hence, from (5.4), we have $s(F, \kappa_1^t) = s(F, \kappa_2^t)$ for *every* graph F, so, by Theorem 2.8 or Theorem 2.9, $\kappa_1^t \sim \kappa_2^t$. Hence, from (2.12), there is a kernel κ and measure-preserving maps $\sigma_1, \sigma_2 : [0, 1] \to [0, 1]$ such that $(\kappa_i^t)^{(\sigma_i)} = \kappa$ almost everywhere, for $i = 1, 2$. Since $(\kappa_i^{(\sigma_i)})^t = (\kappa_i^t)^{(\sigma_i)}$, we thus have $(\kappa_1')^t = (\kappa_2')^t$ almost everywhere for $\kappa_i' = \kappa_i^{(\sigma_i)}$. Since $\kappa_i' \sim \kappa_i$, and our aim is to prove that $\kappa_1 \sim \kappa_2$, it suffices to prove that $\kappa_1' \sim \kappa_2'$. Hence, without loss of generality, we may replace κ_i by κ_i', so $\kappa_1^t = \kappa_2^t$ almost everywhere. It is now straightforward to deduce that $\kappa_1 = \kappa_2$ almost everywhere.

Given a bounded signed kernel, i.e. a bounded function $\kappa : [0, 1]^2 \to \mathbf{R}$ satisfying $\kappa(x, y) = \kappa(y, x)$, let T_κ be the corresponding operator on $L^2([0, 1])$, defined by

$$(T_\kappa f)(x) = \int_0^1 \kappa(x, y) f(y)\, dy. \tag{5.5}$$

From the Cauchy–Schwarz inequality we have

$$\begin{aligned}
\|T_\kappa f\|_2^2 &= \int_0^1 \left(\int_0^1 \kappa(x, y) f(y)\, dy \right)^2 dx \\
&\leq \int_0^1 \left(\int_0^1 \kappa(x, y)^2\, dy \int_0^1 f(y)^2\, dy \right) dx \\
&= \|f\|_2^2 \int\int \kappa(x, y)^2\, dx\, dy = \|f\|_2^2 \|\kappa\|_2^2,
\end{aligned}$$

so the operator norm of T_κ on L^2 satisfies

$$\|T_\kappa\| \leq \|\kappa\|_2 < \infty. \tag{5.6}$$

Now let κ be any bounded kernel, and $\varepsilon > 0$ a real number. By Lemma 4.8 there is some k such that the kernel κ_k obtained by averaging κ over 2^{-k}-by-2^{-k} squares satisfies $\|\kappa - \kappa_k\|_2 \leq \varepsilon$. Writing $T_\kappa = T_{\kappa_k} + T_{\kappa - \kappa_k}$, the first term has finite rank, since $T_{\kappa_k} f$ is constant on intervals of length 2^{-k}. From (5.6), the second term has operator norm at most $\|\kappa - \kappa_k\|_2 \leq \varepsilon$. It follows that the image of the unit ball under T can be covered by a finite number of balls of radius 2ε. Since ε was arbitrary, this shows that T_κ is a compact operator.

Since κ is symmetric, we also have that T_κ self-adjoint. Consequently, T_{κ_1} is a compact self-adjoint operator on the Hilbert space $L^2(0, 1)$, so by standard results (see, for example, Bollobás [5]) there is an orthonormal basis of eigenvectors of T_{κ_1}, and all its eigenvalues are real. It is easy to see that $T_{\kappa_1^t} = (T_{\kappa_1})^t$, so $T_{\kappa_1^t}$ acts on the λ-eigenspace of T_{κ_1} by multiplication by λ^t. Since t is odd (so the map $\lambda \mapsto \lambda^t$ is injective), it follows that $T_{\kappa_1^t}$ has the same eigenspaces as T_{κ_1}. Turning this around, the action of T_{κ_1} on each eigenspace E_λ of $T_{\kappa_1^t}$ with eigenvalue λ is to multiply by $\lambda^{1/t}$. Thus, T_{κ_1} is uniquely determined by $T_{\kappa_1^t}$. In particular, since $\kappa_1^t = \kappa_2^t$ almost everywhere, the operators T_{κ_1} and T_{κ_2} are equal, i.e. $\kappa_1 = \kappa_2$ almost everywhere, as required. □

Note that in Theorem 5.1 the restriction to odd t is essential, as shown by the following example.

Example 5.2 Let κ_1 and κ_2 be the two 2-by-2 'chessboard' kernels defined by

$$\kappa_1(x,y) = \begin{cases} 1 & \text{if } x < 1/2,\ y < 1/2 \text{ or } x \geq 1/2,\ y \geq 1/2 \\ 0 & \text{otherwise,} \end{cases}$$

and

$$\kappa_2(x,y) = \begin{cases} 1 & \text{if } x < 1/2,\ y \geq 1/2 \text{ or } x \geq 1/2,\ y < 1/2 \\ 0 & \text{otherwise.} \end{cases}$$

Thus, in the dense case, κ_1 corresponds to the union of two disjoint complete graphs on $n/2$ vertices, and κ_2 to the complete $n/2$-by-$n/2$ bipartite graph. It is easy to check that for any graph F we have $s(F,\kappa_1) = 2^{1-|F|}$, while $s(F,\kappa_2) = 2^{1-|F|}$ if F is bipartite, and $s(F,\kappa_2) = 0$ otherwise. In particular, $s(F,\kappa_1) = s(F,\kappa_2)$ for all bipartite F, and hence for all $F \in \mathcal{F}_t$, t even.

As we saw from Lemma 4.2 and Corollary 4.7, bounded density is a natural condition to impose on our sequence (G_n) when dealing with d_{cut} for sparse graphs. In the previous sections, when dealing with subgraph counts and d_{sub}, we imposed different conditions, the closest being Assumption 3.2. Let us restate this here in the appropriate form when p need not be as large as $n^{-o(1)}$.

Assumption 5.3 (exponentially bounded admissible subgraph counts) There is a constant C such that, for each fixed $F \in \mathcal{A}$, we have $\limsup s_p(F, G_n) \leq C^{e(F)}$ as $n \to \infty$.

Note that we impose a condition only for $F \in \mathcal{A}$. When comparing d_{cut} and d_{sub}, we need to impose both Assumption 4.1 (bounded density) and Assumption 5.3. In the 'almost dense' case, when we take $\mathcal{A} = \mathcal{F}$, Assumption 5.3 implies Assumption 4.1, with the same constant C. The argument is based on showing that a not-too-small dense part of G_n would contain too many $K_{t,t}$s for some large t. Since the details are very similar to the proof of Lemma 3.5, we omit them.

Unfortunately, in general neither of Assumptions 4.1 and 5.3 implies the other. In one direction, this is easy to see: simply add a complete graph on m vertices, where $m(n)$ is chosen so that $e(K_m) \sim m^2/2 = o(n^2 p)$. This does not affect Assumption 4.1, but, if m is chosen large enough, will create too many copies of any fixed connected graph F with $|F| \geq 3$. For the reverse direction, consider the following example.

Example 5.4 Fix a real number $D > 1$, and let $\kappa = \kappa_D$ be the unbounded kernel defined as follows. First partition $[0,1]$ into intervals $I_1, I_2, \ldots$, so that I_i has length 2^{-i}. Then set $\kappa(x,y) = i^{2/D}$ if $x,y \in I_i$, and $\kappa(x,y) = 0$ otherwise. Let F be a connected graph with average degree at most D. Then, since only terms where all vertices are in the same I_i contribute, we have

$$s(F,\kappa) = \sum_{i=1}^{\infty} 2^{-i|F|} i^{2e(F)/D} \leq \sum_{i=1}^{\infty} \left(2^{-i} i\right)^{|F|} \leq \sum_{i=1}^{\infty} 2^{-i} i = 2.$$

Let $G_n = G_p(n, \kappa)$ be the random graph defined from κ as before. If every component of any admissible graph has average degree at most D, then it is easy to check that with probability 1 the sequence (G_n) satisfies Assumption 5.3 (with $C = 2$). On the other hand, this sequence does not satisfy Assumption 4.1, since, for every i, there will be a subgraph of G_n containing a positive fraction of the vertices with density around $i^{2/D}$.

With the choice of $\mathcal{A}$ made here, whenever $p = p(n)$ does not satisfy $p = n^{-o(1)}$ then only trees and graphs in some $\mathcal{F}_{\geq t}$, $t \geq 2$, are admissible. All such graphs, and all their components, have average degree less than 4, so the example above shows that in this case, Assumption 5.3 does not imply Assumption 4.1.

Example 5.4 also shows that, in contrast to the almost dense case (where all graphs are admissible), in general we cannot tell from the admissible subgraph counts whether a kernel is bounded. For this reason, together with those discussed above, when comparing d_{cut} and d_{sub} we impose both Assumptions 4.1 and 5.3.

5.2 Conjectured equivalence between cut and count convergence

Our main conjecture from Section 3 was that, in the sparse case, if the subgraph counts converge, they converge to those of a kernel. In the present setting, we consider counts for admissible subgraphs. Fix $p(n)$ satisfying (5.1), and a set $\mathcal{A}$ of admissible graphs. By default we take $\mathcal{A} = \mathcal{T} \cup \mathcal{F}_{\geq t}$ as in the previous subsection, although the definitions make sense for other sets $\mathcal{A}$. Throughout we impose Assumptions 4.1 and 5.3 for some fixed constant C. Let $X = [0, \infty)^{\mathcal{A}}$, let $s_p : \mathcal{F} \to X$ be the map defined by

$$s_p(G_n) = (s_p(F, G_n))_{F \in \mathcal{A}} \in X$$

for any graph G_n with n vertices, let d be any metric on X inducing product topology, and define d_{sub} by mapping to X and then applying d; as usual, we suppress the dependence on the normalizing function p. Note that d_{sub} is in general a pseudo-metric rather than a metric: there may be non-isomorphic graphs G, G' with $s_p(F, G) = s_p(F, G')$ for all $F \in \mathcal{A}$. As we only consider questions of convergence for sequences G_n with $|G_n| \to \infty$, this will not be relevant.

Let $\mathcal{L} \subset X$ denote the set of possible limit points of sequences $s_p(G_n)$, where (G_n) satisfies our assumptions.

Recall that we write $\mathcal{K}$ for the space of kernels, that is, symmetric measurable functions $\kappa : [0, 1]^2 \to [0, C]$ quotiented by equivalence. There is a natural map from $\mathcal{K}$ into X given by subgraph counts; we write s for this map, which does not depend on p (except through the choice of $\mathcal{A}$). Since $\mathcal{A}$ always contains some set $\mathcal{F}_{\geq t}$, and hence some $\mathcal{F}_{t'}$ with t' odd, Theorem 5.1 tells us that this map is injective.

Our main conjecture is the following.

Conjecture 5.5 *With the assumptions and definitions above, we have*

$$\mathcal{L} \subset s(\mathcal{K}). \tag{5.7}$$

Note that if $t = 1$ then $p = n^{-o(1)}$ and we recover Conjecture 3.3; Conjecture 5.5 seems to be the natural extension of Theorem 2.1 to functions $p = p(n)$ with $p \to 0$ but $np \geq n^{\alpha}$ for some $\alpha > 0$.

Turning to the equivalent of Theorem 2.4, we believe that in this setting the notions of convergence given by d_{sub} and d_{cut} are equivalent. The most concrete way of saying this is as follows; again we take $\mathcal{A} = \mathcal{T} \cup \mathcal{F}_{\geq t}$ by default, although it might be that the conjecture fails for this $\mathcal{A}$ but holds for some other $\mathcal{A}$.

Conjecture 5.6 *Let (G_n) be a sequence satisfying Assumptions 4.1 and 5.3, and let $\kappa \in \mathcal{K}$. Then $d_{\mathrm{cut}}(G_n, \kappa) \to 0$ if and only if $s_p(G_n) \to s(\kappa)$.*

In this form, the conjecture implies (5.7) (see below). Without assuming (5.7), it still makes sense to compare the notions of Cauchy sequences instead.

Conjecture 5.7 *Let (G_n) be a sequence satisfying Assumptions 4.1 and 5.3. Then (G_n) is Cauchy with respect to d_{cut} if and only if (G_n) is Cauchy with respect to d_{sub}.*

As we shall shortly see, Conjectures 5.6 and 5.7 are equivalent.

Although we cannot prove the conjectures above, we can say something. Conjecture 5.6, for example, asserts two implications. Surprisingly, it is easy to show that, if (5.7) holds, then either of these implications (for all sequences, not just a particular sequence) implies the other! To prove this we shall first show that the random graph $G(n, \kappa)$ behaves 'correctly' with respect to our definition of d_{sub}; the corresponding result for d_{cut} is Lemma 4.10.

Lemma 5.8 *Fix $C > 0$, let $\kappa : [0,1]^2 \to [0,C]$ be a bounded kernel, and let $G_n = G_p(n, \kappa)$. Then, with probability 1, the sequence (G_n) satisfies Assumption 5.3 and we have $s_p(G_n) \to s(\kappa)$.*

Outline proof The first statement follows from the second, since $s(F, \kappa) \leq C^{e(F)}$ holds for every F, and in particular for $F \in \mathcal{A}$.

It is well known that if F is a fixed graph, and $p' = p'(n)$ is a function of n, then the number X_F of subgraphs of $G(n, p')$ isomorphic to F is concentrated about its mean if and only if $\mathbb{E}(X_{F'}) \to \infty$ for every subgraph F' of F. (For early results of this type see Bollobás [3] and Ruciński [35]; for more recent, much stronger, results see Janson [26] and Janson, Oleszkiewicz and Ruciński [28].)

Our choice of the set $\mathcal{A}$ ensures that this holds for every $F \in \mathcal{F}$ with $p' = Cp$ (see (5.2)), proving the result if κ is constant. It is straightforward to adapt this result to finite type κ. It is easy to check that for the F we consider, any $o(n^2 p)$ edges of $G_n \subset G(n, Cp)$ meet $o(n^{|F|} p^{e(F)})$ copies of F. Using this observation, one can approximate the general case by the finite type case as in the proof of Lemma 4.10. We omit the details. □

Lemma 5.8 gives us a sequence tending in d_{sub} to any $\kappa \in \mathcal{K}$. In other words, it shows that $\mathcal{L} \supset s(\mathcal{K})$. Hence, if (5.7) holds,

$$\mathcal{L} = s(\mathcal{K}). \tag{5.8}$$

Let $\mathcal{J} \subset \mathcal{K} \times \mathcal{L}$ denote the set of pairs $(\kappa, \lambda) \in \mathcal{K} \times \mathcal{L}$ such that there is a sequence (G_n) satisfying our assumptions with $d_{\mathrm{cut}}(G_n, \kappa) \to 0$ and $s_p(G_n) \to \lambda$. Together, Lemmas 4.10 and Lemma 5.8 tell us much more than simply that $\mathcal{L} \subset s(\mathcal{K})$: they show that the 'diagonal' $\mathcal{D} = \{(\kappa, s(\kappa)) : \kappa \in \mathcal{K}\}$ is contained in $\mathcal{J}$.

At this point, we have established three basic facts:

Fact 1: Every subsequence of (G_n) has a subsequence converging in d_{sub} to some point of $\mathcal{L}$. This is trivial, since Assumption 5.3 ensures that $s_p(G_n)$ lives in a compact subset of $X = [0, \infty)^{\mathcal{A}}$.

Fact 2: Every subsequence of (G_n) has a subsequence converging in d_{cut} to some kernel $\kappa \in \mathcal{K}$. This is the first part of Corollary 4.7.

Fact 3: The map s is an injection from $\mathcal{K}$ to $\mathcal{L}$. As noted above, this follows from Theorem 5.1.

Facts 1 and 2 tell us that the relationship between the notions of convergence in d_{cut} and d_{sub} is described by the set $\mathcal{J}$. Indeed, any subsequence of (G_n) itself has a subsequence in which we have convergence in both these metrics, to some point of $\mathcal{J}$.

Suppose for the moment that (5.8) holds. There are two possibilities.

If $\mathcal{J}$ is precisely the diagonal $\mathcal{D}$, then the three facts above easily imply that Conjectures 5.6 and 5.7 both hold.

If $\mathcal{J} \neq \mathcal{D}$, then there is some off diagonal point (κ_1, λ) in $\mathcal{J}$. Since we are assuming (5.8), we have $\lambda = s(\kappa_2)$ for some $\kappa_2 \in \mathcal{K}$. From the definition of $\mathcal{J}$ there is a sequence (G_n) satisfying our assumptions, with $d_{\text{cut}}(G_n, \kappa_1) \to 0$ and $s_p(G_n) \to s(\kappa_2)$. Interleaving the sequence G_n with the sequence $G_p(n, \kappa)$, which converges to κ in both d_{cut} and d_{sub}, taking $\kappa = \kappa_1$ or κ_2, we find a sequence which converges in one of d_{cut} or d_{sub} but not in the other. Hence, neither implication in Conjecture 5.6 or 5.7 holds, i.e. these conjectures fail as badly as possible.

In the light of the comments above, Conjecture 5.6 has the following rather vague reformulation as a question.

Question 5.9 *Given a definition of 'suitable' sequences (G_n), let $\mathcal{C}$ be the set of all graphs F with the property that, whenever κ is a bounded kernel and (G_n) is a suitable sequence with $d_{\text{cut}}(G_n, \kappa) \to 0$, then $s_p(F, G_n) \to s(F, \kappa)$. Under what reasonable definition of 'suitable' is the set $\mathcal{C}$ large enough that the counts $s(F, \kappa)$, $F \in \mathcal{C}$, determine a kernel κ up to equivalence?*

The point is that, if $\mathcal{C}$ is large enough, then the three facts above hold with $\mathcal{A} = \mathcal{C}$, and we simply use $\mathcal{C}$ as the set of graphs whose counts we use to define d_{sub}. Then, for our 'suitable' sequences, d_{cut} convergence implies d_{sub} convergence to the same kernel by definition, so $(\kappa, \lambda) \in \mathcal{J}$ implies $\lambda = s(\kappa)$. Thus (5.7) (and hence (5.8)) holds, and $\mathcal{J} = \mathcal{D}$, so d_{sub} convergence also implies d_{cut} convergence. Unfortunately, there is no obvious single choice for the set of suitable sequences. One could hope that sequences with bounded density would do, but this is not the case: by adding a complete graph with many (but still $o(pn^2)$) edges to $G(n, p)$, say, it is easy to check that in this case $\mathcal{C}$ consists only of matchings. Conjecture 5.6 is more specific than Question 5.9, since we define 'suitable' by assuming $s_p(F, G_n)$ bounded for F in some set $\mathcal{A}$, and then require $\mathcal{C} \supset \mathcal{A}$.

If Conjecture 5.5 does not hold, then Conjectures 5.6 and 5.7 cannot hold. Indeed, there is some $\lambda \in \mathcal{L}$ not corresponding to a kernel. Taking G_n converging to λ in d_{sub}, and then a subsequence that converges in d_{cut}, there is some κ with $(\kappa, \lambda) \in \mathcal{J}$. Interleaving a corresponding sequence (G_n) with $G_p(n, \kappa)$, we find a sequence that converges in d_{cut} but not in d_{sub}.

Even if Conjecture 5.5 does not hold, it is still possible that there is some relationship between cut and subgraph convergence: it may be that every sequence that is Cauchy with respect to d_{sub}, and hence converges to some $\lambda \in \mathcal{L}$, is Cauchy with respect to d_{cut}, i.e. converges to some $\kappa \in \mathcal{K}$. This happens if and only if, for every $\lambda \in \mathcal{L}$, there is a unique $\kappa \in \mathcal{K}$ such that $(\kappa, \lambda) \in \mathcal{J}$. This is not as implausible as it may sound. Indeed, suppose Conjecture 5.6 holds for some admissible set $\mathcal{A}_-$, but that the definitions involved make sense for a larger set $\mathcal{A}_+$. It may be that (5.7) fails working with $\mathcal{A}_+$, because we are now allowing as admissible some counts which need not converge to what we expect. However, there is a restriction map from $\mathcal{L}_+$ to $\mathcal{L}_-$ forgetting about the counts outside $\mathcal{A}_-$. Since (5.8) holds for the smaller set of admissible graphs, this would show that for the larger set there is only one κ for each λ, but not vice versa.

In the next section we shall prove a form of Conjecture 5.6. Before doing so, let us briefly compare this conjecture with the corresponding result of Borgs, Chayes, Lovász, Sós and Vesztergombi [15] for the dense case. In the dense setting, as here, Facts 1 and 2 above are easy to prove. That all limiting counts come from kernels was shown by Lovász and Szegedy [34]; this gives (5.8). Surprisingly, the hard part is proving Fact 3, that the counts (now meaning all counts) determine the kernel, up to equivalence as defined in Subsection 2.4. (For us this was easy, since we deduced the sparse equivalent of this statement from the dense result, Theorem 2.8.) Once one knows that the counts determine the kernel, the 'meta-argument' above shows that d_{cut} convergence implies d_{sub} convergence if and only if the reverse implication holds. Since the forward implication is very easy (see Corollary 2.3), the result of [15] that the two metrics are equivalent follows. This gives a proof of this result in which the only non-straightforward step is showing that the counts $s(F, \kappa)$ determine the kernel κ up to the appropriate notion of equivalence. One might expect this uniqueness result to be easy, but this seems to be far from the case. Recently, Borgs, Chayes and Lovász [12] gave a direct proof of this result (which, as noted in Section 2, actually follows from the results of [15]); their proof is far from simple.

5.3 Partial results: embedding lemmas

Our aim in this section is to prove a positive result, that under certain circumstances, if $d_{\text{cut}}(G_n, \kappa) \to 0$, then $s_p(F, G_n) \to s_p(F, \kappa)$ for certain graphs F. In the case where κ is of finite type, this is simply a counting lemma: in this case, $G_n \to \kappa$ says that G_n can be partitioned into (ε, p)-regular pairs with densities given by κ. In the uniform case, Chung and Graham [17] proved such counting lemmas for certain graphs under certain assumptions. The general case turns out to be rather different, but we shall still use several of their ideas.

We start with the simplest case, where F is a path. First we need some definitions. As usual, in the proof it will be easier to consider homomorphisms from F to G_n (i.e. walks in G_n) rather than embeddings. As we shall see later, this makes no difference.

For G_n a graph and $X_0, \ldots, X_\ell$ subsets of $V(G_n)$, let $G_n(X_0, X_1, \ldots, X_\ell)$ denote the number of $(\ell+1)$-tuples (v_i) with $v_i \in X_i$ and $v_i v_{i+1} \in E(G)$ for $0 \le i \le \ell - 1$.

Identifying a subset of $V(G_n)$ with a subset of $[0,1]$ as before, for a kernel κ let

$$\kappa(X_0, X_1, \ldots, X_\ell) = \int_{X_0 \times \cdots \times X_\ell} \kappa(x_0, x_1) \cdots \kappa(x_{\ell-1}, x_\ell) \, dx_0 \cdots dx_\ell.$$

Lemma 5.10 *Let $C > 0$ be constant, let $p(n)$ be any function of n with $np \to \infty$, and let (G_n) be a sequence of graphs with $t_p(T, G_n)$ bounded for each tree T. For every $\varepsilon > 0$ and $\ell \geq 1$ there is a $\delta = \delta_\ell(\varepsilon) > 0$ such that, whenever $\kappa : [0,1]^2 \to [0, C]$ is a kernel with $\|\kappa_{G_n} - \kappa\|_{\mathrm{cut}} \leq \delta$, then*

$$\left| G_n(X_0, X_1, \ldots, X_\ell) - n^{\ell+1} p^\ell \kappa(X_0, X_1, \ldots, X_\ell) \right| \leq \varepsilon n^{\ell+1} p^\ell$$

for any sets $X_0, X_1 \ldots, X_\ell \subset V(G_n)$.

Roughly speaking, the lemma says that if $G_n \to \kappa$ and $t_p(T, G_n)$ is bounded for each T, then $t_p(P_\ell, \kappa) \to s(P_\ell, \kappa)$. The stronger assertion makes it simpler to prove the result by induction.

Proof Renormalizing, we may assume without loss of generality that $C = 1$. Let us do so from now on.

The fact that δ is not allowed to depend on κ allows us to assume without loss of generality that κ is piecewise constant on squares of side $1/n$, i.e. that κ may be interpreted as a (dense) weighted graph with vertex set $V(G_n)$. Indeed, the Frieze–Kannan form of Szemerédi's Lemma shows that there is an integer k such that, given any κ, there is a κ' that is constant on squares of side $1/k$ with $d_{\mathrm{cut}}(\kappa, \kappa') \leq \delta$. Tweaking κ' slightly if k does not divide n, we obtain a kernel κ'' of the required form. Replacing δ by 2δ as appropriate, the result for κ follows from the result for κ''. [Note that we implicitly assumed that n is large here, meaning larger than some n_0 depending on ε and ℓ. We could simply assume this in the statement of the lemma, but it can be achieved by subdividing vertices. In fact, we could work with a kernel instead of a graph throughout the proof.]
Let

$$\Delta(X_0, \ldots, X_\ell) = \frac{G_n(X_0, X_1, \ldots, X_\ell)}{n^{\ell+1} p^\ell} - \kappa(X_0, X_1, \ldots, X_\ell),$$

so our aim is to show that $|\Delta(X_0, \ldots, X_\ell)| \leq \varepsilon$ for all choices of the sets X_i. We shall show much more: let $M = \max_T \sup_n t_p(T, G_n)$, where the maximum is over trees with at most $2\ell + 1$ vertices, noting that $M < \infty$. We shall show that if $d_{\mathrm{cut}}(G_n, \kappa) \leq \delta$, then, for any $1 \leq t \leq \ell$ and any $X_0, \ldots, X_t \subset V(G_n)$ we have

$$|\Delta(X_0, X_1, \ldots, X_t)| \leq \varepsilon_t, \tag{5.9}$$

where $\varepsilon_1 = \delta$, and

$$\varepsilon_t = 7\sqrt{\varepsilon_{t-1}} + \sqrt{M} \varepsilon_{t-1}^{1/4}$$

for $t \geq 2$. Since ε_ℓ tends to zero as $\delta \to 0$, taking δ small enough we have $\delta = \varepsilon_1 \leq \varepsilon_2 \leq \cdots \varepsilon_\ell \leq \varepsilon$, so to complete the proof of the lemma it suffices to prove (5.9) for this choice of δ.

We shall prove (5.9) by induction on t. For $t = 1$, the result is immediate from the definition of the cut norm: indeed, $\Delta(X_0, X_1)$ is one of the quantities appearing in the supremum defining this norm. Suppose now that $2 \le t \le \ell$, and that (5.9) holds with t replaced by $t-1$.

For $v \in V(G)$ and $X_1, \ldots, X_r \subset V(G)$, set

$$\kappa(v, X_1, \ldots, X_r) = \int_{X_1 \times \cdots \times X_r} \kappa(x, x_1) \kappa(x_1, x_2) \cdots \kappa(x_{r-1}, x_r) \, dx_1 \cdots dx_r,$$

where x is any point of the interval of length $1/n$ corresponding to the vertex v, and let

$$\Delta(v, X_1, \ldots, X_r) = \frac{G_n(\{v\}, X_1, \ldots, X_r)}{n^r p^r} - \kappa(v, X_1, \ldots, X_r). \quad (5.10)$$

Note that

$$\Delta(X, X_1, \ldots, X_r) = \frac{1}{n} \sum_{v \in X} \Delta(v, X_1, \ldots, X_r). \quad (5.11)$$

Fix $X_0, \ldots, X_t \subset V(G_n)$, and set $\eta = \sqrt{\varepsilon_{t-1}}$. Let B_1 be the set of $v \in X_1$ with $\Delta(v, X_2, \ldots, X_t) > \eta$. Then, from (5.11), $\Delta(B_1, X_2, \ldots, X_t) \ge \eta |B_1|/n$. But by the induction hypothesis, $\Delta(B_1, X_2, \ldots, X_t) \le \varepsilon_{t-1} = \eta^2$. Hence, $|B_1| \le \eta n$. Arguing similarly, and using $\varepsilon_{t-1} \ge \varepsilon_1$, we see that the set B of vertices $v \in X_1$ for which either $|\Delta(v, X_2, \ldots, X_t)| \ge \eta$ or $|\Delta(v, X_0)| \ge \eta$ holds has size at most $4\eta n$.

If $v \in X_1 \setminus B$, then we have roughly the right number of walks through v, i.e.

$$G_n(X_0, \{v\}, X_2, \ldots, X_t) = G_n(\{v\}, X_0) G_n(\{v\}, X_2, \ldots, X_t)$$

is close to $np\kappa(v, X_0) n^{t-1} p^{t-1} \kappa(v, X_2, \ldots, X_t)$. More precisely, using the fact that κ is pointwise bounded by $C = 1$ to bound the κ terms in the last expression by 1, for $v \in X_1 \setminus B$ we have

$$|\Delta(X_0, v, X_2, \ldots, X_t)| \le 3\eta, \quad (5.12)$$

where the left hand side is defined by analogy with (5.10).

It remains to consider $v \in B$. For $i = 1, 2$, let

$$\sigma_i = \sum_{v \in B} G_n(X_0, \{v\}, X_2, \ldots, X_t)^i,$$

noting that $\sigma_1 \le \sqrt{|B|\sigma_2}$ by the Cauchy–Schwarz inequality. Let T be the tree with $2t$ edges formed by identifying the second vertices of two paths of length t. Then σ_2 counts a subset of the homomorphisms from T into G_n, so

$$\sigma_2 \le \hom(T, G_n) = n^{2t+1} p^{2t} t_p(T, G_n) \le M n^{2t+1} p^{2t}.$$

Since $|B| \le 4\eta n$ it follows that

$$\sigma_1 \le \sqrt{|B|\sigma_2} \le 2\sqrt{M} \eta n^{t+1} p^t.$$

Since κ is bounded by 1, we have $\kappa(X_0, B, X_2, \ldots, X_t) \le \mu(B) \le 4\eta$, so

$$|\Delta(X_0, B, X_2, \ldots, X_t)| \le 2\sqrt{M}\eta + 4\eta.$$

Together with the bound (5.12) for $v \in X_1 \setminus B$ and (the equivalent of) (5.11), this implies that

$$|\Delta(X_0, X_1, \ldots, X_t)| \le 7\eta + 2\sqrt{M}\eta = \varepsilon_t,$$

as required. This completes the proof of (5.9) by induction, and thus the proof of the lemma. □

Note that the argument above works just as well for an arbitrary fixed tree rather than a path: we pick some leaf v to play the role of x_0; the unique neighbour of v then plays the role of x_1. This gives us a counting lemma for trees.

Corollary 5.11 *Let (G_n) be a sequence of graphs with $t_p(T, G_n)$ bounded for every tree T, and suppose that $d_{\mathrm{cut}}(G_n, \kappa) \to 0$, where κ is a bounded kernel. Then for each tree T we have $t_p(T, G_n) \to s(T, \kappa)$ as $n \to \infty$.* □

Chung and Graham [17] proved a version of this result (for paths rather than trees) with κ constant, under the assumption that the maximum degree of G_n is at most Cpn. This maximum degree assumption of course gives $t_p(T, G_n) \le C^{e(T)}$, so it is stronger than the bounded tree counts assumption of Lemma 5.10. In some sense, the maximum degree condition is much stronger, but it turns out that our global assumption is just as good for questions involving subgraph counts. The reason that Lemma 5.10 is more complicated than the corresponding simple result in [17] is that κ is not uniform, not our weaker assumption.

We stated earlier that, in the sparse case, the parameter $s_p(F, \kappa)$ should be preferred to $t_p(F, \kappa)$, even though t_p tends to be easier to work with. Nevertheless, in the case of trees, these parameters are equivalent, as shown by the following observation.

Lemma 5.12 *Let $p(n)$ be any function of n with $np \to \infty$, and let (G_n) be a sequence with $s_p(T, G_n)$ bounded for every tree T. Then, for each tree T, we have $t_p(T, G_n) \sim s_p(T, G_n)$. In particular, $t_p(T, G_n)$ is bounded.*

Proof Fix a tree T with k vertices. It suffices to show that the number N_T of non-injective homomorphisms from T to G_n satisfies $N_T = o(n^k p^{k-1})$ as $n \to \infty$. Now the image of any non-injective homomorphism ϕ from T to G_n is a connected subgraph H of G_n with ℓ vertices, where $1 \le \ell \le k-1$. Any such subgraph contains a tree T' with ℓ vertices, so for each ℓ there are (crudely) at most $\sum_{|T'|=\ell} \mathrm{emb}(T', G_n)$ possibilities for vertex set of H, where the sum is over all trees T' with ℓ vertices. Since there are at most k^ℓ homomorphisms ϕ with image a given set of ℓ vertices, we thus have

$$N_T \le \sum_{\ell=1}^{k-1} k^\ell \sum_{|T'|=\ell} \mathrm{emb}(T', G_n).$$

Since $\mathrm{emb}(T', G_n) = n_{(|T'|)} p^{e(T')} s_p(T', G_n)$, the final term is $O(n^\ell p^{\ell-1})$ by assumption. It follows that $N_T = O(n^{k-1} p^{k-2}) = o(n^k p^{k-1})$, as claimed. □

Lemma 5.12 allows us to restate Corollary 5.11 in terms of the parameter s.

Theorem 5.13 *Let (G_n) be a sequence of graphs with $s_p(T, G_n)$ bounded for every tree T, and suppose that $d_{\mathrm{cut}}(G_n, \kappa) \to 0$, where κ is a bounded kernel. Then for each tree T we have $s_p(T, G_n) \to s(T, \kappa)$ as $n \to \infty$.* □

Theorem 5.13 may be regarded as an embedding lemma for trees. Our next aim is to prove a much more general result. Chung and Graham showed that, in the uniform case, if the number of paths of length $\ell - 1$ between any two vertices is

Metrics for sparse graphs

at most a constant times what it should be, then almost all pairs of vertices are joined by almost the right number of paths of length ℓ, and hence G_n contains asymptotically the expected number of copies of any $F \in \mathcal{F}_\ell$. This result is much harder than the paths result, even in the uniform case. Although we shall use the key idea of Chung and Graham, the proof does not carry over in a simple way. In the following result, we work with t_p rather than s_p for simplicity; we return to this later.

Theorem 5.14 *Let $C > 0$ and $\ell \geq 3$ be fixed, and let $p = p(n)$ be any function of n. Let (G_n) be a sequence of graphs with $\sup_n t_p(F, G_n) < \infty$ for each $F \in \mathcal{T} \cup \mathcal{F}_\ell \cup \{C_{2\ell-2}\}$, and suppose that $d_{\mathrm{cut}}(G_n, \kappa) \to 0$ for some kernel $\kappa : [0,1]^2 \to [0, C]$. Then $t_p(F, G_n) \to s(F, \kappa)$ for each $F \in \mathcal{F}_\ell$.*

Proof Note that by Lemma 4.2, the sequence (G_n) has density bounded by C, i.e. it satisfies Assumption 4.1. Renormalizing, we shall assume without loss of generality that $C = 1$.

Fix $\varepsilon > 0$, and a graph $F_\ell \in \mathcal{F}_\ell$. Let $\eta > 0$ be a small constant to be chosen below (depending on ε, ℓ and F_ℓ). By Lemma 4.5 there is some K such that for n large enough, which we assume from now on, G_n has an (η, p)-regular partition $\Pi = (P_1, \ldots, P_k)$ for some $k = k(n) \leq K$. Passing to a subsequence of (G_n), we may assume that k is constant. As usual, we shall ignore rounding to integers, assuming that each P_i contains exactly n/k vertices.

Passing to a subsequence (again), we may assume that for all i and j the sequence $d_p(P_i, P_j)$ converges to some $\kappa'(P_i, P_j) \in [0, 1]$. Relabelling if necessary so that P_i consists of vertices v with $in/k < v \leq (i+1)n/k$, and identifying vertices with corresponding subsets of $[0, 1]$ as usual, we may view κ' as a kernel on $[0,1]^2$.

If n is large enough, which we assume, then each $d_p(P_i, P_j)$ is within ε of $\kappa(P_i, P_j)$. It follows that $d_{\mathrm{cut}}(G_n/\Pi, \kappa') \leq \|G_n/\Pi - \kappa'\|_1 \leq \varepsilon$. Under our bounded density assumption 4.1, strong regularity implies weak regularity (for suitably transformed parameters), so choosing η small enough we have $d_{\mathrm{cut}}(G_n, G_n/\Pi) \leq \varepsilon$. Hence, choosing n large enough, $d_{\mathrm{cut}}(\kappa, \kappa') \leq d_{\mathrm{cut}}(\kappa, G_n) + d_{\mathrm{cut}}(G_n, G_n/\Pi) + d_{\mathrm{cut}}(G_n/\Pi, \kappa') \leq 3\varepsilon$. Hence, by Lemma 2.2, for any fixed F we have

$$|s(F, \kappa) - s(F, \kappa')| = O(\varepsilon),$$

so it suffices to show that $t_p(F_\ell, G_n)$ is close to $s(F_\ell, \kappa')$ rather than to $s(F_\ell, \kappa)$. To avoid clutter in the notation, from now on we write κ for the finite type kernel κ' defined above; the original κ plays no further role in the proof. Recall that κ (formerly known as κ') is bounded by 1. For $u \in P_i$ and $v \in P_j$ we shall abuse notation by writing $\kappa(u, v) = \kappa(P_i, P_j)$ for the value of κ at any point of $[0,1]^2$ corresponding to (u, v). Recall that $|d_p(P_i, P_j) - \kappa(P_i, P_j)| \leq \varepsilon$ for all i, j.

For $v, w \in V(G_n)$ and $t \geq 1$, let $w_t(v, w)$ denote the number of walks of length t in G_n starting at v and ending at w; we suppress the dependence on G_n in the notation. Let $\kappa^t(v, w)$ denote the normalized 'expected' number of such walks, if G_n behaved like the random graph $G_p(n, \kappa)$. Let $U \subset V^2$ be the set of pairs (v, w) such that $w_\ell(v, w) \leq (\kappa^\ell(v, w) - \varepsilon)n^{\ell-1}p^\ell$. We call the pairs $(v, w) \in U$ *underconnected*, since they are joined by 'too few' walks of length ℓ. We shall show that

$$|U| = \left|\{(v, w) : w_\ell(v, w) \leq (\kappa^\ell(v, w) - \varepsilon)n^{\ell-1}p^\ell\}\right| \leq \varepsilon n^2 \qquad (5.13)$$

if η is chosen suitably, and then n is taken large enough. Before doing so, let us note that this implies the result.

By Lemma 5.10, if we choose η small enough, then the total number of walks of length ℓ in G_n is within $\varepsilon n^{\ell+1} p^\ell$ of the expected number in $G_p(n,\kappa)$, namely $\|\kappa^\ell\|_1 n^{\ell+1} p^\ell$. If (5.13) holds, then if we count only a maximum of $\kappa^\ell(v,w) n^{\ell-1} p^\ell$ walks for each pair (v,w) of endpoints, we still count at least $(1-\varepsilon)(\|\kappa^\ell\|_1 - \varepsilon) n^{\ell+1} p^\ell$ walks, so there are at most $3\varepsilon n^{\ell+1} p^\ell$ walks uncounted, using $\|\kappa^\ell\|_1 \le 1$. Writing W for the set of *overconnected* pairs $(v,w) \in V^2$ with $w_\ell(v,w) \ge (\kappa^\ell(v,w) + \sqrt{\varepsilon}) n^{\ell-1} p^\ell$, it follows that

$$|W| \le 3\sqrt{\varepsilon} n^2. \tag{5.14}$$

In other words, almost all pairs of vertices are joined by almost the right number of walks.

Recall that we fixed a graph $F_\ell \in \mathcal{F}_\ell$. Let F_ℓ be obtained by subdividing the edges of a loopless multi-graph F with vertex set $u_1, \ldots, u_r$, so

$$\hom(F_\ell, G_n) = \sum_{v_1,\ldots,v_r \in V(G_n)} \prod_{u_i u_j \in E(F)} w_\ell(v_i, v_j), \tag{5.15}$$

where the factors in the product corresponding to multiple edges of F are of course repeated. Given $u_i u_j \in E(F)$, let $2F_\ell/E_2$ be the graph formed from two copies of F_ℓ by identifying the vertices corresponding to u_i and identifying the vertices corresponding to u_j. Since $2F_\ell/E_2 \in \mathcal{F}_\ell$, we have $t_p(2F_\ell/E_2, G_n)$ bounded. It follows by the Cauchy–Schwarz inequality that the number of homomorphisms from F_ℓ into G_n mapping u_i and u_j to a pair in $U \cup W$ is small, in fact of order $\varepsilon^{1/4} n^{|F_\ell|} p^{e(F_\ell)}$; the argument is as in the proof of Lemma 5.10.

Since the comment above applies to any edge $u_i u_j$ of F, the contribution to the sum in (5.15) from terms in which one or more pairs (v_i, v_j) fall in $U \cup W$ is small. But in the remaining terms, $w_\ell(v_i, v_j)$ is well approximated by $\kappa^\ell(v_i, v_j) n^{\ell-1} p^\ell$, and it follows that $t_p(F_\ell, G_n)$ is close to $s(F_\ell, \kappa)$: the difference is bounded by some function of $|F_\ell|$ and ε. In short, we have shown that to prove the theorem, it suffices to prove (5.13), i.e. that there are few underconnected pairs.

From now on, we forget the original graph F_ℓ, and aim to prove (5.13), recalling that κ is a fixed finite-type kernel and that G_n/Π is (pointwise) within ε of κ, where $\Pi = (P_1, \ldots, P_k)$ is our (η, p)-regular partition of G_n. It will be convenient to assume that ε is fairly small. In particular, we shall assume that $\varepsilon \le 1/40$.

Recall that all but at most ηk^2 pairs in our partition $(P_i)_1^k$ are (η, p)-regular. Since *all* pairs have density at most $1 + \varepsilon \le 2$, the irregular pairs contain at most $2\eta n^2 p$ edges. By assumption $t_p(T, G_n)$ is bounded for each tree T, and in particular for the trees formed from two paths by identifying an edge from each, so using Cauchy–Schwarz again a small set of edges meets only a small fraction of the walks of length ℓ in G_n. In particular, the number of walks of length ℓ containing one or more edges from irregular pairs is $O(\sqrt{\eta} n^{\ell+1} p^\ell)$. Taking η small enough, we may assume that this quantity is less than $\varepsilon^2 n^{\ell+1} p^\ell / 10$, say. It follows that in proving (5.13), we may delete all edges in irregular pairs, i.e. we may assume that every pair is regular: if (5.13) holds for the resulting graph G'_n and kernel κ' with $\varepsilon/2$ in place of ε, then (5.13) holds for our original graph G_n and kernel κ.

The lower bound in the proof of Lemma 5.10 used only closeness of the graph and kernel in the cut norm, not the bounds on various tree counts. This argument can thus be applied locally to sequences of parts of our partition. Abusing notation, let us write $P_0, P_1, \ldots, P_{\ell-1}$ for an arbitrary sequence of ℓ parts of our partition, with repetition allowed. For any subsets $X_i \subset P_i$, we find that there are at least

$$p^{\ell-1} \prod_{i=0}^{\ell-1} |X_i| \prod_{i=0}^{\ell-2} \kappa(P_i, P_{i+1}) - \gamma \frac{n^\ell}{k^\ell} p^{\ell-1}$$

walks $v_0 v_1 \cdots v_{\ell-1}$ with $v_i \in X_i$, where $\gamma = \gamma(\eta, \ell)$ tends to 0 as $\eta \to 0$. We choose η small enough that $\gamma \le \varepsilon^{12}$. Taking $X_i = P_i$ for $i > 0$, and summing over all choices for the intermediate parts, a consequence of this is that if P_0 and $P_{\ell-1}$ are any two parts, and X_0 is any subset of P_0, then there are at least

$$(\kappa^{\ell-1}(P_0, P_{\ell-1})|X_0|/|P_0| - \gamma)n^\ell p^{\ell-1}/k^2 \tag{5.16}$$

walks of length $\ell - 1$ from X_0 to $P_{\ell-1}$.

Let us call a walk of length $\ell - 1$ in G_n *bad* if there are at least $Mn^{\ell-2}p^{\ell-1}$ walks in G_n with the same endpoints, where M is a constant to be chosen in a moment, depending on ε but not on η; otherwise, the walk is *good*. Each bad walk may be extended to at least $Mn^{\ell-2}p^{\ell-1}$ homomorphic images of $C_{2\ell-2}$. By assumption, $t_p(C_{2\ell-2}, G_n)$ is bounded, so it follows that there are $O(n^\ell p^{\ell-1}/M)$ bad walks. In particular, choosing the constant M large enough, we may assume that there are at most $\varepsilon^9 n^\ell p^{\ell-1}/3$ bad walks.

Suppose for a contradiction that (5.13) does not hold, i.e. the set U of underconnected pairs of vertices has size at least εn^2. Our first aim is to select a pair (P, P') of parts of our partition such that there are many underconnected pairs (u, v) in $P \times P'$, but not too many bad walks start in P. Since $|U| \ge \varepsilon n^2$ by assumption, there are at least $\varepsilon k/2$ parts P with

$$|U \cap (P \times V)| \ge \varepsilon n^2/(2k). \tag{5.17}$$

On the other hand, there are at most $\varepsilon k/3$ parts P with the property that more than $\varepsilon^8 n^\ell p^{\ell-1}/k$ bad walks start in P (otherwise there would be too many bad walks). Hence there exists a part P for which (5.17) holds, with at most $\varepsilon^8 n^\ell p^{\ell-1}/k$ bad walks starting in P. Fix such a P. From (5.17) and averaging, there is a part P' such that

$$|U \cap (P \times P')| \ge \varepsilon n^2/(2k^2) = \varepsilon|P||P'|/2. \tag{5.18}$$

From now on, fix such a P'.

Let us say that a pair (u, P'') with $u \in P$ and P'' a part of our partition is *deficient* if there are fewer than $(\kappa^{\ell-1}(P, P'') - \sqrt{\gamma})n^{\ell-1}p^{\ell-1}/k$ walks of length $\ell - 1$ from u to P'', where γ is as in (5.16). For a given P'', at most $\sqrt{\gamma}n/k$ vertices $u \in P$ form a deficient pair with P'': otherwise, the set X_0 of such vertices would have more than $\gamma n^\ell p^{\ell-1}/k^2$ fewer walks to P'' than it should have, contradicting (5.16). Hence, there are at most $\sqrt{\gamma}n$ deficient pairs. Let $D \subset P$ be the set of vertices u in more than $\gamma^{1/4}k$ deficient pairs. Then $|D| \le \sqrt{\gamma}n/(\gamma^{1/4}k) = \gamma^{1/4}|P|$.

Let us say that a pair (u, P'') with $u \in P$ and P'' a part of our partition is *compromised* if there are more than $\varepsilon^3 n^{\ell-1} p^{\ell-1}/k$ bad walks from u to P''. Since at most $\varepsilon^8 n^\ell p^{\ell-1}/k$ bad walks start in P, there are at most $\varepsilon^5 n$ compromised pairs. Let C be the set of $u \in P$ in more than $\varepsilon^3 k$ compromised pairs; then $|C| \le \varepsilon^2 n/k = \varepsilon^2 |P|$.

Let $S \subset P$ be the set of vertices u for which there are at least $\varepsilon|P'|/4$ vertices $v \in P'$ with $(u,v) \in U$. By (5.18) we have

$$\varepsilon|P||P'|/2 \leq |U \cap (P \times P')| \leq |S||P'| + \varepsilon|P||P'|/4,$$

so $|S| \geq \varepsilon|P|/4 > (\gamma^{1/4} + \varepsilon^2)|P|$. Thus $|S| > |D| + |C|$, and there is some u in $S \setminus (D \cup C)$. Fix such a u for the rest of the proof, and let U_u denote the set of $v \in P'$ for which (u,v) is underconnected.

At this point we have chosen a vertex $u \in P$, a part P', and a set $U_u \subset P'$ with the following properties:

(i) for each $v \in U_u$, there are at most $(\kappa^\ell(u,v) - \varepsilon)n^{\ell-1}p^\ell = (\kappa^\ell(P,P') - \varepsilon)n^{\ell-1}p^\ell$ walks of length ℓ from u to v,

(ii) $|U_u| \geq \varepsilon|P'|/4$,

(iii) there are at most $\gamma^{1/4}k \leq \varepsilon^3 k$ deficient pairs (u, P''),

(iv) there are at most $\varepsilon^3 k$ compromised pairs (u, P'').

From (i) and (ii) above, there are at least $m = \varepsilon^2 n^\ell p^\ell/(4k)$ 'missing walks' from u to U_u: the number of walks of length ℓ from u to U_u falls short of the expected number in $G_p(n,\kappa)$ by at least m. Let P'' be any part of our partition. By a u-U_u walk via P'' we mean a walk of length ℓ from u to U_u whose second last vertex lies in P''; the expected number of such walks is $N_{P''} = \kappa^{\ell-1}(P,P'')\kappa(P'',P')|U_u|n^{\ell-1}p^\ell/k$. Note that $\sum_{P''} N_{P''}$ is simply the expected number of walks from u to U_u. Let $m_{P''}$ be the number of 'missing walks via P''', i.e. the difference between $N_{P''}$ and the number of u-U_u walks via P'', or zero if there are at least $N_{P''}$ such walks. The total number of missing walks is at most the sum of the numbers $m_{P''}$, so

$$\sum_{P''} m_{P''} \geq m \geq \varepsilon^2 n^\ell p^\ell/(4k).$$

Let us say that P'' is *useful* if $m_{P''} \geq \varepsilon^2 n^\ell p^\ell/(8k^2)$, so the contribution to the sum above from non-useful parts P'' is at most half the right hand side. Recalling that we have normalized so that κ is bounded by 1, and that $\varepsilon < 1/40$, for each P'' we have $m_{P''} \leq N_{P''} \leq n^\ell p^\ell/k^2$; it follows that there are at least $\varepsilon^2 k/8 \geq 5\varepsilon^3 k$ useful parts P''.

Using (iii) and (iv) above, it follows that there is a part P'' which is useful, but neither deficient nor compromised. Fix such a part P''.

Recall that a walk of length $\ell - 1$ from u to $w \in P''$ is *good* if it is not bad, i.e. if

$$w_{\ell-1}(u,w) \leq N = Mn^{\ell-2}p^{\ell-1}. \tag{5.19}$$

Since $\gamma^{1/4} \leq \varepsilon^3$, and P'' is neither deficient nor compromised, there are at least

$$(\kappa^{\ell-1}(P,P'') - 2\varepsilon^3)n^{\ell-1}p^{\ell-1}/k$$

good walks from u to P''. On the other hand, there are many missing walks via P''. With this setup, we are finally ready to apply the key idea of Chung and Graham [17], which is to partition the set P'' into subsets according to the approximate number

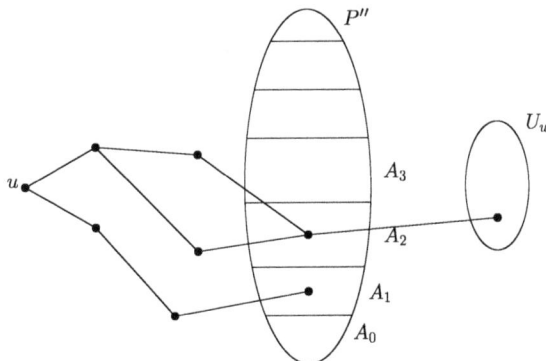

Figure 1: The set P'' is subdivided into sets A_i, with $w_{\ell-1}(u,v) = i$ for each $v \in A_i$. Each edge from A_i to U_u contributes i walks from u to U_u via P''.

of walks from u to the relevant vertex, and then use regularity to show that there are about the right number of walks from U_u to each such subset. In fact, there is a slick way of doing this.

For $i \geq 0$, let A_i be the set of vertices $v \in P''$ with $w_{\ell-1}(u,v) = i$; see Figure 1. Also, let $A_i^+ = \bigcup_{j \geq i} A_j$. Then,

$$w_{\ell-1}(u, P'') = \sum_{i \geq 0} i|A_i| = \sum_{i \geq 1} |A_i^+|.$$

More importantly, $\sum_{i=1}^N |A_i^+|$ is at least the number of *good* walks from u to P'', so

$$\sum_{i=1}^N |A_i^+| \geq (\kappa^{\ell-1}(P, P'') - 2\varepsilon^3) n^{\ell-1} p^{\ell-1}/k. \tag{5.20}$$

Since (P'', P') is (η, p)-regular with (normalized) density $\kappa(P'', P') \leq 1$, if $A \subset P''$ and $B \subset P'$ then $e(A, B) \geq p\kappa(P'', P')|A||B| - \eta p(n/k)^2$ (this is trivially true if one of A or B has size less than $\eta n/k$). Since each edge from U_u to A_i forms the final edge of exactly i walks from u to U_u, the number of walks from u to U_u via P'' is given by

$$\sum_{i \geq 1} ie(A_i, U_u) \geq \sum_{i=1}^N e(A_i^+, U_u)$$
$$\geq \sum_{i=1}^N p\kappa(P'', P')|A_i^+||U_u| - \eta p(n/k)^2$$
$$\geq (\kappa^{\ell-1}(P, P'') - 2\varepsilon^3)\kappa(P'', P')|U_u|n^{\ell-1}p^\ell/k - \eta N p(n/k)^2,$$

where we used (5.20) in the last step. The main term is simply the expected number of walks from u to U_u via P'', so the conclusion is that there are at most

$$2\varepsilon^3 \kappa(P'', P')|U_u|n^{\ell-1}p^\ell/k + \eta N p(n/k)^2 \tag{5.21}$$

missing walks from u to U_u via P''. The two terms above may be bounded above by $2\varepsilon^3 n^\ell p^\ell/k^2$ and, recalling (5.19), $\eta M n^\ell p^\ell/k^2$, respectively. Choosing $\eta \le \varepsilon^3/M$ we thus have at most $3\varepsilon^3 n^\ell p^\ell/k^2$ missing walks via P'', i.e. $m_{P''} \le 3\varepsilon^3 n^\ell p^\ell/k^2$, which contradicts the fact that P'' is useful. This contradiction completes the proof. □

Note that the argument above does not extend to $\ell = 2$, and not only because C_2 makes no sense. The problem is that we cannot define N as in (5.19) (this quantity is now $o(1)$), but must take $N = 1$ instead, and then the second term in (5.21) is too large.

The proof of Theorem 5.14 actually gives rather more with almost no extra work.

Theorem 5.15 *Let $C > 0$ and $\ell \ge 3$ be fixed, and let $p = p(n)$ be any function of n. Let (G_n) be a sequence of graphs with $\sup_n t_p(F, G_n) < \infty$ for each $F \in \mathcal{T} \cup \mathcal{F}_{\ge \ell} \cup \{C_{2\ell-2}\}$, and suppose that $d_{\mathrm{cut}}(G_n, \kappa) \to 0$ for some bounded kernel κ. Then $t_p(F, G_n) \to s(F, \kappa)$ for each $F \in \mathcal{T} \cup \mathcal{F}_{\ge \ell}$.*

Proof The conclusion for $F \in \mathcal{T}$ follows from Corollary 5.11.

Fix $F \in \mathcal{F}_{\ge \ell}$ and $\varepsilon > 0$, and let L be the length of the longest induced path in F. Noting that for $t > \ell$ we have $C_{2t-2} \in \mathcal{F}_{\ge \ell}$, the hypotheses of Theorem 5.14 are satisfied with ℓ replaced by any t in the range $\ell \le t \le L$. The proof of that result thus shows that if η is chosen small enough, then when we take an (η, p)-regular partition of G_n with associated kernel κ', almost all pairs (v, w) of vertices are joined by almost the 'right' number of walks of each length t, $\ell \le t \le L$. More precisely, writing κ for κ' as in the proof of Theorem 5.14, and writing U_t for the set of pairs (v, w) with $w_t(v, w) \le (\kappa^t(v, w) - \varepsilon)n^{t-1}p^t$ and W_t for the set of pairs with $w_t(v, w) \ge (\kappa^t(v, w) + \sqrt{\varepsilon})n^{t-1}p^t$, the proof of Theorem 5.14 shows that $|U_t| \le \varepsilon n^2$ for $\ell \le t \le L$, and (hence) that $|W_t| \le 3\sqrt{\varepsilon}n^2$ for each t in this range. Using the analogue of (5.15) in which each term $w_\ell(\cdot, \cdot)$ is replaced by an appropriate term $w_t(\cdot, \cdot)$, as before we can use the Cauchy–Schwarz inequality to show that the contribution to $t_p(F, G_n)$ from terms with some pair (v_i, v_j) in the small set $\bigcup_t U_t \cup W_t$ is small (of order $\varepsilon^{1/4}$), and it follows as before that if η is small enough, then $|t_p(F, G_n) - s(F, \kappa)|$ is bounded by some function of F and ε, giving the result. □

Let us note for later reference that, in one way, the assumptions of Theorems 5.14 and 5.15 are weaker than they may first appear. Let F be a loopless multigraph with vertex set $u_1, u_2, \ldots, u_k$, and let $F_\ell \in \mathcal{F}_\ell$ be obtained by subdividing each edge of F exactly $\ell - 1$ times. Then (5.15) may be rewritten as

$$\hom(F_\ell, G_n) = n^k \mathbb{E}\left(\prod_{u_i u_j \in E(F)} w_\ell(v_i, v_j)\right),$$

where the expectation is over the uniform choice of $(v_1, v_2, \ldots, v_k) \in V(G_n)^k$. Applying Hölder's inequality, $\mathbb{E}(\prod_{i=1}^r X_i) \le (\prod \mathbb{E}(|X_i|^r))^{1/r}$, with $r = e(F)$, it follows that

$$\hom(F_\ell, G_n)^r \leq n^{kr} \prod_{u_i u_j \in E(F)} \mathbb{E}(w_\ell(v_i, v_j)^r) = n^{kr}\mathbb{E}\big(w_\ell(v_1,v_2)^r\big)^r$$

$$= n^{kr-2r}\hom(H_{r,\ell}, G_n)^r, \quad (5.22)$$

where $H_{r,\ell} \in \mathcal{F}_\ell$ is the 'theta graph' consisting of r internally vertex disjoint paths of length ℓ joining the same pair of vertices. The normalizing factors work out correctly, so we have

$$t_p(F_\ell, G_n) \leq t_p(H_{r,\ell}, G_n). \quad (5.23)$$

Hence, the condition that $t_p(F, G_n)$ remain bounded for every $F \in \mathcal{F}_\ell$ is equivalent to the condition that $t_p(F, G_n)$ is bounded for $F = H_{r,\ell}$, $r = 1, 2, \ldots$.

Arguing similarly, for any $F \in \mathcal{F}_{\geq \ell}$ we may bound $t_p(F, G_n)$ in terms of the quantities $t_p(H_{r,\ell'}, G_n)$, where ℓ' ranges over the lengths of the paths making up F. Hence, to show that $t_p(F, G_n)$ is bounded for all $F \in \mathcal{F}_{\geq \ell}$, it suffices to prove the same condition for the graphs $H_{r,\ell'}$, $r \geq 1$, $\ell' \geq \ell$. Note that these latter conditions are simply moment conditions on the numbers of walks of various lengths joining a random pair of vertices of G_n.

In the case where the limiting kernel κ is of finite type, Theorem 5.15 may be seen as a form of counting lemma. In this case, it is easy to strengthen the result to count homomorphisms from F into G_n with each vertex mapped to a specified part of the partition of G_n corresponding to the finite type kernel κ, obtaining a result similar in form to Lemma 5.10. Such a (strengthened) finite type case of Theorem 5.14 or Theorem 5.15 is very much easier to prove than the general case: there is no need to apply Szemerédi's Lemma, and the proof of the result of Chung and Graham [17] for the uniform case goes through without much modification. One might hope that, using Szemerédi's Lemma, the full generality of Theorem 5.15 would follow easily from the finite type case, but this is not true. The problem is that our assumptions are inescapably global: we assume, for instance, that the number of copies of $C_{2\ell-2}$ in G_n is bounded by a multiple of the expected number of copies. When we take an (ε, p)-regular partition, this gives no useful information about the number of copies of $C_{2\ell-2}$ in each regular pair: we have a bound that is of the form $Mk^{2\ell-2}$ times the expected number of copies, where k is the number of parts. To apply the finite type case, we would need a bound independent of k. For this reason there seems to be no easy way around the work in the proof of Theorem 5.14.

Theorem 5.15 may be seen as some progress towards a proof of some form of Conjecture 5.6. More precisely, it is almost an answer to Question 5.9: the only problem is that for Theorem 5.15 we work with t_p rather than s_p. We shall return to this in detail in a moment. However, even ignoring this, Theorem 5.15 is a little disappointing in some ways. Let $\mathcal{A} = \mathcal{T} \cup \mathcal{F}_{\geq \ell}$. Assuming boundedness of $t_p(F, G_n)$ for $F \in \mathcal{A} \cup \{C_{2\ell-2}\}$, we obtain convergence of the counts $t_p(F, G_n)$ for $F \in \mathcal{A}$. The extra assumption for $F = C_{2\ell-2}$ is somehow annoying. This is perhaps clearest if we consider the range where p is fairly large, say $n^{-o(1)}$. In this case $s_p \sim t_p$, and it makes sense to assume boundedness of all counts $s_p(F, G_n)$. However, since C_2 does not make sense, the smallest value of ℓ for which we can apply Theorem 5.15 is $\ell = 3$, and we obtain convergence of the counts $s_p(F, G_n)$ for $F \in \mathcal{F}_{\geq 3}$. In comparison,

Theorem 3.20 shows that with the counts s_p bounded, and $s_p(C_4, G_n) \to 1$, which should roughly correspond to convergence to the uniform kernel $\kappa = 1$, we obtain $s_p(F, G_n) \to s(F, \kappa) = 1$ for all $F \in \mathcal{F}_{\geq 2}$, rather than just for $F \in \mathcal{F}_{\geq 3}$.

In fact, Theorem 3.20 gives much more: it gives convergence for all F with girth at least 4. Chung and Graham [17] asked whether an analogous result holds for sparse graphs under the appropriate assumptions (what they call 'ℓ-quasi randomness', which corresponds roughly to the assumptions of Theorem 5.14 with κ constant), with girth at least 4 replaced by girth at least 2ℓ. In our language, they asked whether (when $\kappa = 1$) the conclusion of Theorem 5.14 can be extended to all F with girth at least 2ℓ. Unfortunately, the answer is no for a trivial reason, namely that there are graphs F with arbitrarily large girth and arbitrarily large average degree. Taking $p = n^{-\alpha}$ for some $0 < \alpha < 1$, and d large enough, for any graph F with average degree d the expected number of copies of F in $G_n = G(n,p)$ is $o(1)$, so the normalizing constant in the definition of $t_p(F, G_n)$ is $o(1)$. Since $\hom(F, G_n)$ is an integer, we cannot have $t_p(F, G_n) \to 1$ in this case.

5.4 Embeddings or homomorphisms?

In this subsection we return to the use of t_p rather than s_p in Theorems 5.14 and 5.15. Although this simplifies the proof, it is unsatisfactory for a reason we shall now explain. We start by discussing the analogous problem with the corresponding result of Chung and Graham [17], their Theorem 8. We shall use the following fact, proved by Blakley and Roy [2] in a slightly more general form in the context of symmetric matrices.

Theorem 5.16 *Let G be a graph with n vertices and average degree d. Then G contains at least nd^ℓ walks of length ℓ.* □

Recall that we write $w_t(u, v)$ for the number of walks of length t from u to v. Chung and Graham [17] impose the condition that $w_{\ell-1}(u, v) < c_0 p^{\ell-1} n^{\ell-2}$ holds for *every* pair of vertices u, v, where c_0 is a constant: they call this *condition $U(\ell)$*. In other words, the number of walks from u to v is at most a constant times what it should be. Normalizing so that G_n contains exactly $pn^2/2$ edges, Chung and Graham note that $U(\ell)$ can only hold if $p = \Omega(n^{-1+1/(\ell-1)})$: otherwise, the expected number of walks of length $\ell - 1$ from a random u to a random v is much less than 1, so $w_{\ell-1}(u, v)$ must sometimes be much larger than its expectation.

In fact, $U(\ell)$ cannot hold unless p is quite a bit larger, but for the 'wrong' reason: taking ℓ odd for simplicity, let $\ell = 2k+1$. Considering walks of length $\ell - 1$ formed by tracing a walk of length k forwards and then backwards, we see that if G_n has $pn^2/2$ edges, then

$$\sum_v w_{\ell-1}(v, v) \geq \hom(P_k, G_n) \geq n(np)^k, \tag{5.24}$$

where the second inequality is Theorem 5.16. Thus there is some v with $w_{\ell-1}(v,v) \geq (np)^k$, and it follows that $U(\ell)$ can only hold if $p = \Omega(n^{-1+2/(\ell-1)})$, so Theorem 8 of [17] can only be applied for p in this range. Note that this is an essential problem: this result counts homomorphisms (Chung and Graham use the notation #{$H \subset G$}

for hom(H, G)), and the bound on $w_{\ell-1}(u, v)$ is definitely used with $u = v$. Indeed, as we shall see, the conclusion fails if $p = o(n^{-1+2/(\ell-1)})$.

Turning to Theorem 5.14, the condition that $t_p(C_{2\ell-2}, G_n)$ remain bounded corresponds roughly to the condition $U(\ell)$: indeed, the former says exactly that

$$\sum_{u,v} w_{\ell-1}(u, v)^2 = O(n^{2\ell-2} p^{2\ell-2}), \tag{5.25}$$

which follows immediately from $U(\ell)$. It turns out that the problem described above does not arise with (5.25) – in this second moment (rather than uniform) condition, the few pairs with $u = v$ matter less. Indeed, it is easy to check that in $G(n, p)$, for example, (5.25) holds as long as $p = \Omega(n^{-1+1/(\ell-1)})$. (The expected number of homomorphisms from $C_{2\ell-2}$ whose image is a tree with k edges is $O(n(np)^k) = O(n(np)^{\ell-1})$, and the expected number whose image is a graph with k vertices containing a cycle is $O(n^k p^k) = O((np)^{2\ell-2})$.) However, the same problem arises in a different place.

As before, let $H_{k,\ell} \in \mathcal{F}_\ell$ be the 'theta graph' formed by k paths of length ℓ joining the same pair (s, t) of vertices, with the paths internally vertex disjoint. Suppose that ℓ is even. Writing $w_t(v) = w_t(v, V(G_n))$ for the number of walks of length t in G_n starting at v, normalizing still so that $e(G_n) = pn^2/2$, and considering homomorphisms from $H_{k,\ell}$ to G_n mapping s and t to a common vertex v, we have

$$\hom(H_{k,\ell}, G_n) \geq \sum_v w_{\ell/2}(v)^k \geq n \left(\frac{1}{n} \sum_v w_{\ell/2}(v) \right)^k \geq n(np)^{k\ell/2},$$

where the second inequality is from convexity and the last from Theorem 5.16. Since $|H_{k,\ell}| = 2 + k(\ell - 1)$ and $e(H_{k,\ell}) = k\ell$, it follows that $t_p(H_{k,\ell}, G_n) \geq n^{k-1}(np)^{-k\ell/2}$. Suppose that $p \leq n^{-1+2/\ell-\varepsilon}$ for some $\varepsilon > 0$. Then taking k large enough we see that $t_p(H_{k,\ell}, G_n) \to \infty$, so neither the assumptions nor the conclusion of Theorem 5.14 can hold. When ε is small, this value of p is much larger than that above which the number of subgraphs of $G(n, p)$ isomorphic to $H_{k,\ell}$ is well behaved.

The calculations above illustrate the problem with working with t_p: we count certain trees as copies of $H_{k,\ell}$, for example, and the number of these trees exceeds the number of embeddings of $H_{k,\ell}$ in a wide range of densities in which Theorem 5.14 might otherwise apply. For this reason, if we could replace t_p by s_p throughout the statement of the theorem, we would obtain a much stronger and more satisfactory result: not only would it count embeddings, which is what we are really interested in, but it would apply to a much larger family of graphs, for example, to random graphs with much lower densities. Unfortunately, the proof breaks down in various places if we simply replace t_p by s_p. However, the next result is a major step in this direction.

Given vertices v, w of a graph G_n, suppressing the dependence on G_n, let us write $p_\ell(v, w)$ for the number of *paths* of length ℓ from v to w, so $p_\ell(v, w) \leq w_\ell(v, w)$.

Theorem 5.17 *Let $C > 0$ and $\ell \geq 3$ be fixed, and let $p = p(n)$ be any function of n. Let (G_n) be a sequence of graphs satisfying the following three conditions:*

$$\sup_n s_p(F, G_n) < \infty \text{ for each } F \in \mathcal{T}, \tag{5.26}$$

$$\sum_u \sum_{v \neq u} p_{\ell-1}(u,v)^2 = O(n^{2\ell-2} p^{2\ell-2}), \tag{5.27}$$

and

$$\sum_u \sum_{v \neq u} p_\ell(u,v)^k = O(n^{2+k(\ell-1)} p^{k\ell}), \tag{5.28}$$

for each fixed $k \geq 1$. Suppose also that $d_{\mathrm{cut}}(G_n, \kappa) \to 0$ for some kernel $\kappa : [0,1]^2 \to [0,C]$. Then $s_p(F, G_n) \to s(F, \kappa)$ for each $F \in \mathcal{F}_\ell$.

Before turning to the proof of this result, let us make some remarks on the conditions above. Firstly, in (5.26) it makes no difference whether we write s_p or t_p, by Lemma 5.12.

Condition (5.28) is almost the same as the condition $s_p(H_{k,\ell}, G_n) = O(1)$. Indeed, $\mathrm{emb}(H_{k,\ell}, G_n)$ is simply the sum over distinct u and v of the number of k-tuples of *internally vertex disjoint* paths from u to v, so (5.28), which bounds the same sum without the restriction to disjoint paths, is formally stronger than $s_p(H_{k,\ell}, G_n) = O(1)$. Since there are (typically) many paths from u to v in the range of p for which (5.27) may hold, it seems very likely that, assuming the other conditions of Theorem 5.17, $s_p(H_{k,\ell}, G_n) = O(1)$ implies (5.28), so (5.28) could be replaced by this more pleasant condition. However, we do not have a proof of this.

Similarly, condition (5.27) is closely related to $s_p(C_{2\ell-2}, G_n) = O(1)$, and could perhaps be replaced by this weaker condition. This is less clear, however, as Theorem 5.17 can be applied for p small enough that the typical number of paths of length $\ell - 1$ between a given pair of vertices is $O(1)$.

Instead of (5.27) we can always impose the stronger condition $t_p(C_{2\ell-2}, G_n) = O(1)$; these conditions are probably equivalent in the present setting. The corresponding statement for (5.28) and the stronger assumption $t_p(H_{k,\ell}, G_n) = O(1)$ is not true; see the discussion of the behaviour of $t_p(H_{k,\ell}, G_n)$ in the paragraphs preceding Theorem 5.17.

Finally, let us note that (5.28) gives us control over $s_p(F_\ell, G_n)$ for all $F_\ell \in \mathcal{F}_\ell$, not just for $F_\ell = H_{k,\ell}$. Let F_ℓ be obtained by subdividing a graph F with vertex set $u_1, u_2, \ldots, u_k$. Then

$$\mathrm{emb}(F_\ell, G_n) \leq \sum_{v_1, v_2, \ldots, v_k} \prod_{u_i u_j \in E(F)} p_\ell(v_i, v_j),$$

where the sum is over all $n_{(k)}$ k-tuples of distinct vertices of G_n. Applying Hölder's inequality as in the proof (5.22) of (5.23), but in a probability space with $n_{(k)}$ elements rather than n^k, we find that

$$\mathrm{emb}(F_\ell, G_n) \leq n_{(k)} \mathbb{E}(p_\ell(v_1, v_2)^{e(F)}),$$

where the expectation is over the choice of a random pair (v_1, v_2) of *distinct* vertices of G_n. Condition (5.28) bounds the final expectation; as usual the normalizing factors work out, and we see that if (5.28) holds for every k then $s_p(F_\ell, G_n) = O(1)$ for every $F_\ell \in \mathcal{F}_\ell$.

Outline proof of Theorem 5.17 Since the proof is a relatively simple modification of that of Theorem 5.14, we shall give only an outline, concentrating on the differences.

The first change we make is that we work with paths rather than walks, replacing the quantities $w_t(u,v)$, $t = \ell-1, \ell$, appearing in the proof of Theorem 5.14 with the corresponding quantities $p_t(u,v)$. By Lemma 5.12, all but a vanishing fraction of the walks in G_n of a given length are paths, so (5.13), for example, implies the same statement with $w_\ell(v,w)$ replaced by $p_\ell(v,w)$. Of course, (5.13) was proved using the assumption $t_p(C_{2\ell-2}, G_n) = O(1)$, whereas we now have the weaker assumption (5.27). However, following through the proof it is easy to see that if we count paths instead of walks, then (5.27) suffices. (The key point is that (5.27) suffices to bound the number of *bad paths*, i.e. paths between endpoints u, v with $p_{\ell-1}(u,v) > Mn^{\ell-2}p^{\ell-1}$.)

Let us fix (a small) $\varepsilon > 0$ and a graph $F_\ell \in \mathcal{F}_\ell$. We also fix an integer N to be chosen later, depending only on ε and F_ℓ. Finally, let η be a small positive constant depending on ε, F_ℓ and N. For reasons that will become clear later, we first partition $V(G_n)$ into N almost equal parts $Q_1, \ldots, Q_N$. Then we take an (η, p)-regular partition (P_i) with each P_i contained in some Q_j. For the moment we ignore the partition (Q_i).

As before, passing to a subsequence we assume that the densities $d_p(P_i, P_j)$ converge to a finite-type kernel κ. Let $S \subset V \times V$ be the set of pairs of vertices joined by the 'wrong' number of paths of length ℓ:

$$S = \{(v,w) : v \neq w, \ |p_\ell(v,w) - \kappa^\ell(v,w)n^{\ell-1}p^\ell| \geq \varepsilon n^{\ell-1}p^\ell\}.$$

If η is chosen small enough then the proofs of (5.13) and (5.14) carry through, counting paths instead of walks, and (replacing ε by $\varepsilon^2/10$), the equivalents of (5.13) and (5.14) imply that

$$|S| \leq \varepsilon n(n-1). \tag{5.29}$$

We proceed from here to our bound on $s_p(F_\ell, G_n)$ in two steps. First we count something that is not quite an embedding of F_ℓ.

Let F_ℓ be obtained from the loopless multigraph F by subdividing each edge $\ell - 1$ times, and let $u_1, \ldots, u_k$ be the vertices of F, which we also regard as vertices of F_ℓ. By a *semiembedding* of F_ℓ into G_n we mean a homomorphism from F_ℓ into G_n that maps the vertices $u_1, \ldots, u_k$ to distinct vertices of G_n, and each of the $e(F)$ u_i–u_j paths of length ℓ that make up the graph F_ℓ into a *path* in G_n. Clearly, every embedding is a semiembedding; the only additional condition on an embedding is that the paths in G_n are internally vertex disjoint.

Let $\mathrm{emb}^+(F_\ell, G_n) \geq \mathrm{emb}(F_\ell, G_n)$ denote the number of semiembeddings of F_ℓ into G_n. Then, from the definition of a semiembedding, we have

$$\mathrm{emb}^+(F_\ell, G_n) = \sum_{v_1,\ldots,v_k} \prod_{u_i u_j \in E(F)} p_\ell(v_i, v_j), \tag{5.30}$$

where the sum is over all $n_{(k)}$ sequences $(v_1, \ldots, v_k)$ of distinct vertices of G_n and, as usual, any multiple edges in F give rise to multiple factors in the product.

As before, we can rewrite the formula above as an expectation over a random choice of $(v_1, \ldots, v_k)$. Normalizing correctly for a change, let X_{ij} be the random variable $p_\ell(v_i, v_j)/(n^{\ell-1}p^\ell)$, so

$$s_p^+(F_\ell, G_n) = \frac{\mathrm{emb}^+(F_\ell, G_n)}{n_{(|F_\ell|)}p^{e(F_\ell)}} \sim \frac{\mathrm{emb}^+(F_\ell, G_n)}{n_{(|F|)}n^{|F_\ell|-|F|}p^{e(F_\ell)}} = \mathbb{E}\left(\prod_{u_i u_j \in E(F)} X_{ij}\right).$$

Equation (5.29) says, roughly speaking, that each X_{ij} is with high probability close to 'what it should be', which is a random variable depending on κ, the kernel corresponding to the partition $(P_1, \ldots, P_k)$ of G_n. We should like to deduce that the expectation of the product is close to what it should be.

Let Z be the set of k-tuples $(v_1, \ldots, v_k)$ with the v_i distinct such that $(v_i, v_j) \in S$ for some $1 \le i < j \le k$. Regarding Z as an event in our probability space,

$$\mathbb{P}(Z) \le \binom{k}{2} \mathbb{P}((v_1, v_2) \in S) \le \varepsilon \binom{k}{2},$$

from (5.29). Hölder's inequality thus gives

$$\mathbb{E}\left(1_Z \prod_{u_i u_j \in E(F)} X_{ij}\right) \le \left(\mathbb{E}\left(1_Z^{e(F)+1}\right) \prod_{u_i u_j \in E(F)} \mathbb{E}\left(X_{ij}^{e(F)+1}\right)\right)^{1/(e(F)+1)},$$

where 1_Z is the indicator function of the event Z. Now, for each i and j, we have

$$\mathbb{E}(X_{ij}^{e(F)+1}) = \frac{1}{n(n-1)} \sum_u \sum_{v \ne u} \left(\frac{p_\ell(u,v)}{n^{\ell-1} p^\ell}\right)^{e(F)+1},$$

which is $O(1)$ by our assumption (5.28). Also, $\mathbb{E}(1_Z^{e(F)+1}) = \mathbb{E}(1_Z) = \mathbb{P}(Z) \le \varepsilon$. Hence,

$$\mathbb{E}\left(1_Z \prod_{u_i u_j \in E(F)} X_{ij}\right) = O(\varepsilon^{1/(e(F)+1)}). \tag{5.31}$$

In other words, the contribution to (5.30) from semiembeddings mapping some edge of F into a pair $(u,v) \in S$ is negligible. By definition of S, the contribution from all other semiembeddings is 'what it should be', and it follows that

$$|s_p^+(F_\ell, G_n) - s(F_\ell, \kappa)| \le O(\varepsilon^{1/(e(F)+1)}) + O(\varepsilon).$$

Since $\varepsilon > 0$ was arbitrary, we thus have $s_p^+(F_\ell, G_n) \sim s(F_\ell, \kappa)$.

In the end, of course, it is $s_p(F_\ell, G_n)$ that we wish to bound, not $s_p^+(F_\ell, G_n)$. Since $s_p(F_\ell, G_n) \le s_p^+(F_\ell, G_n)$ it remains to show that most semiembeddings are in fact embeddings, i.e. that the paths in G_n making up a typical semiembedding are internally vertex disjoint. For paths corresponding to vertex disjoint edges of F, this is quite easy, using the fact that $s_p(T, G_n)$ is bounded for each tree, which tells us that almost all pairs of paths of length ℓ are vertex disjoint. For paths corresponding to edges of F sharing a vertex, there is a similar argument. We shall not spell these arguments out as there is a third case that cannot be handled in this way, namely paths corresponding to duplicate edges in F. We must allow these, since we include, for example, $C_{2\ell}$ in F_ℓ. It is in handling these paths that our 'crude' partition (Q_i) comes in.

Let us classify paths $w_0 w_1 \ldots w_\ell$ in G_n into $N^{\ell+1}$ *types*, according to which part Q_i each w_i lies in. We say that a pair (u,v) of distinct vertices of G_n is *good* if, for all $N^{\ell-1}$ possible types of u–v path, the number of u–v paths of this type is 'close' to what it should be, i.e. within $\varepsilon |Q_1|^{\ell-1} p^\ell \sim \varepsilon n^{\ell-1} p^\ell / N^{\ell-1}$ of what it should be. As usual, 'what it should be' means the expected number in $G_p(n,\kappa)$, which depends not only on which parts P_i the vertices u and v lie in, but also on the type of path being considered. Let S' be the set of pairs (u,v), $u \neq v$, that are *bad*, i.e. not good.

Since N is fixed before η is chosen, it is not hard to see that the argument giving (5.29) (applied with $\varepsilon/N^{\ell-1}$ in place of ε) also shows that $|S'| \leq \varepsilon n(n-1)$; we omit the details. In other words, almost all pairs of vertices are joined by about the right number of paths of any given type. As before, we break down the set of embeddings of F_ℓ into G_n according to which vertices $v_1,\ldots,v_k$ of G_n the 'branch vertices' $u_1,\ldots,u_k$ are mapped to. Defining Z' analogously to Z, but using S' instead of S, the argument giving (5.31) shows that we may assume that $(v_1,\ldots,v_k) \notin Z'$, i.e. that no pair (v_i,v_j) is in S'. Counting embeddings with $v_1,\ldots,v_k$ fixed, it remains to choose $e(F)$ paths joining the appropriate pairs v_i, v_j. Let us choose these paths one by one. Since the total number of paths joining v_i to v_j is about what it should be, all we must show is that few (say at most $\varepsilon n^{\ell-1}p^\ell$) paths from v_i to v_j meet one of our at most $e(F)-1$ earlier paths. But this is now easy: we must avoid a set X of at most $(e(F)-1)(\ell-1) = O(1)$ vertices, the internal vertices of the previously chosen paths. In fact, we shall do much more, avoiding any part Q_a that meets X! This rules out at most $(\ell-1)|X|N^{\ell-2}$ of the $N^{\ell-1}$ types of v_i–v_j paths. Choosing N large enough (larger than $1/\varepsilon$), this is only a fraction $O(\varepsilon)$ of all possible types. Since $(v_i,v_j) \notin S'$, we have almost the right number of paths of each remaining type, and hence almost the right number of paths in total. This completes our outline proof of Theorem 5.17. □

Of course, there is a variant of Theorem 5.17 which is to Theorem 5.17 as Theorem 5.15 is to Theorem 5.14; we shall not state this separately.

Let us close this section by giving one simple example of a setting in which the conditions of Theorem 5.17 are satisfied. Fix $\ell \geq 3$, and suppose that our sequence (G_n) has the following two properties. Firstly, the maximal degree $\Delta(G_n)$ is not too large:

$$\Delta(G_n) \leq Mpn, \tag{5.32}$$

for some constant M. Secondly,

$$p_{\ell-1}(u,v) \leq Mn^{\ell-2}p^{\ell-1} \tag{5.33}$$

for all $u \neq v \in V(G_n)$. Condition (5.32) is called DEG in Chung and Graham [17]; condition (5.33) is related to their condition $U(\ell)$, but, as noted in the paragraph containing (5.24), is much weaker. In particular, it is easy to check that if $p = n^{-\alpha}$ with $0 < \alpha < 1$ constant, and κ is any bounded kernel, then the random graphs $G_p(n,\kappa)$ satisfy (5.32) and (5.33) with probability 1, as long as $\alpha < 1 - 1/(\ell-1)$. If (5.32) and (5.33) hold then $p_\ell(v,w) \leq M^2 n^{\ell-1}p^\ell$ for all v and w, while $s_p(T,G_n) \leq M^{e(T)}$ for any tree T, so the conditions of Theorem 5.17 are satisfied. Similarly, $p_t(v,w) \leq M^{t-\ell+2}n^{t-1}p^t$ holds for all $t \geq \ell$, so the variant of Theorem 5.17 corresponding to Theorem 5.15 applies.

It follows that conditions (5.32) and (5.33) provide an answer to Question 5.9. Indeed, Theorem 5.17 tells us that, under these conditions, if κ is a bounded kernel, then $d_{\text{cut}}(G_n,\kappa) \to 0$ implies $s_p(F,G_n) \to s(F,\kappa)$ for all $F \in \mathcal{F}_\ell$; its variant gives us $s_p(F,G_n) \to s(F,\kappa)$ for all $F \in \mathcal{F}_{\geq \ell}$. By Theorem 5.1, the counts $s(F,\kappa)$, $F \in \mathcal{F}_{\geq \ell}$, do determine the kernel (up to equivalence), so conditions (5.32) and (5.33) are 'suitable' in the sense of Question 5.9. As noted after Question 5.9, this implies the following result.

Theorem 5.18 *Fix $\ell \geq 3$, let $p = p(n)$ be any function, and let (G_n) be a sequence of graphs satisfying (5.32), (5.33) and the bounded density assumption 4.1. Then, for any bounded kernel κ, we have $d_{\mathrm{cut}}(G_n, \kappa) \to 0$ if and only if $d_{\mathrm{sub}}(G_n, \kappa) \to 0$, where d_{sub} is defined using $\mathcal{A} = \mathcal{T} \cup \mathcal{F}_{\geq \ell}$ for the set of admissible graphs.* □

In this section we discussed how to extend the subgraph (count) metric from dense to sparse graphs, noting that there are various possibilities (depending on the choice of the set $\mathcal{A}$ of admissible graphs), and conjectured that one particular extension is equivalent to the cut metric. In the next section we turn to a different metric on dense graphs, that extends much more easily to sparse graphs.

6 The partition metric

As noted in Section 2, for dense graphs there are many natural metrics that turn out to be equivalent, in the sense of generating the same topology. So far we have focussed on the cut and subgraph (or count) metrics; we now turn to the *partition metric*, introduced by Borgs, Chayes, Lovász, Sós and Vesztergombi [16]. In the dense case, it turns out to be relatively easy to show that the partition and cut metrics are equivalent; in this brief section we show that, under mild assumptions, this equivalence holds also in the sparse setting, as long as $np \to \infty$.

On the one hand, this result (Theorem 6.2, below) shows that for graphs with $\omega(n)$ edges, no new questions arise by considering the partition metric. On the other hand, it reinforces the conclusion that the cut metric remains extremely natural for sparse graphs, and gives a way of considering the cut metric from a very different point of view. There is another, very important, motivation for introducing partition metrics for sparse graphs: when we come to extremely sparse graphs, with $\Theta(n)$ edges, the cut metric turns out to make very little sense, while the partition metric (which is no longer equivalent) remains natural. This is a major topic in its own right and will be discussed in a companion paper [11].

6.1 Partition matrices and the partition metric

Turning to the formal definitions, as in the rest of the paper, let $p = p(n)$ be a normalizing function and G_n a graph with n vertices. Let $k \geq 2$ be fixed. For $n \geq k$ and $\Pi = (P_1, \ldots, P_k)$ a partition of $V(G_n)$ into k non-empty parts, let $M_\Pi(G_n) = (d_p(P_i, P_j))_{1 \leq i,j \leq k}$ be the matrix encoding the normalized densities of edges between the parts of Π (see (4.1)). Since $M_\Pi(G_n)$ is symmetric, we may think of this matrix as an element of $\mathbf{R}^{k(k+1)/2}$. Set

$$\mathcal{M}_k(G_n) = \{M_\Pi(G_n)\} \subset \mathbf{R}^{k(k+1)/2},$$

where Π runs over all balanced partitions of $V(G_n)$ into k parts, i.e. all partitions $(P_1, \ldots, P_k)$ with $|P_i| - |P_j| \leq 1$.

As usual, we assume that G_n has $O(pn^2)$ edges. For definiteness, let us assume that $e(G_n) \leq Cpn^2/2$. Since each part of a balanced partition has size at least

$n/(2k)$, the entries of any $M_\Pi(G_n) \in \mathcal{M}_k(G_n)$ are bounded by $C_k = (2k)^2 C$, say. Thus, $\mathcal{M}_k(G_n)$ is a subset of the compact space $B_k = [0, C_k]^{k(k+1)/2}$.

Let $\mathcal{C}_0(B_k)$ denote the set of non-empty compact subsets of B_k, and let d_H be the Hausdorff metric on $\mathcal{C}_0(B_k)$, defined with respect to the ℓ_∞ distance, say. Thus

$$d_H(X, Y) = \inf\{\varepsilon > 0 : X^{(\varepsilon)} \supset Y, Y^{(\varepsilon)} \supset X\},$$

where $X^{(\varepsilon)}$ denotes the ε-neighbourhood of X in the ℓ_∞ metric. Since (B_k, ℓ_∞) is compact, by standard results (see, for example, Dugundji [20, p. 253]), the space $(\mathcal{C}_0(B_k), d_H)$ is compact. To ensure that the metric we are about to define is a genuine metric, it is convenient to work with $\mathcal{C}(B_k) = \mathcal{C}_0(B_k) \cup \{\emptyset\}$, setting $d_H(\emptyset, X) = C_k$, say, for any $X \in \mathcal{C}(B_k)$, so the empty set is an isolated point in $(\mathcal{C}(B_k), d_H)$.

Let $\mathcal{C} = \prod_{k \geq 2} \mathcal{C}(B_k)$, and let $\mathcal{M} : \mathcal{F} \mapsto \mathcal{C}$ be the map defined by

$$\mathcal{M}(G_n) = (\mathcal{M}_k(G_n))_{k=2}^\infty$$

for every graph G_n on n vertices, noting that $\mathcal{M}_k(G_n)$ is empty if $k > n$. Then we may define the *partition metric* d_{part} by

$$d_{\text{part}}(G, G') = d(\mathcal{M}(G), \mathcal{M}(G')),$$

where d is any metric on $\mathcal{C}$ giving rise to the product topology. Considering the partition of an n vertex graph into n parts shows that d_{part} is a metric on the set $\mathcal{F}$ of isomorphism classes of finite graphs. Recalling that each space $(\mathcal{C}(B_k), d_H)$ is compact, the key property of the partition metric is that (G_n) is Cauchy with respect to d_{part} if and only if there are compact sets $Y_k \subset B_k$ such that $d_H(\mathcal{M}_k(G_n), Y_k) \to 0$ for each k. In particular, convergence in d_{part} is equivalent to convergence of the set of partition matrices for each fixed k. Thus we may always think of k as fixed and n as much larger than k.

In the dense case, a metric equivalent to d_{part} has been introduced independently by Borgs, Chayes, Lovász, Sós and Vesztergombi [16]; the only difference is that in [16], all partitions into k parts are considered, rather than just balanced partitions. Of course, one then needs to take care to ensure that the densities between small parts are counted with an appropriate weight when computing the distance between density matrices $\mathcal{M}_k$. Whether one takes all partitions or just balanced partitions is a matter of taste: it is very easy to see that convergence in either of the resulting metrics implies convergence in the other.

We may extend the map $\mathcal{M} : \mathcal{F} \to \mathcal{C}$, and hence d_{part}, to bounded kernels in a natural way: instead of partitioning the vertex set into k almost equal parts, we partition $[0, 1]$ into k exactly equal parts, and consider the closure of the set of 'density matrices' that may be obtained from κ using such partitions; we omit the details. Note that, as shown by Borgs, Chayes, Lovász, Sós and Vesztergombi [16, Example 4.4], the set of density matrices is not in general closed.

As for the cut metric, it is easy to check that it makes little difference whether we define d_{part} for graphs directly, or by going via kernels. (The corresponding dense result appears in [16]: the sparse case here is slightly more complicated due to the possibility of 'high-degree' vertices.)

Lemma 6.1 *Let $p = p(n)$ satisfy $p \geq 1/n$, and let (G_n) be a sequence of graphs with $e(G_n) = O(pn^2)$ and $\Delta(G_n) = o(pn^2)$. Then $d_{\text{part}}(G_n, \kappa_{G_n}) \to 0$ as $n \to \infty$.*

Proof By definition, we must show that $d_H(\mathcal{M}_k(G_n), \mathcal{M}_k(\kappa_{G_n})) \to 0$ for each $k \geq 1$. Fix k. Since $e(G_n) = O(pn^2)$, there is a constant D such that at most $n/(2k)$ vertices of G_n have degree more than Dpn. Let L denote the set of 'low-degree' vertices, with degree at most Dpn, so $|L| \geq n - n/(2k)$.

We must show that for any density matrix in $\mathcal{M}_k(G_n)$ there is a nearby matrix in $\mathcal{M}_k(\kappa_{G_n})$, and vice versa. The forward implication is trivial: a balanced partition Π of $V(G_n)$ corresponds to a partition of $[0,1]$ into sets whose sizes differ by $O(1/n) = o(1)$. Adjusting these parts slightly, making changes only in subintervals of $[0,1]$ corresponding to low-degree vertices, the entries of the corresponding density matrix change by $o(1)$.

For the reverse implication, let Π be a partition of $[0,1]$ into k parts $P_1, \ldots, P_k$, and let $M \in \mathcal{M}_k(\kappa_{G_n})$ be the corresponding density matrix, with entries m_{ij}. For $v \in V(G_n) = [n]$ and $1 \leq i \leq k$, let $p_{v,i}$ be the fraction of the subinterval of $[0,1]$ corresponding to the vertex v that lies in P_i, noting that $\sum_i p_{v,i} = 1$ for each v, and $\sum_v p_{v,i} = n/k$ for each i. Form a random partition $\Pi' = (P_1', \ldots, P_k')$ as follows: put each vertex v into a random part P_{i_v}' with $\mathbb{P}(i_v = i) = p_{v,i}$, with the choices independent for different vertices v.

It is immediate that $\mathbb{E}(|P_i'|) = n/k$ and $\text{Var}(|P_i'|) \leq n/k$. It follows that for some constant C we have
$$\forall i: ||P_i'| - n/k| \leq C\sqrt{n} \tag{6.1}$$
with probability at least 0.99. Writing $v \sim w$ if $vw \in E(G_n)$, for $1 \leq i, j \leq k$ we have
$$\mathbb{E}(e(P_i', P_j')) = \sum_{(v,w):v \sim w} \mathbb{E}(1_{i_v=i}1_{i_w=j}) = \sum_{(v,w):v \sim w} p_{v,i}p_{w,j}$$
$$= n^2 p \int_{P_i \times P_j} \kappa_{G_n}(x,y)\,dx\,dy,$$
so the expectation of $e(P_i', P_j')/(n^2p)$ is exactly m_{ij}/k^2. For edges vw, $v'w'$ of G_n, the random variables $1_{i_v=i}1_{i_w=j}$ and $1_{i_{v'}=i}1_{i_{w'}=j}$ are independent unless vw and $v'w'$ share a vertex, in which case their covariance is at most one. It follows that $\text{Var}(e(P_i', P_j'))$ is bounded by $2\hom(P_2, G_n)$; the factor 2 arises since we may put the common vertex of two incident edges into P_i or P_j. But $\hom(P_2, G_n) \leq 2e(G_n)\Delta(G_n)$, which is $o(n^4p^2)$ by assumption. Hence, for any ε, the probability that we have
$$\left| \frac{e(P_i', P_j')}{n^2 p} - \frac{m_{ij}}{k^2} \right| \leq \varepsilon \tag{6.2}$$
for every i and j with $1 \leq i, j \leq k$ is at least 0.99, provided n is large enough.

From the comments above, if n is large enough, there is a partition Π' for which both (6.1) and (6.2) hold. Starting from such a partition and moving at most $O(\sqrt{n}) = o(n)$ vertices of L (the set of low-degree vertices) between parts, we may find a balanced partition with almost the same density matrix. In other words, we may find an element of $\mathcal{M}_k(G_n)$ close to M, completing the proof. □

If $np \to \infty$, then the condition of Lemma 6.1 that $\Delta(G_n) = o(n^2p)$ holds trivially, since $\Delta(G_n) \leq n = o(n^2p)$. When np is bounded, this condition is necessary. Taking G_n to be a star, for example, every partition of $V(G_n)$ has the property that there is one part meeting all edges. But the corresponding kernel has partitions which are very far from having this property, namely those in which, roughly speaking, the central vertex of the star has been split between parts.

6.2 The relationship between the cut and partition metrics

We now turn to the main result of this section, showing the equivalence of d_{cut} and d_{part} under mild assumptions. The key idea of the proof is that one can identify the density matrix corresponding to a weakly (ε, p)-regular partition from the set of density matrices.

Theorem 6.2 *Let* $np \to \infty$, *and let* (G_n) *be a sequence of graphs with* $|G_n| = n$ *satisfying the bounded density assumption 4.1. Let* κ *be a bounded kernel. Then* $d_{\text{part}}(G_n, \kappa) \to 0$ *if and only if* $d_{\text{cut}}(G_n, \kappa) \to 0$.

Proof Suppose first that $d_{\text{cut}}(G_n, \kappa) \to 0$, i.e. that $d_{\text{cut}}(\kappa_{G_n}, \kappa) \to 0$. If κ_1 and κ_2 are any kernels with $d_{\text{cut}}(\kappa_1, \kappa_2) < d$, and $M \in \mathcal{M}_k(\kappa_1)$, then there is an $M' \in \mathcal{M}_k(\kappa_2)$ whose entries differ from those of M by at most $k^2 d$: one simply takes the corresponding partition for κ_2, after rearranging so that $\|\kappa_1 - \kappa_2\|_{\text{cut}} < d$. It follows that $d_{\text{H}}(\mathcal{M}_k(\kappa_1), \mathcal{M}_k(\kappa_2)) \leq k^2 d_{\text{cut}}(\kappa_1, \kappa_2)$. Hence, $d_{\text{H}}(\mathcal{M}_k(\kappa_{G_n}), \mathcal{M}_k(\kappa)) \to 0$. Using Lemma 6.1, it follows that $d_{\text{part}}(G_n, \kappa) \to 0$.

Now suppose that $d_{\text{part}}(G_n, \kappa) \to 0$. By the *index* $\text{ind}(M)$ of a density matrix $M = (m_{ij}) \in \mathcal{M}_k(\kappa')$ we mean simply $k^{-2} \sum m_{ij}^2$. Let $f(k, \varepsilon) \geq k$ be a function to be specified later. A k-by-k density matrix $M \in \mathcal{M}_k(\kappa')$ is *locally ε-optimal* for a kernel κ' if

$$\sup_{\ell \leq f(k,\varepsilon)} \sup_{M' \in \mathcal{M}_\ell(\kappa')} \text{ind}(M') \leq \text{ind}(M) + \varepsilon,$$

i.e. if M has almost maximal index among density matrices with not too many parts; the definition of local optimality for $M \in \mathcal{M}_k(G_n)$ is similar.

Fix $\varepsilon > 0$. Since (G_n) has bounded density, whenever n is large enough as a function of k, any density matrix in $\mathcal{M}_k(G_n)$ has index at most some constant C. It follows that there is a $K = K(C, \varepsilon)$ such that, for n large enough, every G_n has some locally optimal density matrix $M_k(n)$ of size at most K. (This statement is a key part of the proof of Szemerédi's Lemma.)

Since $d_{\text{part}}(G_n, \kappa) \to 0$, if n is large enough, there is an $M'_k(n) \in \mathcal{M}_k(\kappa)$ with all entries within $\varepsilon/(10C)$ of those of $M_k(n)$. It follows that $\text{ind}(M'_k(n)) \geq \text{ind}(M_k(n)) - \varepsilon/2$. Similarly, for n large, every $M' \in \cup_{\ell \leq f(k,\varepsilon)} \mathcal{M}_\ell(\kappa)$ has all entries within $\varepsilon/(10C)$ of some $M \in \cup_{\ell \leq f(k,\varepsilon)} \mathcal{M}_\ell(G_n)$, which implies

$$\text{ind}(M') \leq \text{ind}(M) + \varepsilon/2 \leq \text{ind}(M_k(n)) + 3\varepsilon/2 \leq \text{ind}(M'_k(n)) + 2\varepsilon,$$

using the assumption that $M_k(n)$ is locally ε-optimal for G_n for the second inequality. Thus $M'_k(n)$ is locally 2ε-optimal for κ.

Recall that a partition Π of $[0, 1]$ is *weakly (ε, p)-regular* with respect to a kernel κ' if the corresponding averaged kernel κ'/Π satisfies $\|\kappa'/\Pi - \kappa'\|_{\text{cut}} \leq \varepsilon$. The proof of Lemma 4.3 (a sparse form of the Frieze-Kannan form of Szemerédi's Lemma) shows that if (G_n) has bounded density, then there is a function $f(k, \varepsilon)$ such that, if $n \geq n_0(k, \varepsilon)$ and $M \in \mathcal{M}_k(G_n)$ is locally ε-optimal, then the corresponding partition of κ_{G_n} is weakly (ε,p)-regular; the same applies to κ. It follows that for n large, identifying each density matrix with a corresponding kernel, we have $d_{\text{cut}}(\kappa_{G_n}, M_k(n))$, $d_{\text{cut}}(M_k(n), M'_k(n))$ and $d_{\text{cut}}(M'_k(n), \kappa)$ all of order $O(\varepsilon)$. Since ε was arbitrary, it follows that $d_{\text{cut}}(\kappa_{G_n}, \kappa) \to 0$, as required. □

In the light of Corollary 4.7, Theorem 6.2 implies that a sequence (G_n) satisfying Assumption 4.1 is Cauchy with respect to d_{part} if and only if it is Cauchy with respect to d_{cut}.

The bounded density assumption in Theorem 6.2, which is trivially satisfied in the dense case $p = \Theta(1)$, is necessary in general. This can be seen by considering, for example, a graph G_n made up of n/m complete graphs of order m, with $m \sim pn = o(n)$ chosen so that G_n has $pn^2/2$ edges. By compactness, any sequence with $e(G_n) = O(pn^2)$ has a subsequence that is Cauchy with respect to d_{part} (here, in fact, the original sequence is Cauchy). However, it is easy to check that no subsequence of (G_n) is Cauchy with respect to d_{cut}.

The proof of Theorem 6.2 applies just as well to kernels as to graphs (and one can in any case approximate kernels by dense graphs), showing that $d_{\text{part}}(\kappa_n, \kappa) \to 0$ if and only if $d_{\text{cut}}(\kappa_n, \kappa) \to 0$. It follows that d_{part} induces a metric on $\mathcal{K}$, the set of kernels quotiented by equivalence, and that d_{part} and d_{cut} give rise to the same topology on $\mathcal{K}$. This was proved by Borgs, Chayes, Lovász, Sós and Vesztergombi [16] in their study of the dense case, as part of their Theorem 3.5.

7 Discussion and closing remarks

For dense graphs, with $\Theta(n^2)$ edges, the results of Borgs, Chayes, Lovász, Sós and Vesztergombi [15, 16] show that one single metric, say d_{cut}, effectively captures several natural notions of local and global similarity. Indeed, convergence in d_{cut} is equivalent to convergence in the partition metric d_{part} (a natural global notion) and to convergence in d_{sub}, i.e. convergence of all small subgraph counts, a natural local notion. These results apply to all sequences (G_n) of graphs, but if G_n has $o(n^2)$ edges then they become trivial: any such sequence is Cauchy with respect to any of the metrics, and indeed converges to the zero kernel. To make interesting statements about sparse graphs one should adapt the metrics so that, roughly speaking, given an 'edge density function' $p = p(n)$ satisfying $p \to 0$, one compares a graph G_n with $p\binom{n}{2}$ edges to the Erdős–Rényi random graph $G(n,p)$ and its inhomogeneous variants rather than to K_n. Our main aim in this paper has been to introduce such metrics, and to discuss the relationships between them. In this final section we turn to a slightly different question, that of the relationship between metrics and random graph models.

7.1 Models and metrics

In the dense case, there is a very natural correspondence between limit points of sequences converging in d_{cut}, and the inhomogeneous random graph model $G(n, \kappa)$. In general, given any metric, we can ask whether there is a corresponding random graph model: for each metric d on some class of (sparse) graphs satisfying certain restrictions, we can ask the following question.

Question 7.1 *Given a metric d, can we find a 'natural' family of random graph models with the following two properties: (i) for each model, the sequence of random graphs (G_n) generated by the model is Cauchy with respect to d with probability 1, and (ii) for any sequence (G_n) with $|G_n| = n$ that is Cauchy with respect to d, there is a model from the family such that, if we interleave (G_n) with a sequence of random graphs from the model, the resulting sequence is still Cauchy with probability 1.*

In the above question, we are implicitly assuming a coupling between the probability spaces on which the graphs (G_n) are defined. There is of course no need to do so: we can replace 'Cauchy with probability 1' with the less familiar 'Cauchy in probability', which is equivalent to convergence in probability in the completion; see Kallenberg [29, Lemma 4.6].

Although Question 7.1 is rather vague, for $d = d_{\text{cut}}$ the answer is 'yes' in the dense case, since (G_n) is Cauchy if and only if $d_{\text{cut}}(G_n, \kappa) \to 0$ for some kernel κ, while the dense inhomogeneous random graphs $G(n, \kappa)$ converge to κ in d_{cut} with probability 1. Thus our family consists of one model $G(n, \kappa)$ for each kernel κ (to be precise, for each equivalence class of kernels under the relation $\sim$ defined in Subsection 2.4).

In the sparse case we do not have an entirely satisfactory answer for any of the metrics considered in this paper. Assuming that $np \to \infty$, there is an almost completely satisfactory answer for d_{cut}: if we impose the bounded density assumption 4.1, then Corollary 4.7 and Lemma 4.10 show that the sparse inhomogeneous models $G_p(n, \kappa)$ answer Question 7.1. For d_{sub}, defined with respect to certain restricted sets of subgraphs, the results in Section 5 (in particular, Theorem 5.18) show that once again $G_p(n, \kappa)$ answers this question for suitably restricted sequences.

The extremely sparse case, where $p = \Theta(1/n)$, turns out to be even more complicated; we shall discuss this in a forthcoming paper [11].

There is an even vaguer, but perhaps more important, 'mirror image' of Question 7.1. Suppose that we have a random graph model, and we would like to test whether it is appropriate for some network in the real world. Then we would like to have a suitable metric to compare a 'typical' graph from the model with the real-world network. It is too much to hope that one metric will be appropriate in all situations; in particular, taking the simple case in which our model is $G(n, p)$ for some $p = p(n) \to 0$, the unnormalized metrics d_{cut}, d_{sub} or d_{part}, that are very suitable for dense graphs, will declare any graph with $o(n^2)$ edges to be close to the model.

In general, a random graph model (or family of models) may suggest an appropriate metric, or at least properties such a metric should have. For example, the inhomogeneous models $G_p(n, \kappa)$ and the results here suggest the sparse version of d_{cut}. Suppose, however, that we are trying to model a network with rather few edges but high 'clustering', i.e. many triangles and other small subgraphs. One possible model is a denser version of the sparse random graphs with clustering introduced by Bollobás, Janson and Riordan [9]: given, for each fixed graph F, a 'kernel' $\kappa_F : [0,1]^{|F|} \to [0, \infty)$ and a normalizing function $p_F(n)$, we choose vertex types $x_1, \ldots, x_n$ independently and uniformly at random and then, for each F, add each possible copy of F with vertex set $v_1, \ldots, v_k$, $1 \le v_1 < v_2 < \cdots < v_k \le n$, with probability $\kappa_F(x_{v_1}, \ldots, x_{v_k}) p_F(n)$.

In this model, a huge family of normalizations are possible: we can take each p_F to be any function of n bounded by 1. Of course, certain restrictions will be necessary for the model to make much sense; otherwise, for example, the copies of some F_1 added directly may be swamped by copies of F_1 arising as subgraphs of some F_2, in which case there was no point adding any copies of F_1 directly. However, there is no doubt that many different normalizations will be interesting: for example, for any $0 < a \le 4/3$, we can produce graphs with, say, $\Theta(n^{4/3})$ edges and $\Theta(n^a)$ triangles. Indeed, to do so we need only two kernels, one for edges (which we may take to generate a bipartite graph if needed), and one for triangles.

If, for some reason, we are considering graphs with, say, around $n^{4/3}$ edges and $n^{6/5}$ triangles, which is many more triangles than expected in $G(n, n^{-2/3})$, then the triangles are an important part of the structure, so in comparing two such graphs we should certainly compare the number of triangles, normalized by dividing by $n^{6/5}$. This suggests a family of metrics generalizing d_{sub}.

For each $F \in \mathcal{F}$ let $N_F = N_F(n)$ be a normalizing function satisfying $0 < N_F \le \infty$. (We allow infinity to include the possibility of totally ignoring copies of some F. In fact, $N_F = n^{|F|+1}$ will do just as well.) Then we may define a subgraph metric associated to $\mathbf{N} = (N_F)_{F \in \mathcal{F}}$ by modifying the definition of d_{sub} given in Section 3, using the normalized count $\text{emb}(F, G)/N_F(|G|)$ in place of $s_p(F, G)$. This metric will only make sense for suitably restricted families of graphs, but for such families, it will make much better sense than d_{sub}.

7.2 Closing Remarks

The main aim of this paper is to draw attention to the possibility that there is a rich theory of sparse (quasi-)random graphs waiting to be explored. The beginnings of such a theory can be found in the papers of Bollobás, Janson and Riordan [8, 9] in the very sparse case, and of Borgs, Chayes, Lovász, Sós, Szegedy and Vesztergombi [13, 14, 34] in the dense case; it would be desirable to build a theory encompassing these two extreme threads. As we have just shown, this task is unlikely to be easy: there are numerous unexpected difficulties and pitfalls, and much work has to be done even to arrive at concrete problems whose solutions would represent genuine progress in this endeavour. In this paper we have attempted to do some of this groundwork, and have identified some intriguing problems.

Our main focus has been the introduction of normalized versions of the metrics d_{cut}, d_{sub} and d_{part}, adapted to the study of graphs with $\Theta(pn^2)$ edges, where $p = p(n) \to 0$. We have shown in Section 6 that (under a mild assumption) d_{cut} and d_{part} have the same Cauchy sequences, and in Section 4 that (again under a mild assumption) these metrics have the property that any sequence (G_n) contains a subsequence converging to a kernel.

Turning to d_{sub}, things become more difficult. We have conjectured that if our p-normalized subgraph counts are suitably bounded and $p = p(n)$ is not too small then an appropriate Cauchy sequence does converge to a kernel (see Conjectures 3.3 and 3.4). Tantalizingly, we cannot even prove this convergence in just about the simplest case, when we know that the limit *has to be* a constant kernel (Conjecture 3.9).

Section 5 is devoted to the relationship between d_{cut} and d_{sub}. A sound understanding of the relationship between these two metrics, the cut and count metrics, would bring us much closer to a proper theory of sparse inhomogeneous quasi-random graphs. We have conjectured that under some natural and not too restrictive conditions, these two metrics are equivalent in the sense that if (G_n) is a sequence of graphs that are not too 'lumpy' then (G_n) converges to a kernel κ in the p-cut metric if and only if it converges to κ in the p-count metric (see Conjecture 5.6). As one of our main results, we have proved that p-cut convergence does imply p-count convergence for a restricted set of subgraph counts, under a mild assumption on the distribution of paths of certain lengths (see Theorems 5.15 and 5.17).

The case of graphs of bounded average degree turns out to be even more difficult, and will be discussed in a companion paper [11].

Acknowledgements

The first author's research was supported in part by NSF grants CCR-0225610 and DMS-0505550 and ARO grant W911NF-06-1-0076. The authors would like to thank an anonymous referee for many detailed suggestions improving the presentation of the paper.

References

[1] N. Alon, Explicit Ramsey graphs and orthonormal labelings, *Electron. J. Combin.* **1** (1994), R12.

[2] G. R. Blakley & P. Roy, A Hölder type inequality for symmetric matrices with nonnegative entries, *Proc. Amer. Math. Soc.* **16** (1965), 1244–1245.

[3] B. Bollobás, Threshold functions for small subgraphs, *Math. Proc. Cambridge Philos. Soc.* **90** (1981), 197–206.

[4] B. Bollobás, *Modern Graph Theory*, Graduate Texts in Mathematics, 184, Springer, New York (1998).

[5] B. Bollobás, *Linear Analysis*, Second edition, Cambridge University Press, Cambridge (1999).

[6] B. Bollobás, *Random Graphs*, Second edition, Cambridge University Press, Cambridge (2001).

[7] B. Bollobás, C. Borgs, J. T. Chayes & O. Riordan, Percolation on dense graph sequences, preprint, arXiv:0701346.

[8] B. Bollobás, S. Janson & O. Riordan, The phase transition in inhomogeneous random graphs, *Random Structures Algorithms* **31** (2007), 3–122.

[9] B. Bollobás, S. Janson & O. Riordan, Sparse random graphs with clustering, preprint, arXiv:0807:2040.

[10] B. Bollobás, S. Janson & O. Riordan, The cut metric, random graphs, and branching process, preprint, arXiv:0901.2091.

[11] B. Bollobás & O. Riordan, Sparse graphs: metrics and random models, preprint, arXiv:0812.2656.

[12] C. Borgs, J. T. Chayes & L. Lovász, Moments of two-variable functions and the uniqueness of graph limits, preprint, arXiv:0803.1244.

[13] C. Borgs, J. T. Chayes, L. Lovász, V. T. Sós & K. Vesztergombi, Counting graph homomorphisms, in *Topics in Discrete Mathematics* (eds. M. Klazar, J. Kratochvil, M. Loebl, J. Matousek, R. Thomas & P. Valtr), *Algorithms Combin.*, 26, Springer, Berlin (2006), pp. 315–371.

[14] C. Borgs, J. T. Chayes, L. Lovász, V. T. Sós, B. Szegedy & K. Vesztergombi, Graph limits and parameter testing, in *STOC'06: Proceedings of the 38th Annual ACM Symposium on Theory of Computing*, ACM, New York (2006), pp. 261–270.

[15] C. Borgs, J. T. Chayes, L. Lovász, V. T. Sós & K. Vesztergombi, Convergent sequences of dense graphs I: Subgraph frequencies, metric properties and testing, preprint, arXiv:0702004.

[16] C. Borgs, J. T. Chayes, L. Lovász, V. T. Sós & K. Vesztergombi, Convergent sequences of dense graphs II: Multiway cuts and statistical physics, preprint, http://www.cs.elte.hu/~lovasz/ConvRight.pdf

[17] F. Chung & R. Graham, Sparse quasi-random graphs, *Combinatorica* **22** (2002), 217–244.

[18] F. R. K. Chung, R. L. Graham & R. M. Wilson, Quasi-random graphs, *Combinatorica* **9** (1989), 345–362.

[19] P. Diaconis & S. Janson, Graph limits and exchangeable random graphs, *Rend. Mat. Appl., VII. Ser.* **28** (2008), 33–61.

[20] J. Dugundji, *Topology*, Allyn and Bacon, Inc., Boston, Mass. (1966).

[21] P. Erdős & A. Rényi, On random graphs. I, *Publ. Math. Debrecen* **6** (1959), 290–297.

[22] P. Erdős & A. Rényi, On a problem in the theory of graphs, *Magyar Tud. Akad. Mat. Kutató Int. Közl.* **7** (1962), 623–641.

[23] A. Frieze & R. Kannan, Quick approximation to matrices and applications, *Combinatorica* **19** (1999), 175–220.

[24] S. Gerke & A. Steger, The sparse regularity lemma and its applications, in *Surveys in combinatorics 2005*, London Math. Soc. Lecture Notes, 327, Cambridge University Press, Cambridge (2005), pp. 227–258.

[25] E. N. Gilbert, Random graphs, *Ann. Math. Statist.* **30** (1959), 1141–1144.

[26] S. Janson, Poisson approximation for large deviations, *Random Structures Algorithms* **1** (1990), 221–230.

[27] S. Janson, Standard representation of multivariate functions on a general probability space, preprint, arXiv:0801.0196.

[28] S. Janson, K. Oleszkiewicz & A. Ruciński, Upper tails for subgraph counts in random graphs, *Israel J. Math.* **142** (2004), 61–92.

[29] O. Kallenberg, *Foundations of Modern Probability*, Second edition, Springer, New York (2002).

[30] J. H. Kim, The Ramsey number $R(3,t)$ has order of magnitude $t^2/\log t$, *Random Structures Algorithms* **7** (1995), 173–207.

[31] Y. Kohayakawa, Szemerédi's regularity lemma for sparse graphs, in *Foundations of computational mathematics (Rio de Janeiro, 1997)*, Springer, Berlin (1997), pp. 216–230.

[32] Y. Kohayakawa & V. Rödl, Szemerédi's regularity lemma and quasi-randomness, in *Recent advances in algorithms and combinatorics*, CMS Books Math., 11, Springer, New York (2003), pp. 289–351.

[33] L. Lovász & V. T. Sós, Generalized quasirandom graphs, *J. Combin. Theory B* **98** (2008), 146–163.

[34] L. Lovász & B. Szegedy, Limits of dense graph sequences, *J. Combin. Theory B* **96** (2006), 933–957.

[35] A. Ruciński, When are small subgraphs of a random graph normally distributed?, *Probab. Theory Relat. Fields* **78** (1988), 1–10.

[36] E. Szemerédi, Regular partitions of graphs, in *Problèmes combinatoires et théorie des graphes, Colloq. Internat. CNRS, Univ. Orsay, Orsay, 1976*, 260, CNRS, Paris (1978), pp. 399–401.

[37] A. Thomason, Pseudorandom graphs, in *Random Graphs '85 (Poznań, 1985) North-Holland Math. Stud.*, 144, North-Holland, Amsterdam (1987), pp. 307–331.

[38] A. Thomason, Random graphs, strongly regular graphs and pseudorandom graphs, in *Surveys in Combinatorics 1987, London Math. Soc. Lecture Note Ser.*, 123, Cambridge University Press, Cambridge (1987), pp. 173–195.

Department of Pure Mathematics and Mathematical Statistics,
Wilberforce Road, Cambridge CB3 0WB, UK
and
Department of Mathematical Sciences, University of Memphis,
Memphis TN 38152, USA.
b.bollobas@dpmms.cam.ac.uk

Mathematical Institute, University of Oxford,
24–29 St Giles', Oxford OX1 3LB, UK.
riordan@maths.ox.ac.uk

Recent Results on Chromatic and Flow Roots of Graphs and Matroids

Gordon F. Royle

Abstract

This paper surveys recent developments in the study of the real and complex roots of the chromatic and flow polynomials of graphs and matroids.

1 Introduction

The *chromatic polynomial* of a graph G is the polynomial $P(G, q)$ that counts the number of proper q-colourings of G when q is a positive integer. It was introduced by Birkhoff [5] in 1912 in the hope that a quantitative study of the *numbers* of colourings of planar graphs might lead to an analytic proof of the 4-colour conjecture by demonstrating that $P(G, 4) \neq 0$ whenever G is planar. Although originally defined only for positive integers q, it is well known that $P(G, q)$ is a polynomial with integer coefficients (see for example, Read [41]) and although there may be no immediate combinatorial interpretation of the results, we can evaluate it at real or complex arguments and calculate its real or complex zeros. In this vein, Birkhoff and Lewis [4] showed that if G is a planar graph then $P(G, q) > 0$ for all real $q \geq 5$ and conjectured that the same result was true for real $q \geq 4$. It is perhaps embarrassing that the *very first* conjecture on the location of real chromatic roots has still not been resolved more than sixty years later:

Conjecture 1.1 (The Birkhoff-Lewis Conjecture) *If G is a planar graph then $P(G, q) > 0$ for all real $q \geq 4$.*

Despite the lack of progress on this conjecture, a considerable theory has been developed on the location of the *real chromatic roots* of graphs with many fascinating results and open problems. I will discuss recent developments in this theory in Section 7.

The first mention of the *complex* roots of a chromatic polynomial appears to be in a 1965 paper of Hall, Siry and Vanderslice [23]. This paper describes the computation of the chromatic polynomial of the dual of the truncated icosahedron and includes, without further explanation, a list of its real and complex roots. By the early 1970s the increasing availability of electronic computers allowed researchers to calculate greater numbers of complex chromatic roots, and the first results and conjectures on their location started to appear. In a beautiful paper, Sami Beraha and Joseph Kahane [2] proved that certain planar graphs have *complex* chromatic roots arbitrarily close to 4 and therefore wondered "Is the Four-Color Conjecture Almost False?"[1]

Complex chromatic roots are also of considerable interest to statistical physicists because the chromatic polynomial is a special case of the partition function of the q-state Potts model. Loosely speaking, statistical physicists study the behaviour of

---
[1] The delay between submission and publication meant that by the time this paper appeared, it was nearly 3 years after the 4-colour theorem had actually been proved.

the complex chromatic roots of various families of regularly structured graphs (such as lattices) as their size increases. The Lee-Yang theory [67, 35] of phase transitions asserts that a phase transition in the underlying physical system can only occur at a real limit point of the complex zeros of the partition function thereby motivating results identifying chromatic root-free regions or limiting curves of chromatic roots for various families of graphs. As a result there is a very substantial physics literature on techniques for computing chromatic polynomials for families of graphs, their chromatic roots, limiting curves of chromatic roots and so on. In the last decade, the introduction of techniques either directly imported from statistical physics or at least inspired by "physical thinking" has resolved several of the most significant and long standing open questions on the location of chromatic roots. In fact, much more than this is true, because the chromatic polynomial is a special case of the Tutte polynomial which is *equivalent* to the partition function of the q-state Potts model, but for further details on these connections the reader is referred to the survey article by Sokal [54].

The *flow polynomial* $F(G,q)$ of a graph is defined analogously to the chromatic polynomial, except that it counts the *nowhere-zero* $\mathbb{Z}_q$-*flows* on the graph (these are defined in Section 3), and the *flow roots* of a graph are the integer, real or complex roots of the flow polynomial. The flow polynomial is dual to the chromatic polynomial in that if G is a connected planar graph with planar dual G^* then

$$qF(G,q) = P(G^*,q), \qquad (1.1)$$

and hence for planar graphs, the theory of flow roots and chromatic roots coincides. Many of the basic results on the flow roots of general (not-necessarily planar) graphs arise in this way but there appear to be relatively few results relating *strictly* to flow roots. However Jackson [25] has pointed out that in many ways, the flow roots of general graphs behave in a qualitatively similar way to the chromatic roots of planar graphs. Understanding the extent to which this is true and the underlying reasons for this behaviour is an important goal of research in this area.

Rather than viewing chromatic and flow polynomials as dual objects defined on graphs, we can consider them as the *same* object — the *characteristic polynomial* — defined on dual classes of matroids, namely graphic matroids and cographic matroids respectively. From this broader perspective, there is then no particular reason to restrict attention just to graphic and cographic matroids and so we get the following overall goal:

> Given a class of matroids, determine bounds on the location of the characteristic roots of the matroids in this class.

These bounds can either be *absolute* bounds that apply uniformly to the matroids in a particular class, or *parameterized* bounds varying according to the value of some parameter associated with the matroid. (Sokal [54] calls these "hard" and "soft" theorems respectively.)

The topics of this paper have previously been very well surveyed by Jackson [25] and in Chapters 12–14 of Dong, Koh and Teo [16] and so the focus of the current paper will be on the progress that has occurred since these appeared.

2 Graphs and matroids

I assume that the reader is familiar with graphs and just note that in this paper graphs are undirected, but may have both loops and multiple edges. Any specialized or possibly ambiguous terminology will be introduced and/or clarified where it is used. For those completely unfamiliar with graphs, the books by Diestel [10] or Bondy and Murty [6] are just two of many possible texts.

Matroids are far less ubiquitous than graphs and so need more introduction. However in this paper matroids are used only to set the overall context for the results and to provide a unifying language rather than for any deep matroid-theoretic results. Roughly speaking, a matroid is a *combinatorial* generalization of a collection of vectors in a vector space, in that it consists of a set of elements and combinatorially defined concepts of dependence, independence etc. with properties analogous to the same concepts in vector spaces.

There are many equivalent ways to make this description precise, but for definiteness we will take a *matroid* $M = (E, r)$ to consist of a set of *elements* E together with a *rank function* $r : 2^E \to \mathbb{N} \cup \{0\}$ satisfying the following properties:

(R1) If $X \subseteq E$, then $0 \leq r(X) \leq |X|$.

(R2) If $X \subseteq Y \subseteq E$, then $r(X) \leq r(Y)$.

(R3) If $X, Y \subseteq E$, then $r(X \cap Y) + r(X \cup Y) \leq r(X) + r(Y)$.

Matroids were introduced by Whitney [63] in an attempt to axiomatize the properties of linear independence in a vector space and so a fundamental class of examples is derived from linear algebra. In particular, if A is a matrix with entries from a field $\mathbb{F}$ then we can define a matroid $M[A] = (E, r)$ where E is the set of columns of A, and for $X \subseteq E$ the rank $r(X)$ is given by the rank of the submatrix of A with X as its columns. A matroid M is called *representable* over $\mathbb{F}$ if there is a matrix A over $\mathbb{F}$ such that M is isomorphic to $M[A]$, and *binary* if it is representable over $GF(2)$. Row operations and removal of zero rows does not affect the matroid, and column permutations yield an isomorphic matroid so it is often convenient to take a matrix to be in *standard form* $A = [I_r \mid D]$ where r is the rank of A. Multiplication of a column by a non-zero scalar also does not affect $M[A]$ and so we can view the non-zero columns of A as points of the projective space $PG(r-1, \mathbb{F})$. As we are usually only interested in matroids up to isomorphism, it is often useful to view an $\mathbb{F}$-representable matroid as a multiset of points in this projective space.

If $G = (V, E)$ is a undirected graph which may have loops and multiple edges, then the *cycle matroid* $M(G) = (E, r)$ has the edge-set of G as its elements and for $X \subseteq E$

$$r(X) = |V| - k(X) \qquad (2.1)$$

where $k(X)$ is the number of connected components in the graph (V, X). A matroid is called *graphic* if it is isomorphic to the cycle matroid of some graph. If $G = (V, E)$ is a graph then a *vertex-edge incidence matrix* for G is a $\{0, \pm 1\}$-matrix $B = B(G)$ with rows indexed by V and columns by E such that for every non-loop edge $e = \{u, v\}$ we have

$$\{B_{ue}, B_{ve}\} = \{+1, -1\} \qquad (2.2)$$

and all other entries equal to zero. It is a routine exercise to show that $M[B] = M(G)$ for any vertex-incidence matrix B over any field $\mathbb{F}$ and so graphic matroids are representable over all fields.

The basic terminology of matroid theory is largely derived from either linear algebra or graph theory. Thus a set $X \subseteq E$ is *independent* if $r(X) = |X|$ and is a *basis* if $r(X) = |X| = r(E)$. A *circuit* is a dependent set X such that $X - e$ is independent for all $e \in X$. In a graphic matroid, a set of edges is independent if and only if it does not contain the edges of a cycle and the circuits are exactly the edge sets of the cycles. If a one-element subset $\{e\}$ is a circuit then e is called a *loop*, and if $\{e, f\}$ is a circuit then e and f are said to be *parallel*. A matroid is said to be *simple* if it has no loops or pairs of parallel elements. In graphic matroids, loops are indeed loops of the graph and parallel elements are multiple edges.

If $M = (E, r)$ is a matroid then the function

$$r^*(X) = |X| - r(E) + r(E - X) \qquad (2.3)$$

also satisfies (R1), (R2) and (R3) and hence is the rank function of a matroid $M^* = (E, r^*)$ which is called the *dual matroid* to M. If G is a planar graph with planar dual G^* then

$$M(G)^* = M(G^*) \qquad (2.4)$$

and so matroid duality is consistent with, and extends, the corresponding graph-theoretic notion. More generally if A is a matrix in standard form $[I \mid D]$ and we define $A^* = [-D^T \mid I]$ then

$$M[A]^* = M[A^*] \qquad (2.5)$$

and so matroid duality is also consistent with the coding-theoretic concept of dual codes. A *coloop* or *cocircuit* in a matroid is an element or set of elements that forms a loop or circuit respectively in the dual matroid (and similarly for other terms prefixed with "co").

In the class of graphs, only planar graphs have duals, but all graphic matroids have dual matroids. A matroid is called *cographic* if it is isomorphic to the dual of a graphic matroid. The circuits of a cographic matroid $M(G)^*$ are exactly the edge sets of the *bonds* of G, where a bond is a minimal set of edges whose deletion increases the number of components of G, and so $M(G)^*$ is often called the *bond matroid* of G. A matroid is both graphic and cographic if and only if it is the cycle matroid (or equivalently, bond matroid) of a planar graph.

If $M = (E, r)$ is a matroid with $e \in E$ then the matroid $M \backslash e = (E \backslash e, r')$ where r' is given by

$$r'(X) = r(X) \qquad (2.6)$$

for all $X \subseteq E \backslash e$ is the matroid obtained by *deleting* the element e, and the matroid $M/e = (E \backslash e, r'')$ where r'' is given by

$$r''(X) = r(X \cup \{e\}) - r(\{e\}) \qquad (2.7)$$

is the matroid obtained by *contracting* e. These operations generalize the usual operations of deleting and contracting an edge in a graph in that $M(G \backslash e) = M(G) \backslash e$ and $M(G/e) = M(G)/e$. Contraction and deletion are dual operations in that

$$(M \backslash e)^* = (M^*/e). \qquad (2.8)$$

A *minor* of a matroid M is any matroid obtained from M by a sequence of deletions and contractions, and a class $\mathcal{M}$ of matroids is called *minor-closed* if any minor of a matroid $M \in \mathcal{M}$ is also in $\mathcal{M}$.

The standard text for matroid theory is Oxley [40], while Oxley's expository paper [38] is a good choice for a gentle introduction to matroids.

3 The Polynomials

The *chromatic polynomial* $P(G,q)$ of a graph G is defined by the property that when q is a positive integer, the value $P(G,q)$ is the number of proper q-colourings of G. The fact that it *is* a polynomial function can be shown in a number of different ways, usually by appealing to a recurrence involving deletion and contraction of edges. A more direct approach is to express $P(G,q)$ directly as a sum of polynomials as in the following lemma.

Lemma 3.1 *If $G = (V, E)$ is a graph then*

$$P(G,q) = \sum_{X \subseteq E} (-1)^{|X|} q^{k(X)} \qquad (3.1)$$

where $k(X)$ denotes the number of connected components of the graph (V, X).

Proof If G has n vertices then there are q^n ways of assigning a colour from a palette of q colours to each vertex. Most of these assignments are improper colourings because at least one edge is monochromatic, i.e. has the same colour assigned to each end vertex. If we define B_e to be the set of colourings where e is monochromatic, then the total number of *proper* colourings is

$$P(G,q) = q^n - \left| \bigcup_{e \in E} B_e \right|. \qquad (3.2)$$

Using the inclusion-exclusion principle to evaluate the second term of (3.2) and noting that

$$\left| \bigcap_{e \in X} B_e \right| = q^{k(X)}, \qquad (3.3)$$

the result follows immediately. □

If $G = (V, E)$ is a graph then we can associate a pair of oppositely directed *arcs* $\{(e, u), (e, v)\}$ to each edge $e = \{u, v\} \in E(G)$, where (e, u) is viewed as the arc directed away from u. If $A(G)$ denotes the arc set of G then a $\mathbb{Z}_q$-flow on G is an assignment

$$\varphi : A(G) \to \mathbb{Z}_q$$

such that $\varphi(e, u) = -\varphi(e, v)$ for each edge $e = \{u, v\}$ and Kirchhoff's law holds at each vertex, i.e. for each vertex v

$$\sum_{(e,v) \in A(G)} \varphi(e, v) = 0 \qquad (3.4)$$

where all arithmetic is performed in $\mathbb{Z}_q$. A $\mathbb{Z}_q$-flow is called *nowhere-zero* if every arc is assigned a non-zero value.

Now suppose that $G = (V, E)$ is a graph with a spanning tree T. Then any assignment of values from $\mathbb{Z}_q$ to the arcs of $E - T$ can be uniquely extended to a $\mathbb{Z}_q$-flow on G. As shown in the following lemma, this implies both that the number of $\mathbb{Z}_q$-flows is a polynomial in q and that this polynomial is the same if the flows are defined over any abelian group of order q.

Lemma 3.2 *If $G = (V, E)$ is a graph then the number of nowhere-zero $\mathbb{Z}_q$-flows on G is given by*

$$F(G, q) = \sum_{X \subseteq E(G)} (-1)^{|X|} q^{|E-X|-|V|+k(E-X)} \tag{3.5}$$

Proof If G has $k(E)$ components then a spanning forest obtained by taking one spanning tree per component has $|V| - k(E)$ edges. Therefore the total number of $\mathbb{Z}_q$-flows on G is $q^{|E|-|V|+k(E)}$. If $X \subseteq E(G)$ then the number of flows that are zero on every edge in X is simply the total number of flows on the graph $(V, E - X)$ which is $q^{|E-X|-(|V|-k(E-X))}$. Therefore once again the result follows immediately by inclusion-exclusion. □

A *nowhere-zero q-flow* is an integer-valued flow taking values in $\{\pm 1, \pm 2, \ldots, \pm(q-1)\}$. Although Tutte [58] showed that the *existence* of a nowhere-zero q-flow is equivalent to the existence of a nowhere-zero $\mathbb{Z}_q$-flow, their numbers differ (see Kochol [32]).

The similarity between (3.1) and (3.5) arises from their common origin as characteristic polynomials of matroids. If M is a matroid with rank function r, then its *characteristic polynomial*[2] is

$$C(M, q) = \sum_{X \subseteq E} (-1)^{|X|} q^{r(E)-r(X)}. \tag{3.6}$$

If $M = M(G)$ is the cycle matroid of a graph G with c connected components then it follows immediately from (2.1) that

$$C(M, q) = q^{-c} P(G, q) \tag{3.7}$$

and when $M = M(G)^*$ is the bond matroid of a graph G then

$$C(M, q) = F(G, q) \tag{3.8}$$

and so the chromatic and flow polynomials are simply characteristic polynomials of graphic and cographic matroids respectively, with an additional factor of q^c for the chromatic case. The familiar deletion-contraction expression for the chromatic polynomial can equally well be seen in this more general context.

Theorem 3.3 *Let $M = (E, r)$ be a loopless matroid, and let $e \in E$. Then*

$$C(M, q) = \begin{cases} (q-1)C(M \backslash e, q), & \text{if } e \text{ is a coloop;} \\ C(M \backslash e, q) - C(M/e, q), & \text{otherwise.} \end{cases} \tag{3.9}$$

---

[2] The term "characteristic polynomial" is somewhat unfortunate because it is not related to the characteristic polynomial of the adjacency matrix of a graph.

If a matroid has a loop then $C(M,q)$ is identically zero and hence we normally assume that the matroid is loopless. Note however that a cographic matroid $M(G)^*$ is loopless if G has no *coloops* (i.e. bridges).

3.1 The multivariate Tutte polynomial

Suppose now that we associate commuting indeterminates $\{v_e \mid e \in E\}$ to the edges of a graph G. Then the *multivariate Tutte polynomial* $Z_G(q, \{v_e\})$ of G is defined by

$$Z_G(q, \{v_e\}) = \sum_{X \subseteq E} q^{k(X)} \prod_{e \in X} v_e. \tag{3.10}$$

If $v_e = v$ for all edges, then $Z_G(q, v)$ is a 2-variable polynomial equivalent to the "usual" Tutte polynomial $T_G(x, y)$ of G under a simple change of variables:

$$T_G(x, y) = (x-1)^{-k(E)}(y-1)^{-|V|} Z_G((x-1)(y-1), y-1). \tag{3.11}$$

Comparing (3.1) with (3.10) it is immediate that

$$P(G, q) = Z_G(q, -1) \tag{3.12}$$

and so the chromatic polynomial arises by setting all the edge weights equal to -1. The flow polynomial of G is given by

$$F(G, q) = (-1)^{|E(G)|} q^{-|V(G)|} Z_G(q, -q). \tag{3.13}$$

Although at first sight, the multivariate Tutte polynomial appears more complicated than the single variable chromatic polynomial, it is often extremely useful to have the extra freedom to assign different edge weights to each edge, even when the ultimate aim is to prove results for the chromatic polynomial alone. Sokal [54] has eloquently argued in favour of this "multivariate ideology" in great detail, but for our current purposes there are two points of particular importance.

The first of these arises from the observation that to a statistical physicist, the edge weight v_e is a physical parameter determined by the "coupling" between the sites (vertices) joined by e and the ambient temperature. From this viewpoint, the value $v_e = -1$ is the zero-temperature limit of an *anti-ferromagnetic* model i.e. one where adjacent sites tend to have *different* spins (colours). However if a certain property holds at the zero-temperature limit, then it is often the case that it holds throughout the entire antiferromagnetic regime $-1 \leq v_e \leq 0$ for each e (and sometimes further). In these cases the more general result is often easier to prove, and moreover provides additional insight into the special case where all $v_e = -1$.

The second point is that inductive proofs often use local operations on graphs and these operations can easily be handled by changing the edge weight only on the affected edges. If the edge weights are constrained to be equal then even a local operation results in a global change to the edge weights.

The most important of these local operations are *parallel reduction* and *series reduction* which replace edges in parallel or series by single edges with a new weight. More precisely, we say that edges $e, f \in E$ are *parallel* if e and f connect the same pair of vertices x and y. In this case they can be replaced, without changing the value of Z_G, by a single new edge with "effective weight"

$$v_{\text{eff}} = (1+v_e)(1+v_f) - 1. \qquad (3.14)$$

This operation of replacing two parallel edges by a single edge is called *parallel reduction*, and we write $v_e \parallel v_f$ for $(1+v_e)(1+v_f) - 1$.

Two edges $e, f \in E(G)$ are *in series* if there are vertices $x, y, z \in V$ with $x \neq y$, $y \neq z$ such that e connects x and y and f connects y and z and y has degree 2. In this case, replacing them with a single edge of effective weight

$$v_{\text{eff}} = \frac{v_e v_f}{q + v_e + v_f} \qquad (3.15)$$

yields a graph whose multivariate Tutte polynomial — when multiplied by the prefactor $(q + v_e + v_f)$ — is the same as that of the original graph. Naturally this operation is called *series reduction*, and we write $v_e \bowtie_q v_f$ for $v_e v_f / (q + v_e + v_f)$. (Note that this definition of "series" is more restrictive than the matroidal definition of elements in series, but the distinction is not important in our context.)

There is a version of the deletion-contraction formula for the multivariate Tutte polynomial that depends on the weight of the edge involved. The vector of edge weights is denoted by $\mathbf{v}$ and for any edge $e \in E(G)$ the vector obtained from $\mathbf{v}$ by deleting v_e is denoted $\mathbf{v}_{\neq e}$. Then

$$Z_G(q, \mathbf{v}) = Z_{G \setminus e}(q, \mathbf{v}_{\neq e}) + v_e Z_{G/e}(q, \mathbf{v}_{\neq e}). \qquad (3.16)$$

3.2 Examples

It is customary for papers on chromatic polynomials to include some specific chromatic polynomials to illustrate techniques for applying the deletion-contraction recurrence and to provide examples to illustrate subsequent theorems.

3.2.1 Complete graphs

For the complete graphs K_n the chromatic polynomial can be deduced immediately just by counting colourings, yielding

$$P(K_n, q) = q(q-1)\cdots(q-k+1), \qquad (3.17)$$

and we use $q_{(k)}$ to denote this "falling factorial" expression. The expression for the flow polynomial of a complete graph is considerably less simple (see [60] and [34]):

$$(-1)^{\binom{n}{2}} q^n F(K_n, q) =$$
$$\sum_{(b_1, b_2, \ldots, b_n)} \left[\frac{n!}{\prod_i b_i! (i!)^{b_i}} \right] q_{\langle b_1 + b_2 + \cdots + b_n \rangle} (1-q)^{\sum_i \binom{i}{2} b_i} \qquad (3.18)$$

where the summation is over all tuples $(b_1, b_2, \ldots, b_n)$ of non-negative integers such that $\sum_i i b_i = n$. Figure 1 shows the flow roots of K_{10} and K_{30} and they appear to be converging to some algebraic curve — perhaps even $|q - 1| = 1$ — but this has not yet been shown.

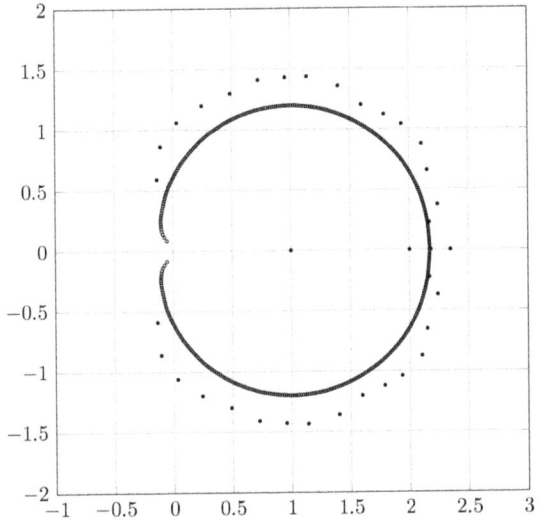

Figure 1: Flow roots of K_{10} and K_{30}

3.2.2 Complete bipartite graphs
For the complete bipartite graphs $K_{m,n}$ another counting argument is used to get the expression

$$P(K_{m,n}, q) = \sum_{k=1}^{m} S(m,k) q_{\langle k \rangle} (q-k)^n \tag{3.19}$$

where $S(m, k)$ is the Stirling number of the second kind. Each term in the sum counts the number of colourings that use exactly k colours on the side of the bipartition containing m vertices.

The flow polynomial of the complete bipartite graphs can be computed symbolically and the flow roots of $K_{10,10}$ and $K_{20,20}$ are shown in Figure 2.

3.2.3 Projective and affine geometries
The complete graph K_n is the maximum sized simple graph with n vertices, and any other graph on n vertices can be obtained from K_n by deleting edges. Similarly the maximum sized simple matroid of rank r that is representable over $GF(q)$ is the *projective geometry* $PG(r-1, q)$ and any other simple $GF(q)$-representable matroid of rank at most r can be obtained from $PG(r-1, q)$ by deleting points.

When discussing finite geometries, the symbol q is universally used to denote the field size. This conflicts with the choice of q for the variable of the characteristic polynomial, which is a choice derived from the connections with the q-state Potts model. As (most) statistical physicists are unlikely to consider q-state Potts models on projective or affine geometries, I will stick to the geometric use of q for this subsection and temporarily use λ as the variable of the characteristic polynomial.

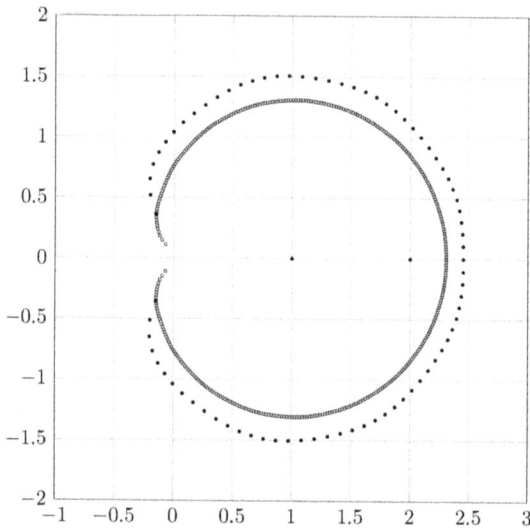

Figure 2: Flow roots of $K_{10,10}$ and $K_{20,20}$

Using this notation, the characteristic polynomial of $PG(r-1,q)$ is

$$C(PG(r-1,q),\lambda) = (\lambda-1)(\lambda-q)(\lambda-q^2)\cdots(\lambda-q^{(r-1)}). \quad (3.20)$$

Thus the characteristic polynomial of the rank-4 binary matroid $PG(3,2)$ is

$$C(PG(3,2),\lambda) = (\lambda-1)(\lambda-2)(\lambda-4)(\lambda-8). \quad (3.21)$$

The *affine geometry* $AG(r-1,q)$ is obtained by deleting a hyperplane from $PG(r-1,q)$. The characteristic polynomial of $AG(r-1,q)$ is (see [8])

$$C(AG(r-1,q),\lambda) = (\lambda-1)\sum_{k=0}^{r-1}(-1)^k\lambda^{r-k+1}(q^{r-1}-1)(q^{r-2}-1)\cdots(q^{r-k}-1). \quad (3.22)$$

Thus the characteristic polynomial of the rank-4 binary matroid $AG(3,2)$ is

$$C(AG(3,2),\lambda) = (\lambda-1)(\lambda^3 - 7\lambda^2 + 21\lambda - 21). \quad (3.23)$$

3.2.4 Wheels and Whirls

The *rank-n wheel* W_n is the $(n+1)$-vertex graph obtained from the cycle C_n by adding a new vertex adjacent to every vertex of C_n. Thus W_3 is isomorphic to the complete graph K_4. The chromatic polynomial of the cycle C_n is well known:

$$P(C_n, q) = (q-1)^n + (-1)^n(q-1). \quad (3.24)$$

and so the chromatic polynomial of the wheel is given by (6.1) and is

$$P(W_n, q) = q\left((q-2)^n + (-1)^n(q-2)\right). \quad (3.25)$$

The n edges that form the *rim* of the wheel (i.e. the original C_n) is a subset of rank $n-1$ in the graphic matroid $M(W_n)$. The *rank-n whirl* $\mathcal{W}_n$ is the matroid obtained from $M(W_n)$ by altering the rank of this one subset to n (and leaving all other values of the rank function unchanged). The characteristic polynomial of $M(W_n)$ is $P(W_n, q)/q$ and by (3.6) we see that the rim contributes $(-1)^n q$ to the characteristic polynomial of the wheel, but $(-1)^n$ to that of the whirl. Thus

$$C(\mathcal{W}_n, q) = (q-2)^n - (-1)^n. \tag{3.26}$$

4 Fundamentals

If G is a disconnected graph with components G_1 and G_2 then $P(G,q) = P(G_1,q)P(G_2,q)$ and so for chromatic roots, it suffices to consider connected graphs. A graph is called *separable* if its edge set can be partitioned into two disjoint nonempty sets E_1 and E_2 such that $|V(E_1) \cap V(E_2)| = 1$ where $V(E_i)$ is the set of vertices incident with an edge in E_i. If G is separable and we set $G_i = (V(E_i), E_i)$ then $P(G,q) = P(G_1,q)P(G_2,q)/q$ and so again for the study of chromatic roots it suffices to consider nonseparable graphs. Note that a graph is nonseparable if it has no loops and is either K_2 or is 2-connected. Similarly, a matroid $M = (E, r)$ is *connected* if there is no partition of its edge set into two disjoint non-empty sets E_1, E_2 such that $r(X) = r(E_1 \cap X) + r(E_2 \cap X)$ for all $X \subseteq E$. With these definitions we have the following basic result:

Theorem 4.1 *Let $M = (E,r)$ be a connected loopless matroid with characteristic polynomial $C(M,q)$. Then*

(1) $C(M,q)$ is a monic polynomial of degree $r(E)$ whose coefficients are non-zero and alternate in sign.

(2) $C(M,q)$ is nonzero with sign $(-1)^{r(E)}$ for $q \in (-\infty, 1)$.

(3) $C(M,q)$ has a simple zero at $q = 1$.

(4) $C(M,q)$ is nonzero with sign $(-1)^{r(E)+1}$ for $q \in (1, \frac{32}{27}]$.

The first three of these are straightforward to prove for graphs, while the fourth was proved for graphs by Jackson [24] and for matroids by Edwards, Hieron and Jackson [18]. In the first of these papers, Jackson proved the existence of the constant 32/27 and also described a family of graphs with real chromatic roots converging to 32/27. Subsequently, Thomassen [57] demonstrated that chromatic roots of graphs are dense in the interval $(32/27, \infty)$ thereby showing that the chromatic root-free intervals of Theorem 4.1 are the best possible.

Recently however, Jackson and Sokal [29] have substantially generalized Jackson's result by considering the multivariate Tutte polynomial with non-constant edge weights $\{v_e\}$ and in the process explained the appearance of the previously-mysterious constant 32/27. In Figure 3, which shows their results for the range $q > 0$ and $-2 \leq v_e \leq 0$, three regions of the diagram are labelled with either "+" or "−". The interpretation of this is that for any fixed q, if the edge weights v_e have the property that all pairs (q, v_e) lie in the labelled region, then $(-1)^{|V|} Z_G(q, \mathbf{v})$ has the stated sign and in particular is non-zero.

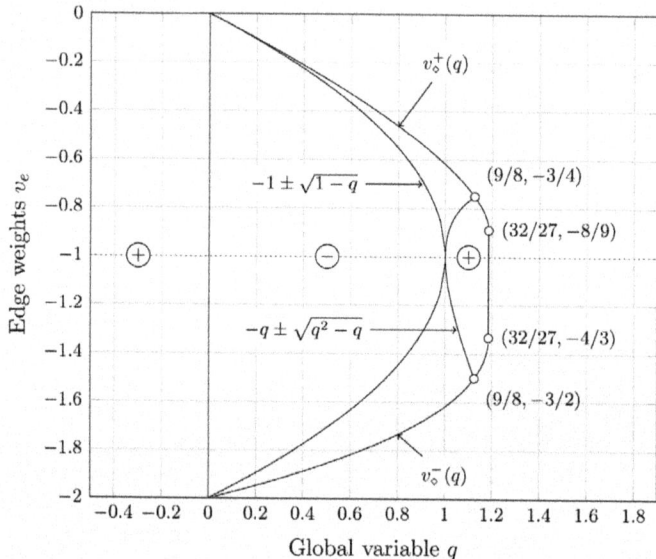

Figure 3: Regions where sign of $(-1)^n Z_G(q, \mathbf{v})$ is known for a loopless 2-connected graph with n vertices

The fundamental idea underlying their proof is surprisingly simple. It is based on trying to find a "good set" $\mathcal{V}(q)$ of edge weights for each fixed value of q so that any graph with edge weights in $\mathcal{V}(q)$ can be reduced to a smaller graph whose edge weights are *also* in $\mathcal{V}(q)$, thereby allowing an inductive proof to start. The base case of the inductive proof is a set of graphs where the sign of $Z_G(q, \mathbf{v})$ is fixed (either positive or negative) whenever $v_e \in \mathcal{V}(q)$.

Theorem 4.2 (Jackson and Sokal [29]) *Fix values $q > 0$, $\delta \in \{0, 1\}$ and $m \geq 2$ and suppose that there is a set of real numbers $\mathcal{V} \subseteq \mathbb{R}$ such that*

(1) $\mathcal{V} \subseteq (-2, -q/2)$,

(2) $\mathcal{V} \parallel \mathcal{V} \subseteq \mathcal{V}$,

(3) $\mathcal{V} \bowtie_q \mathcal{V} \subseteq \mathcal{V}$,

(4) $(-1)^{n+\delta} Z_G(q, \mathbf{v}) > 0$ whenever G is a nonseparable graph on n vertices with exactly m edges and $v_e \in \mathcal{V}$ for all edges of G.

Then $(-1)^{n+\delta} Z_G(q, \mathbf{v}) > 0$ whenever G is a nonseparable graph on n vertices with at least m edges and $v_e \in \mathcal{V}$ for all edges $e \in E(G)$.

Proof Suppose that G is a nonseparable graph on n vertices with more than m edges, and suppose that the result holds for all nonseparable graphs with between m and $|E(G)| - 1$ edges. We divide the argument into three cases according to whether G has any edges in parallel, edges in series or neither.

(Parallel edges) If G has a pair of parallel edges e, f then we can replace them with a single edge of weight $v_e \parallel v_f$ without changing Z_G. As $v_e \parallel v_f \in \mathcal{V}$ and the new graph is nonseparable and has one fewer edge than G, the inductive hypothesis applies and the result follows.

(Series edges) If G has a pair of edges e, f incident with a vertex of degree 2 then $Z_G = (q + v_e + v + f) Z_H$ where H is the graph obtained by replacing e and f with a single edge of weight $v_e \bowtie_q v_f$. As $v_e \bowtie_q v_f \in \mathcal{V}$ and H is nonseparable and has one fewer edge than G, the inductive hypothesis applies. The first condition on $\mathcal{V}$ implies that $q + v_e + v_f < 0$ and as H has $n - 1$ vertices, the signs cancel giving the stated result.

(Neither) If neither of the above cases applies then G is a simple graph with minimum degree 3 in which case it has an edge e such that both $G \backslash e$ and G/e are nonseparable. In this case apply the deletion-contraction recurrence (3.16), and note that because $v_e < 0$ and G/e has $n-1$ vertices, each term has the same sign. □

Armed with this theorem, the next task is to actually find suitable sets $\mathcal{V}(q)$ and here a critical role is played by the "diamond reduction". A *diamond* in a graph consists of two 2-edge paths joining two vertices x and y such that the internal vertices of the paths have degree 2 in G. If G contains a diamond with all edges having weight v, then $Z_G = (q+2v)^2 Z_H$ where H is the graph obtained from G by replacing these four edges with a single edge between x and y of weight

$$\Diamond_q(v) = (v \bowtie_q v) \parallel (v \bowtie_q v) = \frac{v^2(v^2 + 4v + 2q)}{(q+2v)^2}. \tag{4.1}$$

The fixed points of this operation, i.e. those values such that $\Diamond_q(v) = v$, are the zeros of the cubic equation

$$v^3 - 2qv - q^2 = 0. \tag{4.2}$$

The discriminant of this cubic is $q^3(32 - 27q)$ and hence the discriminant is positive if and only if $0 < q < 32/27$. In this region, the cubic has three roots (see Figure 4). If we denote the "middle root" by $v_\Diamond^+(q)$ and let $v_\Diamond^-(q) = q/v_\Diamond^+(q)$ then the "diamond interval" is given by

$$I_\Diamond(q) = [v_\Diamond^-(q), v_\Diamond^+(q)]. \tag{4.3}$$

It is straightforward to see that $I_\Diamond(q)$ satisfies the first three conditions of Theorem 4.2. Jackson and Sokal prove the converse that if there is any set $\mathcal{V}(q)$ satisfying these conditions for $q > 0$ then $q \le 32/27$ and $\mathcal{V}(q)$ is contained in $I_\Diamond(q)$. Therefore taking $\mathcal{V}(q) = I_\Diamond(q)$ gives best possible sets to satisfy the "induction part" of the theorem, and this interval determines the outermost curves shown in Figure 3.

However the "base case" of the induction requires the sets $\mathcal{V}(q)$ to also satisfy the *fourth* condition of Theorem 4.2 and this depends on m. If we take $m = 2$ then the only nonseparable graph with two edges is $K_2^{(2)}$ which consists of two vertices joined by two parallel edges. The multivariate Tutte polynomial of $K_2^{(2)}$ is

$$q^2 + q(v_e + v_f + v_e v_f) = q((q-1) + (1+v_e)(1+v_f)). \tag{4.4}$$

In the range $0 < q < 1$, this expression is negative provided that $(1+v_e)(1+v_f) \le 1 - q$ which is satisfied when $-1 - \sqrt{1-q} \le v_e, v_f \le -1 + \sqrt{1-q}$. This interval

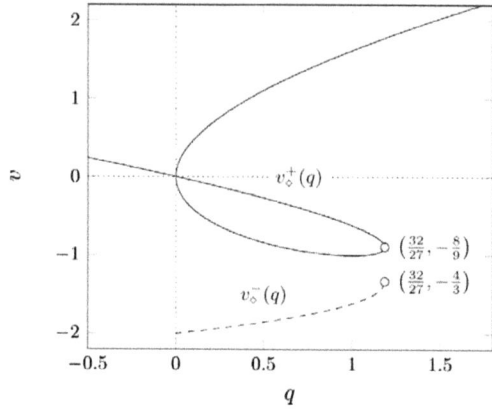

Figure 4: Solutions to the cubic $v^3 - 2qv - q^2 = 0$ where $v_\diamond^+(q)$ is the middle root (for $0 \leq q \leq 32/27$) and $v_\diamond^-(q) = q/v_\diamond^+(q)$ is shown dashed.

satisfies the first three conditions of the theorem and hence to satisfy all four, we must restrict $\mathcal{V}(q)$ to this smaller interval. If $1 < q < 32/27$, then $Z_G > 0$ if the edge weights lie in the interval $(-q - \sqrt{q^2 - q}, -q + \sqrt{q^2 - q})$. When $1 < q < 9/8$ this interval is contained within $I_\diamond(q)$ (and satisfies the first three conditions of the theorem), but when $q > 9/8$ the requirement that $\mathcal{V}(q)$ be contained in $I_\diamond(q)$ is the more restrictive of the conditions.

If we were willing to ignore graphs with only 2 edges and repeat this analysis for graphs with 3 edges, then only the "inner curves" of Figure 3 change. Jackson and Sokal conjecture that as m increases, the conditions imposed by the base cases become less and less restrictive so that in the limit only the outer curves remain.

Jackson and Sokal [29] also show that their proofs can be extended to general matroids with only minor alterations, and so essentially identical results hold in the general matroid case.

5 Integer Roots

The *chromatic number* $\chi = \chi(G)$ of a graph G is the smallest number of colours such that G has a proper χ-colouring. It is then obvious that G has no proper colourings if fewer than χ colours are available, but always has proper colourings if χ or more colours are available. Therefore $P(G, q) = 0$ for all integers $0 \leq q < \chi$ and $P(G, q) > 0$ for all integers $q \geq \chi$, and so the integer chromatic roots of G form a initial segment of the non-negative integers.

Similar considerations for flows show that the integer flow roots of a graph also form a sequence of consecutive integers (though now starting at $q = 1$). However this is not true for the characteristic polynomials of matroids or even binary matroids, as evidenced by the projective geometries (see (3.21)).

The chromatic and flow polynomials of graphs are well-behaved in this respect because they are *counting* colourings and flows respectively, but for general matroids there is no "counting interpretation" for $C(M, q)$. However for *regular matroids*, which are matroids that are representable over *every* field, Lindström gave such an interpretation for $C(M, q)$ thereby proving the following theorem.

Theorem 5.1 (Lindström [36]) *If M is a regular matroid, then there is an integer χ, called the chromatic number of M, such that the integer roots of $C(M,q)$ are precisely the set $\{1, 2, \ldots, \chi - 1\}$.* □

For non-regular matroids, the existence of "gaps" in the sequence of integer roots and positive integers q with $C(M,q) < 0$ means that there are two possible ways to define the "chromatic number" of M — either the smallest value χ such that $C(M, \chi) > 0$ or the smallest value χ such that $C(M,q) > 0$ for all $q \geq \chi$. These are usually denoted $\chi(M)$ and $\pi(M)$ respectively, and Walton and Welsh [62] have investigated $\pi(M)$ for various classes of binary matroids.

In general, the most that can be said is that for matroids representable over a finite field $\mathbb{F}$, there is a counting interpretation for the characteristic polynomial at powers of $|\mathbb{F}|$ (Crapo and Rota [9]). One consequence of this theory is that if a loopless matroid M is represented as a set of points $S \subseteq PG(r-1, \mathbb{F})$ then $C(M, |\mathbb{F}|^k) > 0$ if and only if there is a subspace of codimension at most k disjoint from S. The *critical exponent* of an $\mathbb{F}$-representable matroid is the smallest value of c such that $C(M, |\mathbb{F}|^c) > 0$.

A matroid with critical exponent $c = 1$ is called *affine* and any simple affine matroid of rank at most r can be obtained by deleting points from the affine geometry $AG(r-1, \mathbb{F})$. A graphic matroid $M(G)$ is affine if G is bipartite, and a cographic matroid $M(G)^*$ is affine if G is *eulerian* (that is, every vertex has even degree).

Graphs can have arbitrarily large chromatic number, and so the critical exponent of the class of graphic matroids is unbounded. However the situation is different for cographic matroids. Jaeger's famous 8-flow theorem [31] shows that the critical exponent of a cographic matroid $M(G)^*$ is at most 3, and hence that $C(M(G)^*, 8) = F(G, 8) > 0$. Tutte [59] has conjectured that in fact $F(G, 5) > 0$ for all bridgeless graphs (equivalently, any cographic matroid is affine over $GF(5)$).

Conjecture 5.2 (Tutte's 5-flow conjecture) *A bridgeless graph has a nowhere-zero 5-flow.*

The Petersen graph P has flow polynomial

$$F(P, q) = (q-1)(q-2)(q-3)(q-4)(q^2 - 5q + 10) \tag{5.1}$$

which shows that 5 cannot be replaced by 4, while Seymour [50] has improved Jaeger's result by showing that $F(G, 6) > 0$ for any bridgeless graph G.

5.1 All roots integral

A graph G is *chordal* if every cycle of length at least four has a chord or, equivalently, if G has no induced cycles of length greater than three. There are numerous alternative characterizations of chordal graphs, including the following, which follows directly from results of Dirac [11]. A *simplicial vertex* in a graph G is a vertex whose neighbours induce a complete subgraph (i.e. clique) of G.

Lemma 5.3 *A connected graph G is chordal if it is complete or if it has a simplicial vertex v such that $G - v$ is chordal.*

If G is chordal and v is a simplicial vertex of degree d, then

$$P(G,q) = (q-d)P(G-v,q) \tag{5.2}$$

and so by induction the chromatic polynomial of a chordal graph has only integer roots. The converse is not true as there are many non-chordal graphs with only integer chromatic roots (see [13, 17]).

Turning to flow polynomials, it is immediate that the dual of a planar chordal graph has an integer-rooted flow polynomial. However, no other examples are known and indeed I conjecture that there are no more.

Conjecture 5.4 *If the flow polynomial of a graph G has only integer roots, then G is the dual of a planar chordal graph.*

Joe Kung and I have proved that this conjecture is at least true for 3-connected cubic graphs, i.e. the graphs G for which $M(G)^*$ is a maximum-sized simple cographic matroid. Let $\mathcal{C}$ be the class of cubic planar graphs obtained from K_4 by repeatedly expanding a vertex of degree three into a triangle. If $G \in \mathcal{C}$, then G^* is a chordal planar triangulation.

Theorem 5.5 (Kung and Royle, in preparation) *If G is a 3-connected cubic graph such that $F(G,q)$ has only integer roots, then $G \in \mathcal{C}$.*

The proof of this theorem involves showing that a 3-connected cubic graph with integer flow roots has a large number of 3-edge cutsets that are not vertex stars, and then using this condition to show that $G \in \mathcal{C}$.

6 Complex Roots

The two most important results on the location of complex chromatic roots are both due to Sokal [52, 53]. In the first of these papers, he proved the long-standing conjecture that the chromatic roots of graphs of any fixed maximum degree are bounded. In the second, he showed that complex chromatic roots are dense in the whole complex plane, thereby showing the impossibility of finding any absolute bounds or chromatic root-free regions of the complex plane. These important results are discussed in Section 6.2 and Section 6.1 respectively, while Section 6.3 considers more recent work on bounding chromatic roots in terms of a parameter other than maximum degree, although only for series-parallel graphs.

Almost nothing is known about the location of complex flow roots of graphs, other than results following immediately from results on the complex chromatic roots of planar graphs.

6.1 Absolute Bounds

The earliest computer-generated plots of chromatic roots of small graphs revealed various intriguing patterns, but their most obvious feature was the absence of chromatic roots in the left half-plane (that is, with negative real part) and in

1980, Farrell [19] conjectured that this was always the case. This was subsequently disproved by Read and Royle [42] who found that certain cubic graphs of high girth had some chromatic roots with tiny negative real part.

Now let $\Theta^{(s,p)}$ denote the *generalized theta graph* obtained by connecting two vertices with p internally disjoint paths each with s edges (so $\Theta^{(s,p)}$ has $(s-1)p+2$ vertices). Shrock and Tsai [51] demonstrated that suitably varying s and p gives chromatic roots of arbitrarily large negative real part, thus eliminating any possibility of a lower bound on the real part. In fact something much stronger is true, namely that it is impossible to get *any* absolute bound on the location of chromatic roots of graphs.

Theorem 6.1 (Sokal [53]) *The chromatic roots of the generalized theta graphs $\Theta^{(s,p)}$ are dense in the whole complex plane with the possible exception of the disk $|q-1| < 1$.*

As the generalized theta graphs are series-parallel graphs and hence planar, we get the same result for flow roots also.

Corollary 6.2 *The flow roots of graphs are dense in the whole complex plane, with the possible exception of the disk $|q-1| < 1$.*

The graph $G + K_r$ obtained by completely joining an r-clique K_r to a graph G has chromatic polynomial

$$P(G + K_r, q) = q_{\langle r \rangle} P(G, q - r). \tag{6.1}$$

Thus if z is a chromatic root, then so is $z + r$ for all $r > 0$, and hence by completely joining cliques to suitable generalized theta graphs, we can "fill in" the disk $|q-1| < 1$ and conclude that chromatic roots of graphs are dense in the whole complex plane. As complete joins do not in general preserve planarity, this argument does not apply for flow roots and so the region $|q - 1| < 1$ remains a possible exception to the analogous statement for flow roots.

The question then arises as to whether there are any natural classes of graphs (or indeed matroids) for which one can find absolute bounds on the location of the complex chromatic roots. From a matroidal point of view, the most natural classes are *minor-closed* classes of matroids, but there are only a few very small minor-closed classes (such as outerplanar graphs) that do not include series-parallel graphs. On the other hand, series-parallel graphs have low connectivity as they always have a vertex cutset of size two. It is conceivable that adding a connectivity restriction may constrain complex chromatic roots. For example, although it is known that the roots of 3-connected (and even 4-connected) planar graphs are not bounded, the following question is still open:

Question 6.3 *Are the chromatic roots of 3-connected planar graphs dense in the complex plane?*

6.2 Maximum degree and maxmaxflow

The overall goal in finding parameterized bounds on chromatic or characteristic roots is to determine exactly which properties of a graph or matroid affect the location of its chromatic or characteristic roots. Motivated by Brooks' Theorem and armed with some limited computational evidence, Biggs, Damerell and Sands conjectured that the *degree* of a graph is an important parameter.

Conjecture 6.4 (Biggs, Damerell and Sands [3]) *There exists a function $B(k)$ such that the chromatic roots of any regular graph of degree k lie in the disk $|q| \leq B(k)$.*

Brenti, Royle and Wagner [7] "extended" this conjecture from regular graphs to the case where the maximum degree is k, unfortunately without noticing that the two conjectures are equivalent! Eventually however, an even stronger result showing that *second-largest degree* suffices was proved by Sokal.

Theorem 6.5 (Sokal [52]) *If G is a loopless graph with maximum degree Δ then its chromatic roots lie in the disk $|q| < 7.963907\Delta$. Moreover, if G has second-largest degree Δ_2 then its chromatic roots lie in the disk $|q| < 7.963907\Delta_2 + 1$.*

The graphs $\Theta^{(s,p)}$ show that no bound in terms of *third*-largest degree is possible. The constant 7.963907 is an artifact of the proof and it is likely that the true value is considerably smaller. In fact for many years, I have believed that the actual value is determined by the following conjecture:

Conjecture 6.6 (Royle, c. 1990) *For $\Delta \geq 3$, the complete bipartite graph $K_{\Delta,\Delta}$ has the chromatic root of largest modulus among all graphs of maximum degree Δ (excluding K_4 when $\Delta = 3$).*

In so-far unpublished work, Salas and Sokal [46] have investigated the chromatic roots of the complete bipartite graphs and have identified the limiting curves to which the chromatic roots of $K_{N/2,N/2}$ converge as the total number of vertices (N) increases. Figure 5 shows the scaled chromatic roots of $K_{100,100}$ (thus $N = 200$) together with the end-point of the limiting curve at $q \approx 0.34 \pm 0.72i$. Other than the shortfall at the ends of the curves which do not quite reach the limit point, already the pattern of roots is almost indistinguishable from the theoretical limiting curve depicted in [46].

If the conjecture is true and the pattern of chromatic roots for complete bipartite graphs tends towards the limiting curves without strange outliers, then the limit points would be the chromatic roots of largest modulus, implying that the correct bound is about 1.6Δ.

Despite these results, the parameters Δ and Δ_2 are unsatisfactory in various ways. For example, Δ and Δ_2 can be made arbitrarily large by gluing together graphs at a cut-vertex, yet this operation does not alter the chromatic roots. The underlying reason for this is that the chromatic polynomial is essentially a property of the cycle matroid of the graph, but the maximum degree is not. Therefore it would be of great interest to find a matroidal parameter that could play the role of maximum degree in results of this type. Motivated by some remarks of Shrock and Tsai, Jackson and Sokal [28] considered a graph parameter that they call *maxmaxflow*, defined as follows: If x and y are distinct vertices in a graph G, then let $\lambda_G(x,y)$

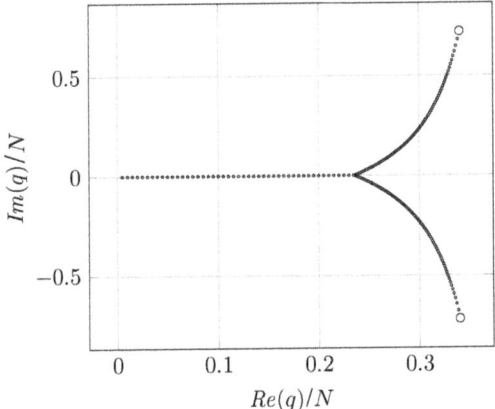

Figure 5: Scaled chromatic roots of $K_{100,100}$ and the limit points $q \approx 0.34 \pm 0.72i$

denote the *maximum flow* from x to y.

$$\lambda_G(x,y) = \text{max. number of edge disjoint paths from } x \text{ to } y$$
$$= \text{min. number of edges separating } x \text{ from } y.$$

Then the maxmaxflow $\Lambda(G)$ is the maximum of these values over all distinct pairs of vertices:

$$\Lambda(G) = \max_{x \neq y} \lambda_G(x,y). \tag{6.2}$$

Although this definition appears to use the non-matroidal concept of a "vertex" in a fundamental way, Jackson and Sokal have shown that there is a surprising alternative definition of the same parameter. If $M = (E, r)$ is a binary matroid then let $V = \mathbb{Z}_2^E$ be the binary vector space with basis E and identify any subset $X \subseteq E$ with a vector $x \in V$ where $x_e = 1$ if and only if $e \in X$. The *cocycle space* of M is the subspace of V spanned by the vectors associated with the cocircuits of M. Jackson and Sokal [28] show that if $M(G)$ is the cycle matroid of G, then

$$\Lambda(G) = \min_{\mathcal{B}} \max_{C \in \mathcal{B}} |C| \tag{6.3}$$

where the min runs over all bases $\mathcal{B}$ of the cocycle space of $M(G)$ and the max runs over all vectors in the basis $\mathcal{B}$. Therefore we can take (6.3) as the *definition* of $\Lambda(M)$ for an arbitrary binary matroid M.

For graphs, $\Lambda(G)$ behaves exactly as one would wish with respect to gluing together blocks at a cut-vertex in that the maxmaxflow of a graph is the maximum of the maxmaxflows of its blocks. Furthermore it is immediate that

$$\Lambda(G) \leq \Delta_2(G). \tag{6.4}$$

There are some serious difficulties in modifying the existing cluster-expansion proof of Theorem 6.5 to get an analogous bound in terms of Λ (Sokal [54]) and even though some progress has been made, a number of obstacles remain. However by using entirely different techniques, Sokal and I [45] have very recently succeeded in obtaining a fairly good bound for the chromatic roots of series-parallel graphs — this is described in the next section.

6.3 Series-parallel graphs and maxmaxflow

In this section, I will sketch a proof of the following result:

Theorem 6.7 (Royle and Sokal [45]) *The chromatic roots of a series-parallel graph of maxmaxflow Λ lie in the disk*

$$|q - 1| \leq (\Lambda - 1)/\log 2 \approx 1.443(\Lambda - 1). \tag{6.5}$$

Although series-parallel graphs form a very small subclass of graphs, they are of great importance in the theory of chromatic roots (as shown by Theorem 6.1) and indeed in many other areas of mathematics and computer science.

First some definitions: A *2-terminal graph* $G = (G, s, t)$ is a graph with two distinguished vertices s and t called the *terminals*. If $G_1 = (G_1, s_1, t_1)$ and $G_2 = (G_2, s_2, t_2)$ are two 2-terminal graphs on disjoint vertex sets, then their *series composition* is the 2-terminal graph

$$G_1 \bowtie G_2 = (H, s_1, t_2) \tag{6.6}$$

where H is obtained from $G_1 \cup G_2$ by identifying t_1 with s_2, and their *parallel composition* is the 2-terminal graph

$$G_1 \parallel G_2 = (H, s_1, t_1) \tag{6.7}$$

where H is obtained from $G_1 \cup G_2$ by identifying s_2 with s_1 and t_2 with t_1.

A *2-terminal series-parallel graph* is a 2-terminal graph that is either K_2 (with both vertices as terminals), or the series or parallel composition of smaller 2-terminal series-parallel graphs. A 2-connected graph G is a *series-parallel graph* if it has distinct vertices s and t such that (G, s, t) is a series-parallel graph, while a graph with cutvertices is a series-parallel graph if each of its blocks (maximal 2-connected subgraphs) is a series-parallel graph. If the graph G is the series or parallel composition of smaller graphs, then the smaller graphs are called the *constituents* of G.

Although we cannot precisely control the maxmaxflow of a 2-terminal series-parallel graph, we *can* control the flow between its terminals.

Lemma 6.8 *Suppose that (G_1, s_1, t_1) and (G_2, s_2, t_2) are 2-terminal graphs (not necessarily series-parallel). Then*

$$\lambda_{G_1 \bowtie G_2}(s, t) = \min\{\lambda_{G_1}(s_1, t_1), \lambda_{G_2}(s_2, t_2)\}, \tag{6.8}$$

$$\lambda_{G_1 \parallel G_2}(s, t) = \lambda_{G_1}(s_1, t_1) + \lambda_{G_2}(s_2, t_2). \tag{6.9}$$

Corollary 6.9 *Suppose that $G = G_1 \parallel G_2$ where $G_1(s_1, t_1)$, $G_2(s_2, t_2)$ are 2-terminal graphs and that $\Lambda = \Lambda(G)$. Then $\lambda_{G_1}(s_1, t_1) < \Lambda$ and $\lambda_{G_2}(s_2, t_2) < \Lambda$.*

This implies that to construct a 2-connected series-parallel graph of maxmaxflow at most Λ through a sequence of series and parallel compositions, only graphs with between-terminal flow strictly less than Λ can be used as constituents.

The multivariate Tutte polynomial of a graph with one (non-loop) edge of weight v_e is $q(q + v_e)$ and so it has roots at $q = 0$ and $q = -v_e$. Given a 2-terminal series-parallel graph G with arbitrary edge weights and a fixed complex number q, we can apply series and parallel reductions until G has been reduced to a single edge with some "effective weight" v_{eff}. If $v_{\text{eff}} \neq \infty$ then none of the prefactors of the form $(q + v_e + v_f)$ generated during the series reductions are zero, and so $Z_G(q, \{v_e\}) = 0$ if and only if $q = 0$ or $v_{\text{eff}} = -q$.

This observation then gives us a strategy for determining root-free regions for the chromatic polynomials of families of series-parallel graphs. For a fixed q in the desired root-free region, we bound the regions of the complex plane where v_{eff} can lie for any graph in the family, and show that $-q$ is not in this region. Although this works, it does not give very good bounds. However, far better bounds can be obtained by changing variables from the edge weights $\{v_e\}$ to the "transmissivities" $\{t_e\}$ which are related to the edge weights as follows:

$$t_e = \frac{v_e}{q + v_e}, \qquad v_e = \frac{qt_e}{1 - t_e}. \tag{6.10}$$

The important points $v_e = -1$, $v_e = \infty$ and $v_e = -q$ correspond to $t_e = 1/(1-q)$, $t_e = 1$ and $t_e = \infty$ respectively. In these variables, the expressions for computing the effective transmissivity after a series and parallel reduction change to

$$t_e \bowtie t_f = t_e t_f, \tag{6.11}$$

$$t_e \|_q t_f = \frac{t_e + t_f + (q-2)t_e t_f}{1 + (q-1)t_e t_f}. \tag{6.12}$$

Theorem 6.10 *Let q be a fixed complex number and Λ a fixed positive integer. Suppose that there are regions $S_1 \subseteq S_2 \subseteq \cdots \subseteq S_{\Lambda-1}$ of the complex t-plane such that*

(1) $1/(1-q) \in S_1$,

(2) $S_k \bowtie S_\ell \subseteq S_{\min(k,\ell)}$,

(3) $S_k \|_q S_\ell \subseteq S_{k+\ell}$ for $k + \ell \leq \Lambda - 1$,

(4) $S_{\Lambda-1} \subseteq \{|t| < 1\}$,

(5) $\infty \notin (S_k \|_q S_\ell)$ for $k + \ell = \Lambda$.

Then q is not a chromatic root of any series-parallel graph of maxmaxflow less than or equal to Λ. □

The main idea here is that during a sequence of series and parallel compositions, every graph constructed should have $t_{\text{eff}} \in S_k$ if it has between-terminals flow equal to k. The first condition covers the base case K_2 which has $t_{\text{eff}} = 1/(1-q)$, while the second and third condition ensure that this property holds during series and parallel compositions. The fourth condition ensures that the effective transmissivity never exceeds 1 in modulus because otherwise (6.11) shows that it could be made arbitrarily large by repeated series composition. Finally the fifth condition ensures that q is not actually a chromatic root of the final graph.

Of course, to apply this theorem it is necessary to actually *exhibit* a suitable collection of regions in the complex plane, and then prove that they are suitably bounded. As $1/(1-q) \in S_1$ (and hence all regions) we initially experimented with taking S_1 to be a disk of radius $\rho = |1/(1-q)|$ (centered at the origin) but this did not lead to very good bounds. Finally, we realized that in fact we should take the *largest* region $S_{\Lambda-1}$ to be a disk of radius ρ and then let every other region be a disconnected region of the form

$$S_i = \{1/(1-q)\} \cup D(r_i) \tag{6.13}$$

where $D(r_i)$ is a disk of radius $r_i < \rho$ centered at the origin. In Royle and Sokal [45] we show that if $r_1, r_2, \ldots, r_{\Lambda-1}$ are determined by the equations

$$r_1 = \rho^2, \tag{6.14}$$

$$r_{s+1} = \frac{r_s(\rho^2 + \rho + 1) + \rho^2}{1 - \rho r_s}, \tag{6.15}$$

and if q is chosen so that $r_{\Lambda-1} < \rho$, then the regions determined by (6.13) for $i < \Lambda - 1$ (and $S_{\Lambda-1} = D(\rho)$) satisfy the conditions of Theorem 6.10. The final step involves diagonalizing the Möbius transformation in (6.15) to get an explicit expression

$$r_k = \rho \frac{(\rho+1)^k - (\rho^2+1)^k}{(\rho^2+1)^k - \rho(\rho+1)^k} \tag{6.16}$$

and then some analysis to show that the condition $|q-1| \geq (\Lambda-1)/\log 2$ is sufficient to ensure that $r_{\Lambda-1} < \rho$.

How good is this bound? On one hand, Sokal (unpublished) has shown that series-parallel graphs with maxmaxflow $k+1$ obtained by taking a complete k-ary tree and then contracting all the leaves to a single vertex have chromatic roots arbitrarily close to the circle $|q-1| = k = \Lambda - 1$, and so the constant is incorrect by a factor of less than $1/\log 2$.

On the other hand, for each fixed q and Λ it is possible to perform computational experiments that approximate the *optimal* regions S_i. Candidate regions S_i can be obtained by starting with $1/(1-q) \in S_1$ and then applying (6.11) and (6.12) in all possible ways. By imposing a fine grid on the disk $|t| < 1$ and only keeping one value per grid cell, this process can be continued until the candidate sets $\{S_i\}$ are closed. If all of the resulting sets are contained in $|t| < 1$ then the approximation labels q as a value that is not the chromatic root of any series-parallel graph of maxmaxflow Λ. Performing this experiment for many different values of q yields a picture of the root-free region for that particular choice of Λ. Figure 6 shows such a picture for $\Lambda = 3$ which suggests that there is still room for improvement in Theorem 6.7.

Of course the main question is whether similar results can be shown to hold for all graphs, not just series-parallel graphs. Certainly this proof will not generalize.

Conjecture 6.11 (Sokal [54]) *There is a function $C(\Lambda)$ such that the chromatic roots of any graph of maxmaxflow Λ lie in the disk $|q| \leq C(\Lambda)$.*

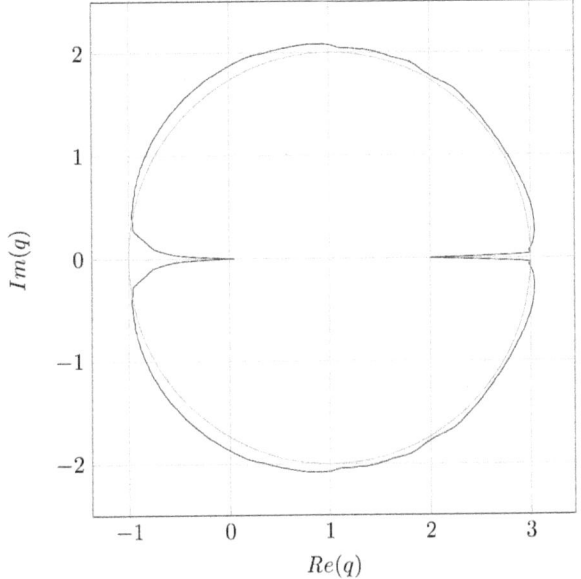

Figure 6: Chromatic root-free region for series-parallel graphs of maxmaxflow $\Lambda = 3$

In fact Sokal has conjectured that $C(\Lambda)$ is a *linear* function and so it is reassuring that this is true at least for series-parallel graphs. More generally we might ask the same question for binary matroids.

Question 6.12 *Is there a function C such that the characteristic roots of a binary matroid M lie in the disk $|q| \leq C(\Lambda(M))$?*

7 Real Roots

Research into real roots is mostly focussed on identifying the extreme values, both large and small, that can be taken by the non-trivial real chromatic or flow roots. At the "large end" lie considerations of which families of graphs have bounded real chromatic roots and real flow roots and analogously for matroids. Section 7.1 discusses the general situation, while Section 7.2 addresses planar graphs and Section 7.3 covers what is known for real flow roots.

At the "small end", there have been several results extending the chromatic root-free interval $(1, 32/27]$ for various classes of graphs. For example, Thomassen [55] has shown that if $\alpha \approx 1.29559$ is the real root of $(t-2)^3 + 4(t-1)^2$ then no graph with a hamiltonian path has a chromatic root in $(1, \alpha]$, and similarly Dong and Koh [15] have shown that graphs with domination number at most 2 have no chromatic roots in the interval $(1, \beta]$ where $\beta \approx 1.31767$. I omit further details in this survey, because Dong and Koh [12] are developing a broader theory giving general conditions under which a large variety of results of this type can be derived.

For certain classes of graphs, the chromatic root-free interval can be stretched to include $(1,2)$. Section 7.4 describes recent results of Dong and Koh [14] on graphs *without* chromatic roots in $(1,2)$ and Royle [44] on graphs *with* chromatic roots in $(1,2)$ which give us further insight into when this is possible.

Section 7.5 gives results for planar triangulations, where the situation is more constrained and the chromatic root-free interval (for non-integer roots) can be pushed beyond 2 and also generalized to cover flow roots of cubic graphs.

7.1 Upper root-free intervals

One of the surprises in the theory of chromatic roots is that there seems to be almost no relationship between the behaviour of integer roots, real roots and complex roots. We have seen that constraining the integer chromatic roots (i.e. chromatic number) does not limit the complex chromatic roots and this is also true for *real* chromatic roots. The chromatic polynomial is strictly increasing on all integer values from the chromatic number onwards, and so one might naively hope that it be increasing between these integer values. At the very least, it seems entirely reasonable that the chromatic polynomial should not become *negative* for non-integer values greater than the chromatic number. However the following result of Woodall shows that this is utterly untrue and that even bipartite graphs have unbounded real chromatic roots.

Theorem 7.1 (Woodall [64]) *If n is sufficiently large compared to m, then the complete bipartite graph $K_{m,n}$ has real chromatic roots arbitrarily close to all integers in $[2, m/2]$.*

Although no bounds on real chromatic roots can be found for the class of bipartite graphs, we can say more when dealing with a *minor-closed* class of graphs. We need a couple of preliminary results.

Theorem 7.2 (Woodall [66]) *If every simple minor of a graph G has a vertex of degree at most d, then $P(G, q) > 0$ for all real $q \in (d, \infty)$.*

Theorem 7.3 (Mader [37]) *There is a function $f(k)$ such that every graph with minimum degree greater than $f(k)$ has K_k as a minor.*

Theorem 7.4 (Woodall [66], Thomassen [57]) *Let $\mathcal{G}$ be a proper minor-closed class of graphs. Then there is an integer $d(\mathcal{G})$ such that no graph in $\mathcal{G}$ has a real chromatic root in the interval $(d(\mathcal{G}), \infty)$.*

Proof As $\mathcal{G}$ is a *proper* minor-closed class of graphs (i.e. not all graphs), there is some complete graph K_k that is not in $\mathcal{G}$. By Theorem 7.3 it follows that every graph in $\mathcal{G}$ has a vertex of degree at most $f(k)$, and hence we can take $d(\mathcal{G}) = f(k)$ and then apply Theorem 7.2. □

A chromatic root-free interval of the form (d, ∞) for a class of graphs is called an *upper root-free interval* and so this result shows that every minor-closed class of graphs has an upper root-free interval. In attempting to extend this result to minor-closed classes of binary matroids, one might first seek extensions of Theorems 7.2 and 7.3 from graphs to matroids. The first of these is readily available.

Theorem 7.5 *If every simple minor of a matroid M has a cocircuit of size at most d, then $C(M,q) > 0$ for all real $q \in (d, \infty)$.*

This theorem was proved for graphs by Woodall [66] with the key step being the observation that if a graph G has a vertex v of degree k then its chromatic polynomial can be expressed as

$$P(G,q) = (q-k)P(G-v,q) + \sum_{H} P(H,q) \qquad (7.1)$$

where the graphs H in the sum range over various minors of G. For matroids, an analogous observation for cocircuits of size k had previously been made by Oxley [39], but Jackson [25] was the first to write down the general theorem.

Unfortunately, there is no analogue of Mader's result for binary matroids because, in general, a large minimum cocircuit size does not force a binary matroid to have a large complete graph as a minor. However it is at least possible to bound the *critical exponent* of any minor-closed class of $GF(q)$-representable matroids that does not include all graphic matroids using the following result.

Theorem 7.6 (Geelen and Whittle [21]) *If q is a prime power and $\mathcal{M}$ is a minor-closed class of $GF(q)$-representable matroids that does not contain all graphic matroids, then there is a constant c such that $|E(M)| \leq cr(M)$ for all simple matroids $M \in \mathcal{M}$.*

But if a class of representable matroids has *linear* growth rate, then we can apply the following result due to Kung.

Theorem 7.7 (Kung [33]) *If q is a prime power and $\mathcal{M}$ is a class of $GF(q)$-representable matroids closed under deletion of elements and every matroid $M \in \mathcal{M}$ satisfies $|E(M)| \leq cr(M)$ then the critical exponent of every matroid in $\mathcal{M}$ is at most c.*

The maximum number of elements in a simple cographic matroid M of rank r occurs when $M = M(G)^*$ for a cubic graph G on $2(r-1)$ vertices, and so $|E(M)| \leq 3(r-1)$. Therefore by Theorem 7.7, cographic matroids have critical exponent at most 3 and we recover Jaeger's 8-flow theorem [31]. However it is still not known if there is any upper bound on the *real* characteristic roots of cographic matroids, which we discuss further in Section 7.3. Conversely there seem to be no known minor-closed classes of $GF(q)$-representable matroids that have *unbounded* real characteristic roots which makes the following conjecture at least slightly plausible:

Conjecture 7.8 *For any finite field $GF(q)$, every minor-closed class of $GF(q)$-representable matroids that does not contain all graphic matroids has an upper characteristic-root-free interval.*

In fact perhaps this conjecture is true for *any* minor-closed class of matroids that does not include all graphic matroids or, equivalently, has linear growth rate.

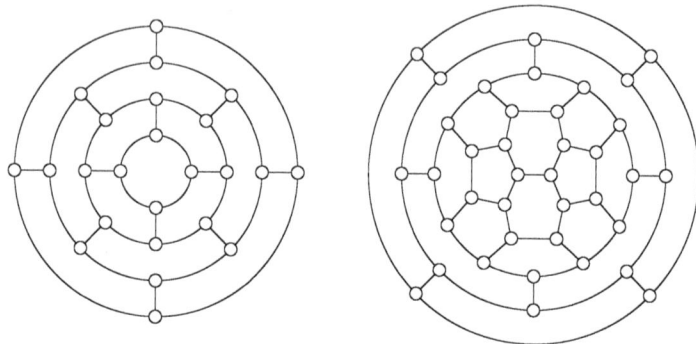

Figure 7: The graph A_3^* and the dual of Woodall's graph

7.2 Upper root-free interval for planar graphs

Exact upper root-free intervals are known for only a few minor-closed classes of graphs (see [66, 57]), but not for the most important family of all, namely planar graphs. Thomassen [57] demonstrated that it is sufficient to consider planar triangulations (i.e. maximal planar graphs) in that the upper root-free intervals for planar graphs and planar triangulations coincide.

The Birkhoff-Lewis conjecture is that the upper chromatic root-free interval for planar graphs is exactly $[4, \infty)$, but until recently planar graphs with real chromatic roots arbitrarily close to 4 were not known. However Beraha and Kahane [2] had found a family of graphs with *complex* chromatic roots arbitrarily close to 4. The dual of a planar triangulation is a planar cubic graph, and as at that time the chromatic polynomial was essentially defined as counting face-colourings of planar maps, researchers described their results in terms of planar cubic graphs. As these tend to be easier to draw and visualize than triangulations, I will do the same in the diagrams (though the chromatic polynomial is still defined as counting vertex-colourings). So let A_n denote the *planar dual* of the planar graph consisting of two 4-faces separated by n rings of four faces (Figure 7 shows the graph A_3^*). Beraha and Kahane computed the limiting curves of the complex chromatic roots of the graphs $\{A_n\}$, and observed that these curves passed through the point $q = 4$.

For *real* chromatic roots of planar graphs, at the time of Jackson's [25] survey, the "record holder" — the planar graph with the largest known real chromatic root — was a 21-vertex graph found by Woodall, with a real chromatic root of approximately 3.82679. Although the reasons why this particular graph should be the record holder were mysterious at the time, drawing the *dual* of Woodall's graph in the right way (see Figure 7) makes it clearer — it is just A_2^* with the inner 4-cycle replaced with a much more complicated graph.

Now A_n is actually a strip of the *periodic triangular lattice* of width 4 and height $n - 1$, but with an additional vertex of degree 4 placed in the top face and the bottom face to make a triangulation. Therefore Woodall's graph is a very short periodic triangular lattice of width 4 and height 2 with a vertex of degree 4 placed

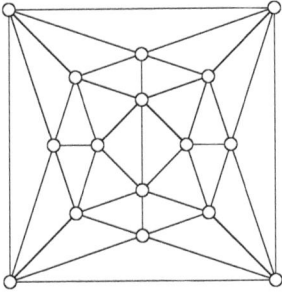

Figure 8: A near-triangulation with a 4-face

in the "bottom" 4-face and the near-triangulation shown in Figure 8 in the "top" 4-face.

From this observation, it was natural to consider inserting the same "end-graphs" into longer and longer strips of the periodic triangular lattice, and indeed this family of graphs has real chromatic roots arbitrarily close to 4 (Royle [43]). There are other similar families of "double-ended lattice strips" using the periodic triangular lattice of width 4 that have chromatic roots arbitrarily close to 4. In fact the near-triangulations of a square can be divided into two classes and any pair of end-graphs consisting of one from each class will suffice.

Theorem 7.9 (Royle [43]) *The triangulations of a square can be divided into two classes such that if a double-ended strip of the periodic triangular lattice of width 4 has one end-graph from each class, then it has real chromatic roots arbitrarily close to 4 as its length increases.*

For a natural number n, the Beraha number B_n is given by the equation

$$B_n = 2 + 2\cos(2\pi/n). \qquad (7.2)$$

The first few Beraha numbers $B_1 = 4$, $B_2 = 0$, $B_3 = 1$, $B_4 = 2$, $B_5 = \tau^2$, $B_6 = 3$, ... where $\tau = (1 + \sqrt{5})/2$ (the golden ratio) and then an increasing sequence tending to 4. Beraha noticed that planar triangulations tend to have roots close to these numbers, and they arise naturally in Tutte's theory of "chromatic sums" [61]. (See also Saleur [49] and Fendley and Krushkal [20].)

Conjecture 7.10 (Beraha's Conjecture) *For all $n \in \mathbb{N}$ and $\epsilon > 0$ there is a planar triangulation with a real chromatic root in $(B_n - \epsilon, B_n + \epsilon)$.*

This is obviously true for the integer values where a planar graph can actually *have* a chromatic root, namely 0, 1, 2 and 3, and had previously been demonstrated for B_5 and B_7 (Beraha, Kahane and Weiss [1]), and the results of this section have added $B_1 = B_\infty = 4$ to the list. Despite the conjecture, none of the Beraha numbers can actually *be* chromatic roots — with the possible exception of B_{10} (Salas and Sokal [48]). This will occur if and only if there is a chromatic polynomial with $q^2 - 5q + 5$ as one of its quadratic factors. Further results and conjectures on planar triangulations and near-triangulations are discussed below (Section 7.5).

Of course any subset of a minor-closed class of graphs will have an upper chromatic root-free interval even if the subset is not itself minor-closed. So one could ask for the upper chromatic root-free interval for, say, *bipartite* planar graphs. Salas and Sokal have found families of graphs with real chromatic roots above τ^2 and have made the following conjecture:

Conjecture 7.11 (Salas and Sokal [47]) *The upper root-free interval for bipartite planar graphs is* $[3, \infty)$.

7.3 Upper flow-root-free interval for graphs

A long-standing open problem is to determine the upper root-free interval for the *flow roots* of graphs or, equivalently, the characteristic roots of *cographic* matroids. Indeed, in the absence of an analogue of Theorem 7.4, it would be significant to demonstrate the mere *existence* of an upper root-free interval. Welsh made the following conjecture (unpublished, but see Jackson[25]), which is essentially a dual version of the Birkhoff-Lewis conjecture.

Conjecture 7.12 (Dominic Welsh) *If G is a bridgeless graph, then $F(G, q) > 0$ for all $q \in (4, \infty)$.*

Maximal cographic matroids arise from cubic graphs and as cubic graphs with high girth seem to exhibit qualitatively extremal behaviour (this is a deliberately imprecise statement) with respect to their *chromatic* roots, these are a natural class to study. Using a recently-completed implementation of an algorithm for Tutte polynomials that uses isomorphism testing to eliminate duplicate portions of the computation tree (see Haggard, Pearce and Royle [22]), we could compute flow polynomials for large numbers of high girth cubic graphs with 20–32 vertices, and smaller numbers of 34–50 vertex graphs.

To our surprise, counterexamples to Welsh's conjecture were almost immediately found, with an explicit example being the generalized Petersen graph $P(16, 6)$ which is a 32-vertex cubic graph of girth 7 shown in Figure 9 with flow polynomial $(q - 1)(q - 2)(q - 3)G(q)$ where

$$\begin{aligned} G(q) &= q^{14} - 42\, q^{13} + 833\, q^{12} - 10358\, q^{11} + 90393\, q^{10} - 587074\, q^9 \\ &+ 2934917\, q^8 - 11515364\, q^7 + 35798907\, q^6 - 88275860\, q^5 \\ &+ 171273551\, q^4 - 256034548\, q^3 + 282089291\, q^2 \\ &- 207662412\, q + 77876944. \end{aligned}$$

This has real roots at two values $q_1 \approx 4.0252205$ and $q_2 \approx 4.2331455$ demonstrating that 4 is not the upper limit for flow roots.

Many more examples were found on 28, 32 and 36 vertices but none on 30 or 34 vertices. The common features of the examples found are that the flow polynomial is a polynomial of reasonably high *odd* degree. In very loose terms, the polynomial is increasing at $q = 1$, decreasing at $q = 2$, increasing at $q = 3$ and then decreasing so fast at $q = 4$ that it has a root just after 4 and then has to have another root in $(4, 5)$

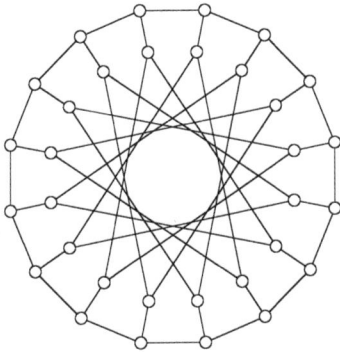

Figure 9: The generalized Petersen graph $P(16,6)$

in order to be positive at $q = 5$. Empirical observations show that the higher degree flow polynomials belonging to the bigger graphs have more pronounced oscillations with the smaller of the two roots in $(4,5)$ being closer to 4 and the larger closer to 5. The logical conclusion of this non-rigorous argument is that taking larger graphs of higher girth should lead to roots arbitrarily close to 5, but no larger, leading to the following modification of Welsh's conjecture,

Conjecture 7.13 (Welsh, modified) *If G is a bridgeless graph with flow polynomial $F(G,q)$, then $F(G,q) > 0$ for all $q \in [5, \infty)$.*

The case $q = 5$ is simply Tutte's 5-flow conjecture and so the truth of this conjecture (and the Birkhoff-Lewis conjecture) would give an appealing parallel between the conjectured results and theorems for the flow roots of general graphs and chromatic roots of planar graphs.

All Flow Roots	Planar Chromatic Roots
No roots in $[5, \infty)$	No roots in $[4, \infty)$
5-flow conjecture	4-colour theorem
Roots arbitrarily close to 5	Roots arbitrarily close to 4

Studying flow polynomials of very large cubic graphs with very high girth is impossible with a direct computational approach. The results for planar graphs described in the previous section were obtained using *transfer matrix* methods where the polynomial is expressed in the form

$$u^T T^n v \qquad (7.3)$$

where T is a matrix, u and v are vectors representing boundary conditions and n is the number of layers in the lattice. To apply these methods, it is necessary to have graphs with some sort of repeating or "layered" structure and then the transfer matrix describes the effect of adding an extra layer. Jacobsen and Salas [30] have described how to use transfer matrix methods with cyclic boundary conditions in the longitudinal direction (i.e. the "long" direction).

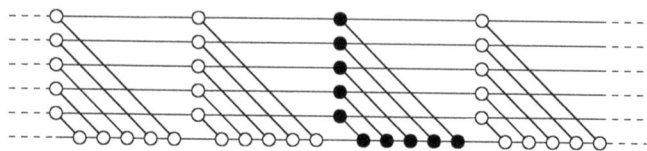

Figure 10: The graph $P(20,5)$ with one "layer" highlighted

To apply these techniques we need to find a family of high girth cubic graphs with a suitable layered structure. The *generalized Petersen graph* $P(n,k)$ is a cubic graph with $2n$ vertices where

$$V(P(n,k)) = \{v_0, v_1, \ldots, v_{n-1}\} \cup \{w_0, w_1, \ldots, w_{n-1}\} \quad (7.4)$$
$$E(P(n,k)) = \{\{v_i, w_i\}, \{v_i, v_{i+1}\}, \{w_i, w_{i+k}\} \mid 0 \leq i < n\} \quad (7.5)$$

with all subscripts taken mod n. Thus $P(5,2)$ is the usual Petersen graph. By drawing $P(mk, k)$ in a non-standard fashion it can be seen to have a suitable repeating structure with cyclic boundary conditions, as shown in Figure 10.

Jacobsen and Salas (personal communication) have applied their transfer matrix techniques to computing the limiting curves of the flow roots of the generalized Petersen graphs of the form $P(mk, k)$ for $k \leq 5$. The limiting curves for $P(5k, 5)$ are shown in Figure 11 along with the flow roots for $P(120, 5)$. The limiting curve crosses the real axis at $q_c \approx 4.9029018$ and $P(120, 5)$ (the largest graph for which an explicit root has been calculated) has a real flow root of $q \approx 4.8679668$.

Intriguingly, it appears that for $k = 6$ there is a limiting curve crossing the real axis at a value $q_c > 5$, but it remains to be seen whether this will lead to *real* roots greater than 5.

7.4 Graphs with chromatic roots in $(1,2)$

The graphs with roots close to $32/27$ and the other families described in the previous section all have many vertex-cutsets of size 2 and hence are far from 3-connected. However there are some 3-connected graphs that are "forced" to have chromatic roots in $(1,2)$, namely bipartite graphs on an odd number of vertices. In this situation $P(G, q)$ is positive for $q \in (0, 1)$ and has a simple root at $q = 1$ and so is negative in some interval of the form $(1, 1+\alpha)$ where $\alpha > 0$. However $P(G, 2) = 2$ and so it is positive at $q = 2$ and hence is forced to have a root in $(1, 2)$. Similarly a 3-connected eulerian graph with $|E| - |V| + 1$ even necessarily has a flow root in $(1, 2)$. Jackson [25] conjectured that these forced examples were the only ones.

Conjecture 7.14 *(1) If G is a 3-connected graph that is not bipartite of odd order, then $P(G, q) \neq 0$ for $q \in (1, 2)$.*

(2) If G is a 3-connected graph that is not eulerian with $|E| - |V| + 1$ even, then $F(G, q) \neq 0$ for $q \in (1, 2)$.

The first conjecture was recently disproved with the discovery of several infinite families of 3-connected graphs that do have chromatic roots in $(1, 2)$ (Royle [44]).

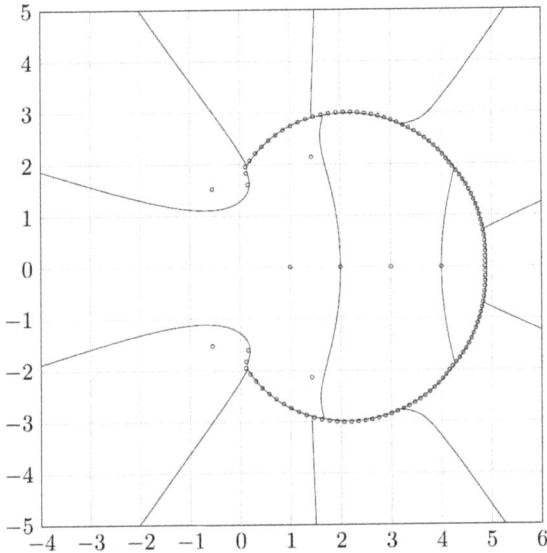

Figure 11: Flow roots of $P(120,5)$ and limiting curves for flow roots of $P(5k,5)$

The simplest of these families has a smallest member with only 11 vertices as shown in Figure 12. It is clear by inspection that $X(3,3)$ is 3-connected and not bipartite and it is sufficiently small that its chromatic polynomial can be easily computed. The result of this computation is that $P(X(3,3), q) = q(q-1)(q-2)Q(q)$, where $Q(q)$ is

$$\left(q^8 - 17\,q^7 + 137\,q^6 - 677\,q^5 + 2228\,q^4 - 4969\,q^3 + 7284\,q^2 - 6363\,q + 2509\right),$$

which has real roots at $q_1 \approx 1.90263148$ and $q_2 \approx 2.42196189$. Therefore we conclude that Jackson's conjecture is false.

Now let $X(s,t)$ be the graph obtained by replacing the two independent sets of size three labelled S and T in Figure 12 by independent sets of size $s \geq 3$ and $t \geq 3$ respectively. If s and t have the right parity, then $P(X(s,t), q)$ goes from positive-to-negative as q increases past 2.

Lemma 7.15 (Royle [44]) *If $s, t > 1$ then the derivative of $P(X(s,t), q)$ evaluated at $q = 2$ is*

$$P'(X(s,t), 2) = 2\left((-1)^s + (-1)^t + (-1)^{s+t}\right). \tag{7.6}$$

In particular, when $s, t \geq 3$ are both odd, $P'(X(s,t), 2) = -2$. □

Theorem 7.16 (Royle [44]) *The graph $X(s,t)$ is 3-connected and not bipartite and if $s, t \geq 3$ are both odd, then it has a chromatic root in $(1,2)$.*

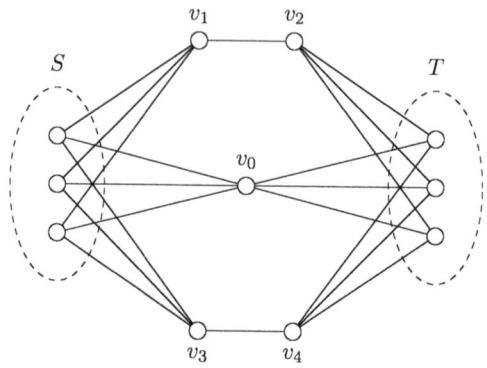

Figure 12: The graph $X(3,3)$

Proof Clearly $X(s,t)$ is 3-connected and bipartite. If s, $t \geq 3$ are both odd then $X(s,t)$ has an odd number of vertices and so its chromatic polynomial is positive on $(0,1)$, has a simple root at $q = 1$ and hence is negative as q increases past 1. By Lemma 7.15 it is positive as q approaches 2, and so there is a chromatic root somewhere in $(1,2)$. □

There are many similar examples, but all of them contain at least one induced $K_{3,3}$. Therefore all these known examples are non-planar and cannot be used to resolve the second of Jackson's two conjectures which remains open.

Although 3-connectivity it not enough in itself to guarantee that $(1,2)$ is free of chromatic roots, Dong and Koh [14] found a somewhat complicated condition that is sufficient. A *double-link ordering* of a graph G is an ordering of its vertices $(v_1, v_2, \ldots, v_n)$ such that v_2 is adjacent to v_1 and for each $i > 2$, the vertex v_i has at least two neighbours in the set $V_{i-1} := \{v_1, v_2, \ldots, v_{i-1}\}$. Let $N(v)$ denote the neighbours of v.

Theorem 7.17 (Dong and Koh [14]) *Let $(v_1, v_2, \ldots, v_n)$ be a double-link ordering of G, and let V' denote the set of vertices v_i ($i > 4$) such that $N(v_i) \cap V_{i-1}$ is independent. If*

$$|U| < \left| \bigcup_{v_i \in U} (N(v_i) \cap V_{i-1}) \right| \qquad (7.7)$$

for every nonempty subset $U \subseteq V'$, then G has no chromatic roots in $(1,2)$. □

Following Dong and Koh, we let Γ denote the set of graphs satisfying the conditions of Theorem 7.17. Although it appears hard to get a simple characterization of the graphs in Γ, several important families of graphs are known to be in Γ and several sufficient conditions are known. For example, if there is a double-link ordering of a graph where none of the "back-neighbourhoods" $N(v_i) \cap V_{i-1}$ are independent, then the graph is automatically in Γ. It is easily seen that 2-connected near triangulations have such an ordering, thereby recovering a result of Birkhoff and Lewis (also see Theorem 7.22).

Recall that a graph G is said to be *1-tough* if there is no nonempty subset S of its vertices such that $G - S$ has more than $|S|$ components. We will call a graph α-1-tough if it has no nonempty *independent* set S of vertices such that $G - S$ has more than $|S|$ components. (The term α-tough has also been used previously, but it has been pointed out that this may cause confusion with the graph-theoretic concept of t-tough, so I have substituted the slightly more unwieldy term.) Dong and Koh showed that the graphs in Γ are all α-1-tough and made the following conjecture:

Conjecture 7.18 (Dong and Koh [14]) *An α-1-tough graph has no chromatic roots in $(1, 2)$.*

In order to make sense, Jackson's original conjecture had to *explicitly* exclude the bipartite graphs of odd order, but this conjecture is more appealing because it gives a structural criterion that *intrinsically* excludes them. Also, it has survived its first "test" because the graphs $X(s,t)$, which were not known at the time the conjecture was made, are also not α-1-tough (removing $\{v_0, v_1, v_3\}$ leaves $s + 1 > 3$ components).

Furthermore, the conjecture extends a conjecture of Thomassen [56] that *hamiltonian* graphs have no chromatic roots in $(1, 2)$. It is well known that hamiltonian graphs are 1-tough and hence α-1-tough. Perhaps then, it is the *toughness* of hamiltonian graphs, rather than the Hamilton cycle, that is actually important. Proving either result by induction appears to require proving (the appropriate) one of the following conjectures:

Conjecture 7.19 (Thomassen [56]) *A hamiltonian graph G with minimum degree at least 3 has an edge e such that both $G\backslash e$ and G/e are hamiltonian.*

Conjecture 7.20 (Dong and Koh [12]) *An α-1-tough graph G with minimum degree at least 3 has an edge e such that both $G\backslash e$ and G/e are α-1-tough.*

7.5 Cubic graphs and planar triangulations

Planar triangulations play a central role in the theory of graph colouring as many questions on colouring or the chromatic roots of planar graphs can be reduced through combinatorial arguments to the case of planar triangulations. Similarly, many questions about flows and flow roots can be reduced to the cubic graph case.

The dual of a planar triangulation is a cubic *planar* graph and so any result on the flow roots of general cubic graphs is automatically a result on the chromatic roots of planar triangulations. Conversely, given any result on the chromatic roots of triangulations, one can try to determine whether it is the restriction of a more general result on flow roots of cubic graphs or whether the planarity condition is essential to the result. As an example, Jackson has proved the following theorem which extends the results proved for planar triangulations by Birkhoff and Lewis [4] and Woodall [65].

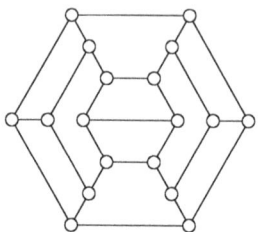

Figure 13: The exceptional graph in Conjecture 7.23

Theorem 7.21 (Jackson [26]) *If G is a bridgeless cubic graph, then it has no flow roots in $(1,2)$ or $(2,\delta)$ where $\delta \approx 2.546$ is the unique flow root of the cube in $(2,3)$.*

The classes of planar triangulations and cubic graphs are not closed under the operations of deletion or contraction, and so researchers often consider the larger classes of *near triangulations* and *near-cubic* graphs. Near triangulations are planar graphs where at most one face is permitted to have size $k > 3$, while near-cubic graphs are graphs where at most one vertex is permitted to have degree $k > 3$. Again extending results of Birkhoff and Lewis, Jackson [27] has proved an analogous result for the flow roots of near cubic graphs.

Theorem 7.22 (Jackson [27]) *If G is a bridgeless near-cubic graph, then it has no flow roots in $(1,2)$ or $(2,\sigma)$ where $\sigma \approx 2.225$ is the real zero in $(2,3)$ of the polynomial $q^4 - 8q^3 + 22q^2 - 28q + 17$. If G is 3-connected, then 2 is a simple zero of $F(G,q)$.*

However the following conjecture cannot be extended to flow roots.

Conjecture 7.23 (Woodall, personal communication) *All 4-connected planar triangulations except for the 11-vertex dual of the graph shown in Figure 13 have a chromatic root in the interval $(2.546602\ldots, 2.677814\ldots)$ where the two bounds are the unique chromatic roots of the octahedron $(2K_1 + C_4)$ and $2K_1 + C_5$ in $(2,3)$ respectively.*

Jackson [25] describes a variety of additional conjectures, also mostly due to Woodall, for planar triangulations of higher connectivity.

7.6 Final Questions

Although much is known about real chromatic roots, there are still some quite fundamental questions that are unanswered. In particular, we still have no good general parameterized upper bound on the largest real chromatic root of a graph, although there are several conjectured bounds.

Conjecture 7.24 (Sokal [54]) *If G is a loopless graph of maxmaxflow Λ then $P(G,q) > 0$ for $q > \Lambda$.*

The analogous (weaker) conjectures with Λ replaced by second maximum degree Δ_2 or even maximum degree Δ are also still unresolved. Rather than strictly real chromatic roots, one might ask similar questions about the *real part* of possibly complex chromatic roots. In this case, I have found examples showing that it is possible for chromatic roots to satisfy $\text{Re}(q) > \Lambda$ and $\text{Re}(q) > \Delta_2$. Let W be the "Wheatstone bridge" which is the 2-terminal graph $K_4 \backslash e$ with the two vertices of degree 2 as the terminals and define a sequence of graphs as follows:

$$G_8 = (W \bowtie K_2) \parallel (W \bowtie K_2) \tag{7.8}$$
$$G_{16} = (G_8 \bowtie K_2) \parallel (G_8 \bowtie K_2) \tag{7.9}$$
$$G_{47} = (G_{16} \bowtie K_2) \parallel (G_{16} \bowtie K_2) \parallel (G_{16} \bowtie K_2) \tag{7.10}$$

Then G_{47} is a 47-vertex graph with $\Lambda = \Delta_2 = 3$ but it has chromatic roots at $q \approx 3.012995071 \pm 0.808962864i$.

We finish with one final conjecture from Sokal:

Conjecture 7.25 (Sokal, personal communication, 2007) *Let G be a bipartite graph with bipartition $V(G) = A \cup B$, let Δ_A and Δ_B be the maximum degrees over the vertices in A and B, and let $\Delta_{\min} = \min(\Delta_A, \Delta_B)$. Then $P_G(q) > 0$ for $q > \Delta_{\min}$.*

The binary matroid version of Conjecture 7.24, with Λ given by (6.3), is essentially unexplored at this time.

Acknowledgements

I would like to thank the following people (in alphabetical order) for valuable conversations and correspondence regarding the topics of this paper: Norman Biggs, Fengming Dong, Bill Jackson, Jesper Jacobsen, Joseph Kung, Dillon Mayhew, Jésus Salas, Robert Shrock, Alan Sokal, Geoff Whittle and Douglas Woodall.

I also thank the Isaac Newton Institute for Mathematical Sciences, University of Cambridge, for support during the programme on Combinatorics and Statistical Mechanics (January-June 2008) where many of these discussions took place.

References

[1] S. Beraha, J. Kahane & N. J. Weiss, Limits of chromatic zeros of some families of maps, *J. Combin. Theory Ser. B* **28** (1980), 52–65.

[2] S. Beraha & J. Kahane, Is the four-color conjecture almost false?, *J. Combin. Theory Ser. B* **27** (1979), 1–12.

[3] N. L. Biggs, R. M. Damerell & D. A. Sands, Recursive families of graphs, *J. Combin. Theory Ser. B* **12** (1972), 123–131.

[4] G. D. Birkhoff & D. C. Lewis, Chromatic polynomials, *Trans. Amer. Math. Soc.* **60** (1946), 355–451.

[5] G. D. Birkhoff, A determinant formula for the number of ways of coloring a map, *Ann. of Math. (2)* **14** (1912/13), 42–46.

[6] J. A. Bondy & U. S. R. Murty, *Graph theory*, Graduate Texts in Mathematics, 244, Springer, New York (2008).

[7] F. Brenti, G. F. Royle & D. G. Wagner, Location of zeros of chromatic and related polynomials of graphs, *Canad. J. Math.* **46** (1994), 55–80.

[8] T. Brylawski & J. Oxley, The Tutte polynomial and its applications, in *Matroid applications*, Encyclopedia Math. Appl., 40, Cambridge University Press, Cambridge (1992), pp. 123–225.

[9] H. H. Crapo & G.-C. Rota, *On the foundations of combinatorial theory: Combinatorial geometries*, The M.I.T. Press, Cambridge, Mass.-London, Preliminary edition, (1970).

[10] R. Diestel, *Graph theory*, Third edition, Graduate Texts in Mathematics, 173, Springer-Verlag, Berlin (2005).

[11] G. A. Dirac, On rigid circuit graphs, *Abh. Math. Sem. Univ. Hamburg* **25** (1961), 71–76.

[12] F. M. Dong & K. M. Koh, On zero-free intervals in $(1,2)$ of chromatic polynomials of some families of graphs, submitted.

[13] F. M. Dong & K. M. Koh, Non-chordal graphs having integral-root chromatic polynomials, *Bull. Inst. Combin. Appl.* **22** (1998), 67–77.

[14] F. M. Dong & K. M. Koh, On graphs having no chromatic zeros in $(1,2)$, *SIAM J. Discrete Math.* **20** (2006), 799–810.

[15] F. M. Dong & K. M. Koh, Domination numbers and zeros of chromatic polynomials, *Discrete Math.* **308** (2008), 1930–1940.

[16] F. M. Dong, K. M. Koh & K. L. Teo, *Chromatic polynomials and chromaticity of graphs*, World Scientific Publishing Co. Pte. Ltd., Hackensack, NJ (2005).

[17] F. M. Dong, K. L. Teo, K. M. Koh & M. D. Hendy, Non-chordal graphs having integral-root chromatic polynomials. II, *Discrete Math.* **245** (2002), 247–253.

[18] H. Edwards, R. Hierons & B. Jackson, The zero-free intervals for characteristic polynomials of matroids, *Combin. Probab. Comput.* **7** (1998), 153–165.

[19] E. J. Farrell, Chromatic roots — some observations and conjectures, *Discrete Math.* **29** (1980), 161–167.

[20] P. Fendley & V. Krushkal, Tutte chromatic identities from the Temperley-Lieb algebra, preprint, http://arxiv.org/abs/0711.0016.

[21] J. Geelen & G. Whittle, Cliques in dense $GF(q)$-representable matroids, *J. Combin. Theory Ser. B* **87** (2003), 264–269.

[22] G. Haggard, D. J. Pearce & G. F. Royle, Computing Tutte polynomials, submitted.

[23] D. W. Hall, J. W. Siry & B. R. Vanderslice, The chromatic polynomial of the truncated icosahedron, *Proc. Amer. Math. Soc.* **16** (1965), 620–628.

[24] B. Jackson, A zero-free interval for chromatic polynomials of graphs, *Combin. Probab. Comput.* **2** (1993), 325–336.

[25] B. Jackson, Zeros of chromatic and flow polynomials of graphs, *J. Geom.* **76** (2003), 95–109.

[26] B. Jackson, A zero-free interval for flow polynomials of cubic graphs, *J. Combin. Theory Ser. B* **97** (2007), 127–143.

[27] B. Jackson, Zero-free intervals for flow polynomials of near-cubic graphs, *Combin. Probab. Comput.* **16** (2007), 85–108.

[28] B. Jackson & A. D. Sokal, Maxmaxflow and counting subgraphs, preprint, http://arxiv.org/abs/math/0703585.

[29] B. Jackson & A. D. Sokal, Zero-free regions for multivariate Tutte polynomials (alias Potts-model partition functions) of graphs and matroids, preprint, http://www.arxiv.org/abs/08063249.

[30] J. L. Jacobsen & J. Salas, Transfer matrices and partition-function zeros for antiferromagnetic Potts models. IV. Chromatic polynomial with cyclic boundary conditions, *J. Statist. Phys.* **122** (2006), 705–760.

[31] F. Jaeger, Flows and generalized coloring theorems in graphs, *J. Combin. Theory Ser. B* **26** (1979), 205–216.

[32] M. Kochol, Polynomials associated with nowhere-zero flows, *J. Combin. Theory Ser. B* **84** (2002), 260–269.

[33] J. P. S. Kung, Extremal matroid theory, in *Graph structure theory (Seattle, WA, 1991)*, Contemp. Math., 147, Amer. Math. Soc., Providence, RI (1993), pp. 21–61.

[34] J. P. S. Kung, Critical problems, in *Matroid theory (Seattle, WA, 1995)*, Contemp. Math., 197, Amer. Math. Soc., Providence, RI (1996), pp. 1–127.

[35] T. D. Lee & C. N. Yang, Statistical theory of equations of state and phase transitions. II. Lattice gas and Ising model, *Phys. Rev. (2)* **87** (1952), 410–419.

[36] B. Lindström, On the chromatic number of regular matroids, *J. Combin. Theory Ser. B* **24** (1978), 367–369.

[37] W. Mader, Homomorphieeigenschaften und mittlere Kantendichte von Graphen, *Math. Ann.* **174** (1967), 265–268.

[38] J. Oxley, What is a matroid?, *Cubo Mat. Educ.* **5** (2003), 179–218.

[39] J. G. Oxley, Colouring, packing and the critical problem, *Quart. J. Math. Oxford Ser. (2)* **29** (1978), 11–22.

[40] J. G. Oxley, *Matroid theory*, Oxford Science Publications,, The Clarendon Press Oxford University Press, New York (1992).

[41] R. C. Read, An introduction to chromatic polynomials, *J. Combin. Theory* **4** (1968), 52–71.

[42] R. C. Read & G. F. Royle, Chromatic roots of families of graphs, in *Graph theory, combinatorics, and applications Vol. 2 (Kalamazoo, MI, 1988)*, Wiley-Intersci. Publ., Wiley, New York (1991), pp. 1009–1029.

[43] G. F. Royle, Planar triangulations with real chromatic roots arbitrarily close to 4, *Ann. Comb.* **12** (2008), 195–210.

[44] G. F. Royle, Graphs with chromatic roots in the interval $(1,2)$, *Electron. J. Combin.* **14** (2007), R18.

[45] G. F. Royle & A. D. Sokal, Linear bounds in terms of maxmaxflow for the chromatic roots of series-parallel graphs, in preparation.

[46] J. Salas & A. D. Sokal, Chromatic roots of the complete bipartite graphs, in preparation.

[47] J. Salas & A. D. Sokal, Transfer matrices and partition-function zeros for antiferromagnetic Potts models. V. Further results for the square-lattice chromatic polynomial, preprint, http://arxiv.org/abs/0711.1738.

[48] J. Salas & A. D. Sokal, Transfer matrices and partition-function zeros for antiferromagnetic Potts models. I. General theory and square-lattice chromatic polynomial, *J. Statist. Phys.* **104** (2001), 609–699.

[49] H. Saleur, Zeroes of chromatic polynomials: a new approach to Beraha conjecture using quantum groups, *Comm. Math. Phys.* **132** (1990), 657–679.

[50] P. D. Seymour, Nowhere-zero 6-flows, *J. Combin. Theory Ser. B* **30** (1981), 130–135.

[51] R. Shrock & S.-H. Tsai, Ground-state degeneracy of Potts antiferromagnets: cases with noncompact w boundaries having multiple points at $1/q = 0$, *J. Phys. A* **31** (1998), 9641–9665.

[52] A. D. Sokal, Bounds on the complex zeros of (di)chromatic polynomials and Potts-model partition functions, *Combin. Probab. Comput.* **10** (2001), 41–77. See http://www.arxiv.org:cond-mat/9904146.

[53] A. D. Sokal, Chromatic roots are dense in the whole complex plane, *Combin. Probab. Comput.* **13** (2004), 221–261.

[54] A. D. Sokal, The multivariate Tutte polynomial (alias Potts model) for graphs and matroids, in *Surveys in Combinatorics 2005*, London Math. Soc. Lecture Note Ser., 327, Cambridge University Press, Cambridge (2005), pp. 173–226.

[55] C. Thomassen, Chromatic roots and Hamiltonian paths, *J. Combin. Theory Ser. B* **80** (2000), 218–224.